Universitext

Universitext

Universitext is a series of textbooks that presents material from a wide variety of mathematical disciplines at master's level and beyond. The books, often well class-tested by their author, may have an informal, personal even experimental approach to their subject matter. Some of the most successful and established books in the series have evolved through several editions, always following the evolution of teaching curricula, to very polished texts.

Thus as research topics trickle down into graduate-level teaching, first textbooks written for new, cutting-edge courses may make their way into *Universitext.*

For further volumes:
www.springer.com/series/223

Marc Hindry

Arithmetics

Prof. Marc Hindry
Institut de Mathématiques de Jussieu
Université Paris 7 Denis Diderot
Paris 75013
France
hindry@math.jussieu.fr

Translation from the French language edition:
'Arithmétique' by Marc Hindry
Copyright © 2008 Calvage et Mounet, France
http://www.calvage-et-mounet.fr/

ISSN 0172-5939 e-ISSN 2191-6675
Universitext
ISBN 978-1-4471-2130-5 e-ISBN 978-1-4471-2131-2
DOI 10.1007/978-1-4471-2131-2
Springer London Dordrecht Heidelberg New York

British Library Cataloguing in Publication Data
A catalogue record for this book is available from the British Library

Library of Congress Control Number: 2011935339

Mathematics Subject Classification: 11A, 11D, 11G, 11H, 11J, 11M, 11N, 11R, 11T, 11Y

Cover design: VTeX UAB, Lithuania

Printed on acid-free paper

Springer is part of Springer Science+Business Media (www.springer.com)

To the memory of my father, Geoffrey

à la mémoire de ma mère, Nicole

Preface to the English edition

This book is a translation from the French *Arithmétique* originally published by Calvage & Mounet. Apart from minor corrections and a couple of examples added in Chap. 3, I have left the book unchanged. I wish to thank all the people from Springer for showing interest and making this new version possible. The book is already dedicated to my parents but I cannot avoid thinking my father would have been very happy to see me publish this book in English. Finally my heartiest thanks go to Sarah Carr, who showed both enthusiasm and expertise in translating the text, she even made me forgive her american spelling.

Preface to the French edition

Amis lecteurs qui ce livre lisez,
Despouillez vous de toute affection,
Et, le lisant ne vous scandalisez.
Il ne contient mal ny infection.
Vray est qu'icy peu de perfection
Vous apprendrez, si non en cas de rire :
Aultre argument ne peut mon cueur eslire.
Voyant le dueil qui vous mine et consomme,
Mieulx est de ris que de larmes escripre
Pour ce que rire est le propre de l'homme.

François Rabelais (Gargantua)

Arithmetic is certainly the oldest mathematical activity. The use of the concept of a *whole number*, numeral systems and the operations of addition, multiplication and division can be found in all civilizations. The invention of *zero* appears to have come from India. Traces of arithmetical operations have been identified on bones dating back to the Paleolithic Era, on Mesopotamian clay tablets, on Chinese turtle shells and on Egyptian papyrus; the Incas, who did not—so it seems—have writing, did develop an evolved numeral system based on knots in strings, called *quipus*.

In our times, number theory is a branch of mathematics which draws its vitality from its rich history. We cite Pythagoras, Euclid, Diophantus, Fermat, Euler, Lagrange, Legendre, Gauss, Abel, Jacobi, Dirichlet, Galois, Riemann, Hilbert, stopping here at the XIXth century. It is also traditionally nourished through interactions with other domains, such as algebra, algebraic geometry, topology, complex analysis, harmonic analysis, etc. More recently, it has made a spectacular appearance in theoretical computer science and in questions of communication, cryptography and error-correcting codes.

The notion of a *number* has actually been progressively extended and enriched throughout history. All of the civilizations considered first the whole numbers (with or without zero). Since the work of Dedekind and Peano at the end of the XIXth century, we consider the *set* of natural numbers, traditionally denoted by $\mathbf{N}$. Advances in logic, calculation techniques and algebra then led us to add negative integers and obtain the set traditionally denoted by $\mathbf{Z}$ and to introduce fractions (the Greeks spoke of proportions) and obtain the set now denoted by $\mathbf{Q}$. Very early on, the necessity of considering even more extraordinary numbers, such as π (the proportion of the circumference of a circle to its diameter) or $\sqrt{2}$ (the proportion of the length of the diagonal of a square to one of its sides) appeared, but it was only very much later that the notions of a *real number* and a *complex number* were clarified. The set of real numbers is today denoted by $\mathbf{R}$ and that of complex number by $\mathbf{C}$. The latter were called *imaginary numbers* for a long time. The two concepts—real and complex numbers—were only rigorously defined in the XIXth century. The first rational approximations of the number π—computed by Archimedes and others—can be viewed as the first chapter in the history of Diophantine approximations. We will cite one more development, even if it did not become what its inventor—Hamilton—had wished it to become, but which has nevertheless proven very useful: the *quaternions*, the set traditionally known as $\mathbf{H}$.

Some other fundamental objects, known at least since the time of Euclid and Pythagoras, are *prime numbers*, traditionally denoted with p—they are so important that in number theory classes, we do not even bother to specify that a number p is prime—and polynomial equations (constructed using the laws of arithmetic and multiplication) or *Diophantine* equations. "Fermat's little theorem", which we write today as $a^p \equiv a \bmod p$, can be considered to be a turning point in the history of number theory in the XVIIth century. The great problem left by Fermat, somewhat accidentally, to mathematicians of subsequent centuries also left its mark on history, culminating in the solution given by Wiles (1995); it can be stated by saying that if $n \geqslant 3$, then there are no non-zero integers x, y, z such that $x^n + y^n = z^n$. Interest in considering subfields and subrings, such as the *Gaussian integers*, $\mathbf{Z}[i]$, or Kummer's *cyclotomic integers*, $\mathbf{Z}[\exp(2\pi i/n)]$, came about little by little, and they were developed as a consequence of the theory of algebraic numbers. Modern arithmetic—it could be more prudent to say *contemporary* arithmetic—is also enriched by the study of finite quotients such as the congruence rings $\mathbf{Z}/N\mathbf{Z}$ and the finite fields $\mathbf{F}_q$. A result dating back to the XIXth century could be considered as a key ingredient in these developments: the *quadratic reciprocity law.* Stated by Legendre and proved by Gauss, it says that "if p and q are odd primes and p or q is congruent to 1 modulo 4 (resp. p and q congruent to 3 modulo 4), then p is a square modulo q if and only if q is (resp. is not) a square modulo p". In another direction, *p-adic numbers*, the set traditionally denoted by $\mathbf{Q}_p$, were invented by Hensel at the end of the XIXth century; we can view the fields $\mathbf{Q}_p$ as ultrametric completions of the field of rationals $\mathbf{Q}$.

This book is naturally a mix of these various notions of numbers. It offers a basic number theory course, followed by an initiation to some contemporary research areas. It is written at the level of an advanced undergraduate course fading into a first-year graduate course, including results which are more advanced but which can be appreciated without having to rely on "heavy" background knowledge. This book is thus divided into two parts which have a gradually different tone.

a) The first part (Chaps. 1 to 4) corresponds to an advanced undergraduate course. All of the statements given in this part are of course accompanied by their proofs, with perhaps the exception of some results appearing at the end of the chapters.

b) The second part (Chaps. 5 and 6 and the appendices) is of a higher level and is relevant for the first year of graduate school. It contains an introduction to elliptic curves and a chapter entitled "Developments and Open Problems", which introduces and brings together various themes oriented toward ongoing mathematical research. Many of the statements about elliptic curves, often coming from courses given at l'Université Paris 7 and the magistère de la rue d'Ulm, are proven, but the panorama proposed in Chap. 6 contains more statements without proof or which are conjectural than proven statements.

On this note, the first four chapters end with a copious list of exercises of varying difficulty; some of them are direct applications of material from the book and others, while not necessarily more difficult, require or develop some aspect not found in the book.

Number theory is a multifaceted and flourishing subject. Every author/ number theorist is condemned to choose between the many themes of this discipline. Our guiding principles in developing the present book were:

- the wish to give some idea of the very large variety of mathematics useful for studying numbers;
- the "necessity" to look at deep and classical themes, such as Gauss sums, Diophantine equations, the distribution of prime numbers and the Riemann zeta function;
- the will to introduce the principle, "arithmetic plays a role in modern applied mathematics". Cryptography and error-correcting codes are introduced and used as a motivation for concepts such as cyclotomic polynomials and the cyclicity of $(\mathbf{Z}/p^m\mathbf{Z})^*$ and $\mathbf{F}_q^*$;
- the effort to include some recent proofs. The polynomial primality algorithm (Agrawal-Kayal-Saxena, 2002) is presented, and its correctness is proven in detail (moreover, the proof is on the level of an advanced

undergraduate course!). The proof that we give of the prime number theorem is essentially due to Newman (1980) and modified by Zagier (1997);

- the desire to approach subjects of contemporary research: elliptic curves, rational points on algebraic varieties, "zeta" and "L" functions, etc.;
- and obviously the incomparable beauty of arithmetic (everybody knows that the others are all jealous of it).

The prerequisites for this text are very modest, at least for the first four chapters: undergraduate algebra is assumed (linear algebra, abelian groups, rings and divisibility), as well as a little topology of $\mathbf{R}^n$ for Chap. 3. In addition to elementary real analysis, Chap. 4 is also based on the theory of complex analysis (holomorphic functions, power series, the Cauchy formula, the residue formula, the complex logarithm) of which we will give a brief overview as a reminder. The first four sections on elliptic curves (Chap. 5) are relatively elementary, even if the material is a little denser than before, and use only simple properties of the projective plane outlined at the beginning of Appendix B. The last section of Chap. 5 and all of Chap. 6 are less accommodating and recall or allude to various more-advanced notions.

We will now finish with a brief description of the individual chapters, which are largely independent of each other.

The 1st chapter, "Finite Structures", provides a systematic study of the congruence groups and rings $\mathbf{Z}/N\mathbf{Z}$ and finite fields $\mathbf{F}_q$, as well as their groups of invertible elements $(\mathbf{Z}/N\mathbf{Z})^*$ and $\mathbf{F}_q^*$. We also confirm the ubiquity of Gauss sums, studied first in their own right, then used to prove the quadratic reciprocity law and to count the number of solutions of diagonal equations over a finite field.

The 2nd chapter, "Applications: Algorithms, Primality and Factorization, Codes", begins with the study of the complexity of basic arithmetic operations (addition, multiplication, computation of the gcd, inversion modulo N, exponentiation, calculations in finite fields). We then briefly introduce the RSA system—the star of public key cryptography procedures—which governs credit cards, internet transactions, etc. This is the motivation for the core of this chapter: the study of algorithms which determine whether an integer is prime or composite. The mathematical prerequisites are those of Chap. 1, plus an elementary statement coming from analytic number theory (which is proven in the first section of Chap. 4). We will also introduce error-correcting codes—used in compact disc technology and the transmission of data—which are another industrial application of number theory and serve as a motivation for the study of the decomposition of cyclotomic polynomials over a finite field.

The 3rd chapter, "Algebra and Diophantine Equations", is an initiation to the study of some classical problems, such as which numbers are expressible as the sum of (two, three or four) squares, integer solutions to Pell's equation $x^2 - dy^2 = 1$ and integer solutions to Fermat's equation $x^n + y^n = z^n$ (done here for $n = 3$ and 4). We then move on to algebraic number theory: number fields, rings of algebraic integers, decomposition of ideals into prime ideals, the group of units and the finiteness of the ideal class group. In addition to commutative algebra, the tools used are a little bit of the geometry of numbers (lattices, Minkowski's theorem) and Diophantine approximations (Dirichlet's theorem and continued fractions).

The 4th chapter, "Analytic Number Theory", is dedicated to the study of the distribution of prime numbers; the two main theorems are the prime number theorem: "the number of prime numbers smaller than x is asymptotically equivalent to $x/\log x$" and the theorem on arithmetic progressions: "there are infinitely many prime numbers congruent to m modulo n when m and n are relatively prime". Apart from some elementary statements (comparison of series and integrals, etc.), the fundamental tool that we use is complex analysis. A brief summary of the necessary tools is included in the chapter. This chapter also introduces a fundamental mathematical object, the "Riemann zeta function", and closes with an introduction to the Riemann hypothesis, which is probably the most important open problem in mathematics.

The 5th chapter, "Elliptic Curves", is an introduction to the rich theory of equations of the type $y^2 = x^3 + ax + b$. We will give you a little bit of projective geometry and examine the group law on a cubic, the theory of heights, notably the Néron-Tate height, before proving the Mordell-Weil theorem: "the group of rational solutions of this equation is a finitely generated abelian group". In the following section, we will prove (modulo a result proved in Chap. 6) Siegel's theorem, "the set of solutions where x and y are integers is finite". We will finish by connecting this to the theory of elliptic functions and by formulating both the extraordinary theorem of Wiles (1995), "Every elliptic curve defined over $\mathbf{Q}$ is modular", and the famous Birch & Swinnerton-Dyer conjecture, which relates the rank of the group of rational solutions to the behavior at $s = 1$ of the Dirichlet series associated to an elliptic curve.

The 6th chapter, "Developments and Open Problems", goes back to some of the subjects of the previous chapters, and pushes them to the level of current research; in particular, each section contains at least one unsolved problem. Of course, some statements must be given without proof, and the prerequisites to read this chapter are more advanced, even though we did make an effort to give all of the necessary definitions and some essential ideas. The six themes that we chose for this chapter are:

- the Weil conjectures, or the computation, which we already started in Chap. 1, of the number of points on an algebraic variety over a finite field. We obtain a precise description of the zeta function of a variety over a finite field and, at the same time, a first glimpse of the connections between arithmetic, geometry and topology;
- the conjectural dictionary, proposed by Serge Lang, between the qualitative properties of the set of rational points on an algebraic variety over a number field and the geometric properties of the variety, as well as the properties of the associated analytic complex variety. For algebraic curves, this dictionary is a theorem and gives us the following trichotomy of curves: curves of genus 0 (conics and the projective line), curves of genus 1 (elliptic curves) and curves of genus $\geqslant 2$ (the others!); however, very little is known for varieties of dimensions at least two;
- an introduction to p-adic numbers, with the goal determining when it is appropriate to apply the "Hasse principle", which, in its most elementary form, asks if an equation $f(x_1, \ldots, x_n) = 0$ has an integral solution whenever it has an integral solution modulo N for every integer N. A key tool in this context, is "Hensel's lemma", which we can consider to be an analogue of Newton's method for finding real-valued solutions of equations. We will also sketch the beginnings of the theory of *adeles* and *ideles*: the global additive and multiplicative groups constructed starting with the local fields $\mathbf{Q}_p$ and $\mathbf{R}$;
- a presentation of the fundamental results of Roth (1955) on rational approximations of algebraic numbers and Baker (1966) on the transcendence of linear combinations of logarithms of algebraic numbers. We will then give the details of the proof of Thue's theorem (a predecessor to and weaker than Roth's theorem) and the method of applying Baker's theorem to Diophantine equations. This provides an opportunity to carefully examine so-called "transcendence" methods and to introduce the reader to problems of computational effectiveness;
- the "a, b, c" conjecture, which is a totally elementary statement and whose proof would have some remarkable consequences, is briefly introduced. Its connections to elliptic curve theory are also presented. This allows us to elaborate on its possible links to the great theorem of Wiles;
- zeta functions associated to algebraic varieties (the function associated to an elliptic curve is described in detail in Chap. 5) are introduced, as well as their connection to the theory of representations of groups. Modular forms and Galois representations also make an appearance. The end of this section touches the tip of the iceberg of Grothendieck's theory of motives and the Langlands program.

Appendix A, entitled "Factorization", follows up on the themes introduced in Chap. 2, but relies on Chaps. 3 (number fields) and 5 (elliptic curves) to describe two recent factorization algorithms for an integer N: Lenstra's algorithm (1986) which uses elliptic curves and an algorithm of Pollard, Lenstra *et al* (1993) called the "number field sieve". We will also briefly discuss the problem of factoring polynomials over a finite field or $\mathbf{Z}$.

Appendix B on "Elementary Projective Geometry" is an introduction to algebraic projective geometry. Some elementary statements on lines, conics and cubics are proven and used in Chap. 5 to construct the group law on a projective plane cubic. We will also describe Hilbert's Nullstellensatz in detail and prove Bézout's theorem: two projective plane curves of degrees d_1 and d_2 with no common components intersect at exactly $d_1 d_2$ points (counted with appropriate multiplicities).

Appendix C, entitled "Galois Theory", is an attempt to fill an intentionally-made gap. We actually avoided relying on any Galois theory in this text (except in the last section of Chap. 6) since it is either absent from the classical university curriculum or taught in the first year of graduate school. It is however such an important tool in modern number theory, that it seemed to us to be necessary to include as a supplement, if only a brief one. We will namely explain how Chebotarev's theorem brings together analytic number theory and Galois groups by generalizing the theorem on arithmetic progressions. This appendix, in particular the description of the concept of a *Galois representation*, is a prerequisite for reading the last section of Chap. 6.

The bibliography is composed of two parts: the first one gives nine reference books which can be read in parallel with this one, as well as commentaries on them; the second part is a more copious list of references to original articles and historical and more advanced books. In [28], you can find a relatively complete overview of the history of number theory up to the beginning of the XXth century. The reference [34] contains numerous open but relatively elementary problems.

Many people were kind enough to take the time to read parts of this book and then bring to my attention remarks on the content as well as the editing. To that effect, I would first like to thank Dominique Bernardi for his careful reading of the entire text. He thus saved the eyes of the happy readers from many misprints and more than one mistake. If there are more of them, only I can take responsibility for them. Olivier Bordellès, Nicolas Ratazzi, Marie-France Vigneras and Michel Waldschmidt suggested some improvements and pointed out some insufficiencies. I owe a large part of Chap. 2 to numerous discussions with Sinnou David and Jean-François Mestre. It would have been very difficult for me to complete this text without the encouragement and suggestions of Alberto Arabia and Rached Mneimné. My mathematical and arithmetical education was nourished for many years by Monday morning lectures by Jean-Pierre Serre at Collège de France.

With this, I thank you all heartily.

Last but not least, this book would not exist without my students, whose listening, reactions, moments of silence and questions often motivated and reorientated me while I was teaching them in front of a blackboard smeared with chalk, the beauties of arithmetic.

Contents

Chapter 1

Finite Structures

"I hope good luck lies in odd numbers. Away! go.
They say there is divinity in odd numbers, either in nativity, chance or death. Away!"
William Shakespeare (The Merry Wives of Windsor)

In this chapter, the theory of congruences will lead into the study of the ring $\mathbf{Z}/n\mathbf{Z}$ *for* $n \geqslant 2$*, as well as the group* $(\mathbf{Z}/n\mathbf{Z})^*$ *of its invertible elements with respect to multiplication. Furthermore, for every power of a prime number,* $q = p^f$*, there exists a unique finite field, up to isomorphism, of cardinality* q*, denoted* $\mathbf{F}_q$*. We will review the construction of these objects and state their main properties. In the following sections, we expand on some structures and applications, notably Gauss sums, Legendre and Jacobi symbols and the number of solutions of congruences.*

1. Review of $\mathbf{Z}/n\mathbf{Z}$, $(\mathbf{Z}/n\mathbf{Z})^*$, $\mathbf{F}_q$ and $\mathbf{F}_q^*$

The group $\mathbf{Z}$ is, up to isomorphism, the only group which is cyclic (generated by one element) and infinite. All of its subgroups are of the type $m\mathbf{Z}$, for $m \geqslant 0$. The set $\mathbf{Z}$ is also equipped with a multiplication which makes it a commutative ring. In this ring, we have the notions of divisibility and of GCD and LCM (greatest common divisor and least common multiple). In the case of $\mathbf{Z}$, the notion of an *ideal* coincides with that of a subgroup. From this, we can easily deduce the following theorem.

1.1. Theorem. (Bézout's lemma) *Let* $m, n \in \mathbf{Z}$ *and let* d *be their GCD. Then there exist* $u, v \in \mathbf{Z}$ *such that*

$$d = um + vn.$$

Proof. The set $H := m\mathbf{Z} + n\mathbf{Z} = \{um + vn \mid u, v \in \mathbf{Z}\}$ is clearly a subgroup, therefore it is of the form $d'\mathbf{Z}$, and there exist u and v such that

M. Hindry, *Arithmetics*, Universitext,
DOI 10.1007/978-1-4471-2131-2_1,

$d' = um + vn$. Since d divides m and n, we see that d divides $um + vn = d'$. But m and n are elements of H, so d' divides m and n, and therefore d' also divides d. It follows then that $d = d'$ (assuming both of them are positive). □

The group $\mathbf{Z}/n\mathbf{Z}$ is, up to isomorphism, the unique cyclic group with n elements, i.e., generated by one element of order n. We will now study the generators of this group.

1.2. Proposition. *Let $m \in \mathbf{Z}$ and let $\bar{m}$ denote its class in $\mathbf{Z}/n\mathbf{Z}$. The following three properties are equivalent.*

i) The element $\bar{m}$ is a generator of $\mathbf{Z}/n\mathbf{Z}$.
ii) The integers m and n are relatively prime.
iii) The integer m is invertible modulo n, in other words, there exists $m' \in \mathbf{Z}$ such that $mm' \equiv 1 \bmod n$ or equivalently $\bar{m}\bar{m}' = 1 \in \mathbf{Z}/n\mathbf{Z}$.

Proof. If $\bar{m}$ generates $\mathbf{Z}/n\mathbf{Z}$, then there exists $m' \in \mathbf{Z}$ such that $m'\bar{m} = 1 \in \mathbf{Z}/n\mathbf{Z}$; hence $mm' \equiv 1 \bmod n$, which means that m is invertible modulo n. If $mm' \equiv 1 \bmod n$, then $mm' = 1 + an$, and therefore m is relatively prime to n. If m is relatively prime to n, then by Bézout's lemma, there exist a and b such that $am + bn = 1$, hence $a\bar{m} = 1 \in \mathbf{Z}/n\mathbf{Z}$, and therefore $\bar{m}$ generates $\mathbf{Z}/n\mathbf{Z}$. □

The group of invertible elements of the ring $\mathbf{Z}/n\mathbf{Z}$ is therefore equal to

$$\begin{aligned}(\mathbf{Z}/n\mathbf{Z})^* &= \{\bar{m} \in \mathbf{Z}/n\mathbf{Z} \mid m \text{ is relatively prime to } n\} \\ &= \{\text{generators of } \mathbf{Z}/n\mathbf{Z}\}.\end{aligned}$$

1.3. Definition. We denote by $\phi(n) := \operatorname{card}(\mathbf{Z}/n\mathbf{Z})^*$ *the Euler totient* of the integer n.

By noticing that $\gcd(m, p^r) = \gcd(m, p)$, we can easily deduce that if p is prime, $\phi(p^r) = p^r - p^{r-1} = (p-1)p^{r-1}$. In general, to calculate $\phi(n)$, we make use of the following classical lemma.

1.4. Proposition. (Chinese remainder theorem) *Let $m, n \in \mathbf{Z}$, and suppose that m and n are relatively prime. Then the groups $\mathbf{Z}/mn\mathbf{Z}$ and $\mathbf{Z}/m\mathbf{Z} \times \mathbf{Z}/n\mathbf{Z}$ are naturally isomorphic. Furthermore, this isomorphism is also a ring isomorphism and consequently induces an isomorphism of $(\mathbf{Z}/mn\mathbf{Z})^*$ and $(\mathbf{Z}/m\mathbf{Z})^* \times (\mathbf{Z}/n\mathbf{Z})^*$. In particular, $\phi(mn) = \phi(m)\phi(n)$.*

Proof. Consider the map $f : \mathbf{Z} \to \mathbf{Z}/m\mathbf{Z} \times \mathbf{Z}/n\mathbf{Z}$ given by $x \mapsto (x \bmod m, x \bmod n)$. It is a group homomorphism with kernel $\operatorname{lcm}(m, n)\mathbf{Z}$, hence we have the injective map

$$\hat{f} : \mathbf{Z}/\operatorname{lcm}(m,n)\mathbf{Z} \hookrightarrow \mathbf{Z}/m\mathbf{Z} \times \mathbf{Z}/n\mathbf{Z}.$$

Since m and n are relatively prime, $\operatorname{lcm}(m,n) = mn$, and by considering the cardinalities the two groups, then homomorphism $\hat{f}$ must be an isomorphism. For any rings A and B, we have $(A \times B)^* = A^* \times B^*$, hence the second assertion. □

1.5. Remark. A function $f : \mathbf{N}^* \to \mathbf{C}$ is generally known as an *arithmetic function*. We say that an arithmetic function $f : \mathbf{N}^* \to \mathbf{C}$ is *multiplicative* (resp. *completely multiplicative*) if $f(mn) = f(m)f(n)$ for all m, n which are relatively prime (resp. for all m, n). Thus the Euler totient ϕ is multiplicative but not completely multiplicative; notice however that $\phi(mn)$ is always greater than or equal to $\phi(m)\phi(n)$.

The description of the subgroups of $\mathbf{Z}/n\mathbf{Z}$ is fairly simple.

1.6. Proposition. *For any integer $d \geqslant 1$ which divides n, there exists a unique subgroup of $\mathbf{Z}/n\mathbf{Z}$ of order d: namely, the cyclic subgroup generated by the class of n/d in $\mathbf{Z}/n\mathbf{Z}$.*

Proof. Assume $n = dd'$. The element $x = \bar{d'} \in \mathbf{Z}/n\mathbf{Z}$ is therefore of order d since obviously $dx = 0$, and if $cx = 0$, then n divides cd', so d divides c. Now let H be a subgroup of $\mathbf{Z}/n\mathbf{Z}$ of order d. Let s be the canonical surjection $s : \mathbf{Z} \to \mathbf{Z}/n\mathbf{Z}$. We know that $s^{-1}(H) = m\mathbf{Z}$ is generated by m, hence H is generated by $\bar{m} \in \mathbf{Z}/n\mathbf{Z}$. We then have $d\bar{m} = 0$, hence n divides dm, and therefore d' divides m, so the subgroup H is contained in the subgroup generated by $\bar{d'}$ and is therefore equal to this subgroup. □

An application of this proposition is the following formula (that we will use further down):

$$n = \sum_{d \mid n} \phi(d). \tag{1.1}$$

To see why this is true, we write $\mathbf{Z}/n\mathbf{Z}$ as the disjoint union of sets, where each set contains the elements of order d and d divides n. The number of elements in each set is the number of generators of the unique subgroup of order d, and since the latter is isomorphic to $\mathbf{Z}/d\mathbf{Z}$, the number of such generators is $\phi(d)$.

A finite field k necessarily has finite characteristic equal to a prime number p and therefore contains $\mathbf{Z}/p\mathbf{Z} = \mathbf{F}_p$ (the homomorphism $\mathbf{Z} \to k$ has kernel $n\mathbf{Z}$ with $n > 0$, and since $\mathbf{Z}/n\mathbf{Z} \hookrightarrow k$, n must be prime). The dimension of k, viewed as a vector space over $\mathbf{F}_p$, is finite, say f, and therefore $\mathrm{card}(k) = p^f$. We know that $\mathrm{card}(k^*) = p^f - 1$, so all of the elements of k^* satisfy $x^{p^f-1} = 1$, and therefore all of the elements of k satisfy $x^{p^f} = x$. Conversely, we can construct a finite field with cardinality p^f as follows: we consider an extension K of $\mathbf{F}_p = \mathbf{Z}/p\mathbf{Z}$ in which the polynomial $P = X^{p^f} - X$ splits completely into p^f linear factors. We then set $k := \{x \in K \mid P(x) = 0\}$. Since $P'(X) = -1$, the roots of P are simple and $\mathrm{card}(k) = \deg(P) = p^f$; furthermore, k is a subfield of K because in characteristic p, the "Frobenius map" given by $\phi(x) = x^p$ is a homomorphism of fields; the same holds true for ϕ^f. In other words we have:

$$(xy)^p = x^p y^p \qquad \text{and} \qquad (x+y)^p = x^p + y^p.$$

From general field theory, we know that the field k of order p^f is unique, up to isomorphism, and is denoted by $\mathbf{F}_{p^f}$. The following statement summarizes these notions.

1.7. Theorem. *Let p be a prime number, $f \geqslant 1$ and $q = p^f$. There exists a unique finite field, up to isomorphism, of order q. The elements of $\mathbf{F}_q$ are the roots of the polynomial $X^q - X \in \mathbf{Z}/p\mathbf{Z}[X]$.*

1.8. Corollary.[1] *Let $q = p^f$ and $\mathbf{F}_q$ the field defined above. The subfields of $\mathbf{F}_q$ are isomorphic to $\mathbf{F}_{p^d}$, where d divides f. Conversely, if d divides f, there exists a unique subfield of $\mathbf{F}_q$ isomorphic to $\mathbf{F}_{p^d}$: it is exactly the set of elements which satisfy $x^{p^d} = x$.*

Proof. If $\mathbf{F}_p \subset k \subset \mathbf{F}_q$, then $\mathrm{card}(k) = p^d$ with $d = [k : \mathbf{F}_p]$, and $k \cong \mathbf{F}_{p^d}$. Furthermore, $f = [\mathbf{F}_q : \mathbf{F}_p] = [\mathbf{F}_q : k][k : \mathbf{F}_p]$, and therefore d divides f. Conversely, if d divides f, $f = ed$, then every element (in an extension of $\mathbf{F}_p$) which satisfies $x^{p^d} = x$ also satisfies $x^{p^f} = x^{p^{ed}} = x$ and is therefore in $\mathbf{F}_q$. These elements form a subfield isomorphic to $\mathbf{F}_{p^d}$. □

In practice, we construct the fields $\mathbf{F}_{p^f}$ as follows: we choose an irreducible, monic polynomial of degree f, say $P \in \mathbf{F}_p[X]$ (the existence of such a polynomial is equivalent to the existence of an element $\alpha \in \mathbf{F}_{p^f}$ such that $\mathbf{F}_{p^f} = \mathbf{F}_p(\alpha)$ and is guaranteed by invoking, for example, Lemma 1-2.1 below) and we represent $\mathbf{F}_{p^f}$ as $\mathbf{F}_p[X]/P\mathbf{F}_p[X]$. An element of $\mathbf{F}_{p^f}$ can be seen as a polynomial of degree $\leqslant f-1$ with coefficients in $\mathbf{Z}/p\mathbf{Z}$. Addition is the obvious addition, and the multiplication rule is simply polynomial multiplication, followed by taking the remainder gotten from the division

[1] This statement can be reinterpreted in terms of Galois theory (see Appendix C).

algorithm. For example,

$$\mathbf{F}_4 = \mathbf{F}_2[X]/(X^2 + X + 1)\mathbf{F}_2[X],\ \mathbf{F}_8 = \mathbf{F}_2[X]/(X^3 + X + 1)\mathbf{F}_2[X],$$
$$\mathbf{F}_{16} = \mathbf{F}_2[X]/(X^4 + X^3 + X^2 + X + 1)\mathbf{F}_2[X].$$

2. The Group Structure of $(\mathbf{Z}/n\mathbf{Z})^*$ and $\mathbf{F}_q^*$

In order to describe the structure of these groups, we start by proving the following lemma, which is interesting in and of itself.

2.1. Lemma. *Let k be a field and G a finite subgroup of k^*. Then G is cyclic. In particular, $(\mathbf{Z}/p\mathbf{Z})^*$ or more generally $\mathbf{F}_q^*$ is cyclic.*

Proof. Set $n := \mathrm{card}(G)$, and let $\psi(d)$ be the number of elements of order d in G. It is clear that $n = \sum_{d\mid n}\psi(d)$. Let d be an integer which divides n: either there are no elements of order d in G in which case $\psi(d) = 0$, or there exists one which generates a cyclic subgroup H of order d. All of the elements of H are solutions to the equation $X^d = 1$, but since k is a field, such an equation has at most d roots in k; all of the elements of order d are therefore in H, and there are $\phi(d)$ of them because $H \cong \mathbf{Z}/d\mathbf{Z}$. Hence $\psi(d)$ is either zero or $\phi(d)$, but since $n = \sum_{d\mid n}\psi(d) = \sum_{d\mid n}\phi(d)$ (by (1.1)), we see that $\psi(d) = \phi(d)$ for every d which divides n. In particular, $\psi(n) = \phi(n) \geqslant 1$, which implies that G is cyclic. □

From what we have seen, if $n = p_1^{\alpha_1}\cdots p_s^{\alpha_s}$, then

$$(\mathbf{Z}/n\mathbf{Z})^* \cong (\mathbf{Z}/p_1^{\alpha_1}\mathbf{Z})^* \times \cdots \times (\mathbf{Z}/p_s^{\alpha_s}\mathbf{Z})^*,$$

and in particular

$$\phi(n) = \phi(p_1^{\alpha_1})\cdots\phi(p_s^{\alpha_s}) = \prod_{i=1}^{s}\left(p_i^{\alpha_i} - p_i^{\alpha_i-1}\right) = n\prod_{i=1}^{s}\left(1 - \frac{1}{p_i}\right). \qquad (1.2)$$

We will now describe the structure of the groups $(\mathbf{Z}/p^\alpha\mathbf{Z})^*$.

2.2. Proposition. *Let p be prime and $\alpha \geqslant 1$.*

i) If p is odd, then $(\mathbf{Z}/p^\alpha\mathbf{Z})^*$ is cyclic.
ii) If $p = 2$ and $\alpha \geqslant 3$, then $(\mathbf{Z}/2^\alpha\mathbf{Z})^* \cong \mathbf{Z}/2^{\alpha-2}\mathbf{Z} \times \mathbf{Z}/2\mathbf{Z}$, which is not cyclic. However, $(\mathbf{Z}/2\mathbf{Z})^* = \{1\}$ and $(\mathbf{Z}/4\mathbf{Z})^* \cong \mathbf{Z}/2\mathbf{Z}$ are cyclic.

Proof. If $\alpha = 1$, we have seen that $(\mathbf{Z}/p\mathbf{Z})^* = \mathbf{F}_p^*$ is cyclic. When $\alpha > 1$, we use the element $p + 1$.

2.3. Lemma. *Let p be an odd prime. The class of $p+1$ in $(\mathbf{Z}/p^\alpha\mathbf{Z})^*$ has order $p^{\alpha-1}$.*

Proof. (of Lemma 1-2.3) We first prove the congruence

$$(p+1)^{p^k} \equiv 1 + p^{k+1} \bmod p^{k+2}$$

by induction. For $k = 0$, the congruence is trivial. For $k = 1$, we have $(p+1)^p \equiv 1 + \binom{p}{1}p + \binom{p}{2}p^2 \equiv 1 + p^2 + p^3(p-1)/2 \bmod p^3$, and the latter is of course congruent to $1 + p^2$ if p is odd (notice however that $3^2 \not\equiv 1 + 2^2 \bmod 2^3$). Assume now that $k \geqslant 1$ and $(p+1)^{p^{k-1}} = 1 + p^k + ap^{k+1}$. Then $(p+1)^{p^k} = \left(1 + p^k + ap^{k+1}\right)^p \equiv 1 + p(p^k + ap^{k+1}) \equiv 1 + p^{k+1} \bmod p^{k+2}$ since $1 + 2k \geqslant k + 2$. In particular, we see that $(p+1)^{p^{\alpha-1}} \equiv 1 \bmod p^\alpha$, but $(p+1)^{p^{\alpha-2}} \equiv 1 + p^{\alpha-1} \not\equiv 1 \bmod p^\alpha$, which implies that $p+1$ has order $p^{\alpha-1}$ in $(\mathbf{Z}/p^\alpha\mathbf{Z})^*$. □

We can now finish the proof of the proposition for p odd. Let $x \in \mathbf{Z}$ such that x modulo p generates $(\mathbf{Z}/p\mathbf{Z})^*$, i.e., has order $p-1$ in $(\mathbf{Z}/p\mathbf{Z})^*$. Therefore $\bar{x}$ has order $m(p-1)$ in $(\mathbf{Z}/p^\alpha\mathbf{Z})^*$, and hence $y = \bar{x}^m$ has order exactly $p-1$ in $(\mathbf{Z}/p^\alpha\mathbf{Z})^*$. The element $u := y(p+1)$ therefore has order $p^{\alpha-1}(p-1)$ because $p^{\alpha-1}$ and $p-1$ are relatively prime, which gives us that u is a generator of $(\mathbf{Z}/p^\alpha\mathbf{Z})^*$.

2.4. Lemma. *Let $\alpha \geqslant 3$. The class of 5 in $(\mathbf{Z}/2^\alpha\mathbf{Z})^*$ has order $2^{\alpha-2}$. Furthermore, the class of -1 does not belong to the subgroup generated by the class of 5.*

Proof. (of Lemma 1-2.4) We first show by induction that

$$5^{2^k} \equiv 1 + 2^{k+2} \bmod 2^{k+3}.$$

The congruence is trivial for $k = 0$, and for $k = 1$ we check that $25 = 5^2 \equiv 1 + 2^3 = 9 \bmod 2^4$. Therefore, we can assume that $5^{2^{k-1}} = 1 + 2^{k+1} + a2^{k+2}$. Then $5^{2^k} = (1 + 2^{k+1} + a2^{k+2})^2 = 1 + 2(2^{k+1} + a2^{k+2}) + 2^{2(k+1)}(1+2a)^2 \equiv 1 + 2^{k+2} \bmod 2^{k+3}$. In particular, $5^{2^{\alpha-2}} \equiv 1 \bmod 2^\alpha$, but $5^{2^{\alpha-3}} \equiv 1 + 2^{\alpha-1} \not\equiv 1 \bmod 2^\alpha$, so 5 has order $2^{\alpha-2}$. For the second assertion, observe that for every integer m, we have $5^m \equiv 1 \not\equiv -1 \bmod 4$. □

For the proof of the second part of the proposition, we can assume that $\alpha \geqslant 3$ (actually, we see immediately how to calculate $(\mathbf{Z}/2\mathbf{Z})^*$ and $(\mathbf{Z}/4\mathbf{Z})^*$). The class of 5 therefore generates a subgroup isomorphic to $\mathbf{Z}/2^{\alpha-2}\mathbf{Z}$, and -1 generates a subgroup of order 2 not contained in the former. Therefore, $(\mathbf{Z}/2^\alpha\mathbf{Z})^* = \langle 5 \rangle \oplus \langle -1 \rangle \cong \mathbf{Z}/2^{\alpha-2}\mathbf{Z} \times \mathbf{Z}/2\mathbf{Z}$.

2.5. Remark. The quaternion subgroup $H_8 = \{\pm 1, \pm i, \pm j, \pm k\}$ is a finite subgroup of the multiplicative group of the division ring $\mathbf{H}$ but is not cyclic (which does not contradict Lemma 1-2.1 because $\mathbf{H}$ is not commutative).

Applications. The previous statements allow us to find the number of solutions to the equation $x^m = 1$ in $\mathbf{F}_q^*$ or $(\mathbf{Z}/N\mathbf{Z})^*$, as well as the number of mth powers. This is true because in a cyclic group of order n, say $G = \mathbf{Z}/n\mathbf{Z}$, the number of elements which satisfy $mx = 0$ is equal to $d := \gcd(m,n)$: by making use of Bézout's lemma, we can show that $\{x \in \mathbf{Z}/n\mathbf{Z} \mid mx = 0\}$ is equal to $\{x \in \mathbf{Z}/n\mathbf{Z} \mid dx = 0\}$, and since d divides n, the latter set is the cyclic subgroup of order d in $\mathbf{Z}/n\mathbf{Z}$. By applying this to $G = \mathbf{F}_q^*$ or $G = (\mathbf{Z}/p^\alpha\mathbf{Z})^*$, we get the first part of the following proposition.

2.6. Proposition. *Let m be an integer $\geqslant 1$.*

1) We have the following formulas:

- $\operatorname{card}\{x \in \mathbf{F}_q^* \mid x^m = 1\} = \gcd(m, q-1)$*;*
- $\operatorname{card}\{x \in (\mathbf{Z}/p^\alpha\mathbf{Z})^* \mid x^m = 1\} = \gcd(m, (p-1)p^{\alpha-1})$ *(for p odd).*

2) More generally, if $N = p_1^{\alpha_1} \cdots p_r^{\alpha_r}$ is odd,

$$\operatorname{card}\{x \in (\mathbf{Z}/N\mathbf{Z})^* \mid x^m = 1\} = \prod_{i=1}^{r} \gcd(m, (p_i - 1)p_i^{\alpha_i - 1}).$$

Proof. The formulas in part *1)* follow from the previous discussion and from the fact that $\mathbf{F}_q^*$ and $(\mathbf{Z}/p^\alpha\mathbf{Z})^*$ are cyclic. Formula *2)* follows from the previous formula and from the Chinese remainder theorem. This is because for all $x \in \mathbf{Z}$, $x^m \equiv 1 \bmod N$ is equivalent to $x^m \equiv 1 \bmod p_i^{\alpha_i}$ for $1 \leqslant i \leqslant r$. □

2.7. Remark. By considering the homomorphism $x \mapsto x^m$, we can easily see that

$$\operatorname{card} \mathbf{F}_q^{*m} = \operatorname{card}\{x \in \mathbf{F}_q^* \mid \exists y \in \mathbf{F}_q^*,\ x = y^m\} = \frac{q-1}{\gcd(m, q-1)}.$$

For example, if q is odd, we have $(\mathbf{F}_q^* : \mathbf{F}_q^{*2}) = 2$.

3. Jacobi and Legendre Symbols

In this section, we mainly concentrate on the study of squares, i.e., the case $m = 2$ of the preceding section.

We begin with a remark. The map $x \mapsto x^2$ is an isomorphism from $\mathbf{F}_2$ to $\mathbf{F}_2$ or more generally from $\mathbf{F}_{2^f}$ to $\mathbf{F}_{2^f}$; in order to study squares, it is therefore natural to assume $p \neq 2$, and that is what we do.

3.1. Definition. We define the *Legendre symbol* for $a \in \mathbf{Z}$ and $p \neq 2$ as follows:

$$\left(\frac{a}{p}\right) := \begin{cases} 0 & \text{if } a \equiv 0 \bmod p, \\ +1 & \text{if } a \text{ is a non-zero square} \bmod p, \\ -1 & \text{if } a \text{ is not a square} \bmod p. \end{cases}$$

3.2. Remark. It is clear that $\left(\frac{a}{p}\right)$ only depends on $a \bmod p$, thus we will continue to use the same notation whenever $a \in \mathbf{F}_p$. If $\left(\frac{a}{p}\right) = +1$, we say that a is a *quadratic residue*; if $\left(\frac{a}{p}\right) = -1$, we say that a is a *quadratic nonresidue*.

3.3. Theorem. *The Legendre symbol satisfies the following properties.*

i) *For any* $a, b \in \mathbf{Z}$,

$$\left(\frac{ab}{p}\right) = \left(\frac{a}{p}\right)\left(\frac{b}{p}\right).$$

ii) *For every* $a \in \mathbf{Z}$,

$$a^{(p-1)/2} \equiv \left(\frac{a}{p}\right) \bmod p.$$

iii) *For every* $p \neq 2$,

$$\left(\frac{-1}{p}\right) = (-1)^{(p-1)/2} \qquad \textit{and} \qquad \left(\frac{2}{p}\right) = (-1)^{(p^2-1)/8}.$$

In particular, -1 *is a square modulo* p *(resp. not a square) if* $p \equiv 1 \bmod 4$ *(resp.* $p \equiv 3 \bmod 4$*), and* 2 *is a square modulo* p *(resp. not a square) if* $p \equiv \pm 1 \bmod 8$ *(resp.* $p \equiv \pm 3 \bmod 8$*).*

iv) (Quadratic reciprocity law) *Let* p *and* q *be two distinct prime numbers. Then we have*

$$\left(\frac{q}{p}\right)\left(\frac{p}{q}\right) = (-1)^{\frac{(p-1)(q-1)}{4}}.$$

Proof. The multiplicativity in part *i)* is clear if p divides a or b, since then the two terms are 0. If $a, b \in \mathbf{F}_p^*$, the formula comes from the fact that $\mathbf{F}_p^*/\mathbf{F}_p^{*2}$ is of order 2, so the product of the two quadratic nonresidues is a quadratic residue.

To prove *ii)*, we observe that since $(a^{(p-1)/2})^2 = a^{p-1} = 1$, we always have $a^{(p-1)/2} = \pm 1$, and by Proposition 1-2.6, the subgroup H of elements satisfying $a^{(p-1)/2} = 1$ is of order $(p-1)/2$. In addition, the set of squares is a subgroup of order $(p-1)/2$. Furthermore, if $a = b^2$, we can deduce that $a^{(p-1)/2} = b^{p-1} = 1$, hence $\mathbf{F}_p^{*2} \subset H$, and we have the desired equality.

The first part of *iii)* follows from equality *ii)*. For the second part, we introduce α, a root of $X^4 + 1 = 0$; it is an 8th primitive root of unity in an algebraic extension of $\mathbf{F}_p$, in other words $\alpha^8 = 1$ but $\alpha^4 \neq 1$, which is equivalent to $\alpha^4 = -1$ and also $\alpha^2 = -\alpha^{-2}$. If we set $\beta := \alpha + \alpha^{-1}$, then $\beta^2 = \alpha^2 + 2 + \alpha^{-2} = 2$; thus we see that 2 is a square in $\mathbf{F}_p$ if and only if $\beta \in \mathbf{F}_p$. We know that $\beta \in \mathbf{F}_p$ is equivalent to $\beta^p = \beta$, so we want to compute $\beta^p = \alpha^p + \alpha^{-p}$. By using the fact that $\alpha^8 = 1$ and $\alpha^4 = -1$, we see that if $p \equiv \pm 1 \bmod 8$, then $\beta^p = \beta$ so $\beta \in \mathbf{F}_p$, whereas if $p \equiv \pm 3 \bmod 8$, we have $\beta^p = -\beta$ and therefore $\beta \notin \mathbf{F}_p$. We will postpone the proof of the quadratic reciprocity law *iv)* until the next section. □

3.4. Remark. To see where the choice of "$\beta = \sqrt{2}$" comes from, notice that if $\zeta := \exp(2\pi i/8) \in \mathbf{C}$, then ζ is an 8th root of unity and $\zeta = \frac{\sqrt{2}}{2} + i\frac{\sqrt{2}}{2}$, so $\zeta + \zeta^{-1} = \zeta + \bar{\zeta} = \sqrt{2}$.

The *Jacobi symbol* is a generalization for odd $N = p_1^{\alpha_1} \cdots p_r^{\alpha_r}$ and is given by

$$\left(\frac{a}{N}\right) := \left(\frac{a}{p_1}\right)^{\alpha_1} \cdots \left(\frac{a}{p_r}\right)^{\alpha_r} \tag{1.3}$$

Its main properties are stated in the following lemma.

3.5. Lemma. *For N, M odd:*

i) $\left(\frac{ab}{N}\right) = \left(\frac{a}{N}\right)\left(\frac{b}{N}\right)$ *and* $\left(\frac{a}{N}\right) = 0$ *if and only if* $\gcd(a, N) > 1$;

ii) $\left(\frac{-1}{N}\right) = (-1)^{\frac{N-1}{2}}$ *and* $\left(\frac{2}{N}\right) = (-1)^{\frac{N^2-1}{8}}$;

iii) $\left(\frac{M}{N}\right) = (-1)^{\frac{(N-1)(M-1)}{4}}\left(\frac{N}{M}\right)$.

Proof. These formulas can be deduced from the analogous formulas for prime numbers M and N. Statement *i)* is clearly true. To prove *ii)* and *iii)*, we write $N = p_1 \cdots p_r$ (with possible repetitions), so that

$$\left(\frac{-1}{N}\right) = \prod_{i=1}^{r}\left(\frac{-1}{p_i}\right) = (-1)^{\sum_{i=1}^{r}(p_i-1)/2} = (-1)^h,$$

with h being equal to the number of indices i where $p_i \equiv 3 \bmod 4$. Furthermore, $N \equiv 3^h \bmod 4$, hence $N \equiv 3 \bmod 4$ if h is odd and $N \equiv 1 \bmod 4$ if h is even; thus we have $\frac{N-1}{2} \equiv h \bmod 2$. Likewise,

$$\left(\frac{2}{N}\right) = \prod_{i=1}^{r}\left(\frac{2}{p_i}\right) = (-1)^{\sum_{i=1}^{r}(p_i^2-1)/8} = (-1)^h,$$

where h is now the number of indices i with $p_i \equiv \pm 3 \bmod 8$. In this case, we have $N \equiv \pm 3^h \bmod 8$. Therefore, $\dfrac{N^2-1}{8} \equiv \dfrac{9^h-1}{8} \equiv h \bmod 2$, which proves the second formula of *ii*).

In order to prove assertion *iii*), we write $M = q_1 \cdots q_s$ and $N = p_1 \cdots p_r$ (with possible repetitions). If h (resp. k) is the number of indices i such that $p_i \equiv 3 \bmod 4$ (resp. $q_j \equiv 3 \bmod 4$), then $\dfrac{N-1}{2}$ is odd if h is odd and $\dfrac{N-1}{2}$ is even if h is even (resp. $\dfrac{M-1}{2}$ is odd if k is odd and $\dfrac{M-1}{2}$ is even if k is even). In other words, $\dfrac{N-1}{2} \equiv h \bmod 2$ and $\dfrac{M-1}{2} \equiv k \bmod 2$. We can deduce from this that

$$\left(\frac{M}{N}\right) = \prod_{i=1}^{r}\prod_{j=1}^{s}\left(\frac{q_j}{p_i}\right) = \prod_{i=1}^{r}\prod_{j=1}^{s}(-1)^{(p_i-1)(q_j-1)/4}\left(\frac{p_i}{q_j}\right)$$
$$= (-1)^{hk}\left(\frac{N}{M}\right) = (-1)^{\frac{(N-1)(M-1)}{4}}\left(\frac{N}{M}\right). \qquad \square$$

Statement *ii*) is *Jacobi's reciprocity law.* The two properties provide an algorithm for calculating the Jacobi symbol. Pay attention however to the fact that the Jacobi symbol does not characterize squares modulo N (if a is relatively prime to N and a square modulo N, then $\left(\frac{a}{N}\right) = 1$, but the converse is not true when N is not prime).

As a first application of the quadratic reciprocity law, we will prove that if d is a square-free integer, the prime numbers which can be written in the form $p = x^2 + dy^2$ satisfy certain congruences modulo $4d$.

To be more precise, if $d = \epsilon p_1 \cdots p_k$ (where $\epsilon = \pm 1$) and $p = x^2 + dy^2$, then p does not divide y because if so p would also divide x, and we could then conclude that p^2 divides p. Therefore, we know that $-d = (xy^{-1})^2 \bmod p$, and if d is odd, then

$$1 = \left(\frac{-d}{p}\right) = (-\epsilon)^{\frac{p-1}{2}}(-1)^{\sum_{i=1}^{k}(p_i-1)(p-1)/4}\left(\frac{p}{p_1}\right)\cdots\left(\frac{p}{p_k}\right).$$

We therefore obtain congruences for p modulo $4p_1 \cdots p_k$. If d is even, we set $p_1 = 2$ and separately calculate $\left(\frac{2}{p}\right) = (-1)^{\frac{p^2-1}{8}}$, thus obtaining congruences for p modulo $8p_2 \cdots p_k$.

3.6. Example. If a prime number can be written as $p = x^2 - 6y^2$, with $x, y \in \mathbf{Z}$, then $\left(\frac{6}{p}\right) = 1$ and also $1 = (-1)^{(p^2-1)/8}(-1)^{(p-1)/2}\left(\frac{p}{3}\right)$, which is equivalent to $p \equiv 1$ or $3 \bmod 8$ and $p \equiv 1 \bmod 3$, or also to $p \equiv -1$

or $-3 \bmod 8$ and $p \equiv -1 \bmod 3$. By the Chinese remainder theorem, we can then conclude that $p \equiv 1, 5, 19$ or $23 \bmod 24$. Thus there is no prime number $p \equiv 7, 11, 13$ or $17 \bmod 24$ which can be written $p = x^2 - 6y^2$.

4. Gauss Sums

Gauss sums are important in arithmetic; we are going to use them to give a proof (due to Gauss of course) of the quadratic reciprocity law. In the following section, we will use them to calculate the number of solutions modulo p of a quadratic equation.

Observe that $\exp\left(\frac{2\pi i a}{p}\right)$ only depends on $a \bmod p$, hence this expression is well-defined for $a \in \mathbf{F}_p$. We will use the following formulas, and leave the proof of them as an instructive exercise.

$$\sum_{x\in\mathbf{F}_p}\left(\frac{x}{p}\right) = \sum_{x\in\mathbf{F}_p^*}\left(\frac{x}{p}\right) = 0 \quad \text{and}$$

$$\sum_{x=0}^{n-1}\exp\left(\frac{2\pi i x y}{n}\right) = \begin{cases} n & \text{if } n \text{ divides } y, \\ 0 & \text{if } n \text{ does not divide } y. \end{cases}$$

The first example of a Gauss sum that we will look at is the following, where p is an odd prime and a is relatively prime to p:

$$\tau(a) := \sum_{x=0}^{p-1}\exp\left(\frac{2\pi i a x^2}{p}\right).$$

4.1. Proposition. *The sums $\tau(a)$ satisfy the following formulas.*

i) $\tau(a) = \left(\frac{a}{p}\right)\tau(1)$.

ii) $|\tau(a)|^2 = p$.

iii) $\tau(1)^2 = \left(\frac{-1}{p}\right)p$.

Proof. If a is a square, then $a\mathbf{F}_p^{*2} = \mathbf{F}_p^{*2}$ hence $\tau(a) = \tau(1)$. Let a be a quadratic residue and b a quadratic nonresidue modulo p.

$$\begin{aligned}
\tau(a) + \tau(b) &= \sum_{x=0}^{p-1}\exp\left(\frac{2\pi i a x^2}{p}\right) + \sum_{x=0}^{p-1}\exp\left(\frac{2\pi i b x^2}{p}\right) \\
&= 2 + 2\sum_{u\in a\mathbf{F}_p^{*2}}\exp\left(\frac{2\pi i u}{p}\right) + 2\sum_{u\in b\mathbf{F}_p^{*2}}\exp\left(\frac{2\pi i u}{p}\right) \\
&= 2\sum_{u\in\mathbf{F}_p}\exp\left(\frac{2\pi i u}{p}\right) = 0.
\end{aligned}$$

Hence we have $\tau(b) = -\tau(a) = -\tau(1)$, which proves *i*). For the second formula, we can do the calculation in two ways: $\sum_{a=1}^{p-1} |\tau(a)|^2 = (p-1)|\tau(1)|^2$ which is also equal to

$$\sum_{a=1}^{p-1} \sum_{x,y \in \mathbf{F}_p} \exp\left(\frac{2\pi i a(x^2-y^2)}{p}\right) = \sum_{a=1}^{p-1} \sum_{u,v \in \mathbf{F}_p} \exp\left(\frac{2\pi i a u v}{p}\right)$$
$$= \sum_{a=1}^{p-1} p = p(p-1),$$

and so we have formula *ii*). Finally, we know that $\overline{\tau(1)} = \tau(-1) = \left(\frac{-1}{p}\right)\tau(1)$, hence $\tau(1)^2 = \left(\frac{-1}{p}\right)|\tau(1)|^2 = \left(\frac{-1}{p}\right)p$. □

4.2. Remark. Formula *iii*) allows us to deduce that if $p \equiv 1 \bmod 4$, then $\tau(1) = \pm\sqrt{p}$, whereas if $p \equiv 3 \bmod 4$, $\tau(1) = \pm i\sqrt{p}$. We can actually show (the proof is a little tricky, see Exercise 1-6.13) that it is always positive. For example,

$$\tau_3(1) = \sum_{x=0}^{2} \exp\left(\frac{2\pi i x^2}{3}\right) = 1+2\exp\left(\frac{2\pi i}{3}\right) = 1+2\left(-\frac{1}{2}+i\frac{\sqrt{3}}{2}\right) = i\sqrt{3},$$

$$\tau_5(1) = \sum_{x=0}^{4} \exp\left(\frac{2\pi i x^2}{5}\right) = 1+2\exp\left(\frac{2\pi i}{5}\right) + 2\exp\left(\frac{-2\pi i}{5}\right)$$
$$= 1+4\cos\left(\frac{2\pi}{5}\right) = 1+4\left(-\frac{1}{4}+\frac{\sqrt{5}}{4}\right) = \sqrt{5}.$$

We can express the sums in another way by proving the following lemma.

4.3. Lemma. *The following equality holds:*

$$\tau(a) = \sum_{x \in \mathbf{F}_p} \left(\frac{x}{p}\right) \exp\left(\frac{2\pi i a x}{p}\right) = \sum_{x \in \mathbf{F}_p^*} \left(\frac{x}{p}\right) \exp\left(\frac{2\pi i a x}{p}\right).$$

Proof. Notice that $1 + \left(\frac{x}{p}\right)$ is equal to the number of solutions in $\mathbf{F}_p$ to the equation $y^2 = x$. This gives us:

$$\sum_{x \in \mathbf{F}_p} \left(\frac{x}{p}\right) \exp\left(\frac{2\pi i a x}{p}\right) = \sum_{x \in \mathbf{F}_p} \left(1+\left(\frac{x}{p}\right)\right) \exp\left(\frac{2\pi i a x}{p}\right)$$
$$= \sum_{y \in \mathbf{F}_p} \exp\left(\frac{2\pi i a y^2}{p}\right) = \tau(a),$$

as desired. □

This leads into the first generalization. We define a *character* as a homomorphism $\chi : \mathbf{F}_p^* \to \mathbf{C}^*$. We generally refer to the constant function, equal to 1, as a *unitary* character, a *principal* character or even a *trivial* character; it is denoted by χ_0. By convention, we extend each character to all of $\mathbf{F}_p$ by $\chi(0) := 0$ if $\chi \neq \chi_0$ and $\chi_0(0) = 1$.
Therefore, for a relatively prime to p, we let

$$G(\chi, a) = \sum_{x\in\mathbf{F}_p} \chi(x) \exp\left(\frac{2\pi i a x}{p}\right) = \sum_{x\in\mathbf{F}_p^*} \chi(x) \exp\left(\frac{2\pi i a x}{p}\right)$$

and prove the following.

4.4. Proposition. *The sums $G(\chi, a)$ satisfy the following formulas.*

i) $G(\chi, a) = \bar{\chi}(a) G(\chi, 1)$.
ii) $|G(\chi, a)|^2 = p$ *(if χ is not a trivial character).*
iii) $\overline{G(\chi, 1)} = \chi(-1) G(\bar{\chi}, 1)$.

Proof. For the first formula, notice that $\chi(a)$ is a root of unity (since $\chi(a)^{p-1} = \chi(a^{p-1}) = 1$), and therefore $\chi(a^{-1}) = \chi(a)^{-1} = \bar{\chi}(a)$. This yields

$$\begin{aligned} G(\chi, a) &= \sum_{x\in\mathbf{F}_p^*} \chi(x) \exp\left(\frac{2\pi i a x}{p}\right) \\ &= \chi(a^{-1}) \sum_{x\in\mathbf{F}_p^*} \chi(ax) \exp\left(\frac{2\pi i a x}{p}\right) = \chi(a^{-1}) G(\chi, 1). \end{aligned}$$

For the second formula, $\sum_{a=1}^{p-1} |G(\chi, a)|^2 = (p-1)|G(\chi, 1)|^2$ and is also equal to

$$\begin{aligned} &\sum_{a=1}^{p-1} \sum_{x,y\in\mathbf{F}_p} \chi(x)\bar{\chi}(y) \exp\left(\frac{2\pi i a(x-y)}{p}\right) \\ &= \sum_{a=0}^{p-1} \sum_{x,y\in\mathbf{F}_p} \chi(x)\bar{\chi}(y) \exp\left(\frac{2\pi i a(x-y)}{p}\right) - \sum_{x,y\in\mathbf{F}_p} \chi(x)\bar{\chi}(y) \\ &= p \sum_{x\in\mathbf{F}_p} \chi(x)\bar{\chi}(x) = p(p-1). \end{aligned}$$

The last formula can be deduced from the equation:

$$\overline{G(\chi, 1)} = G(\bar{\chi}, -1) = \chi(-1) G(\bar{\chi}, 1). \qquad \square$$

To prove the quadratic reciprocity law, we will introduce the analogue of these sums in finite characteristic. More precisely, if p and q are two distinct odd prime numbers, we choose a primitive pth root of unity, α, in an extension of $\mathbf{F}_q$; namely α is a root of the equation

$$\alpha^{p-1} + \alpha^{p-2} + \cdots + \alpha + 1 = 0.$$

We then define the "Gauss sum" in $\mathbf{F}_q(\alpha)$ by

$$\tau := \sum_{x \in \mathbf{F}_p} \left(\frac{x}{p}\right) \alpha^x$$

and prove the following lemma.

4.5. Lemma. *Let τ be the element of $\mathbf{F}_q(\alpha)$ as above. Then*

1) $\tau^2 = \left(\frac{-1}{p}\right) p$;

2) $\tau^{q-1} = \left(\frac{q}{p}\right) \in \mathbf{F}_q(\alpha)$.

Proof. We calculate

$$\tau^2 = \sum_{x,y \in \mathbf{F}_p} \left(\frac{xy}{p}\right) \alpha^{x+y} = \sum_{u \in \mathbf{F}_p} S(u) \alpha^u,$$

where $S(u) := \sum_{x+y=u} \left(\frac{xy}{p}\right) = \sum_{x \in \mathbf{F}_p} \left(\frac{x(u-x)}{p}\right)$. For $u = 0$, we have $S(0) = \sum_{x \in \mathbf{F}_p} \left(\frac{-x^2}{p}\right) = \left(\frac{-1}{p}\right)(p-1)$. For $u \in \mathbf{F}_p^*$, the sum $S(u)$ equals

$$\begin{aligned} \sum_{x \in \mathbf{F}_p} \left(\frac{x(u-x)}{p}\right) &= \sum_{x \in \mathbf{F}_p^*} \left(\frac{-x^2(1-ux^{-1})}{p}\right) \\ &= \left(\frac{-1}{p}\right) \sum_{x \in \mathbf{F}_p^*} \left(\frac{1-ux^{-1}}{p}\right) \\ &= \left(\frac{-1}{p}\right) \left\{ \sum_{y \in \mathbf{F}_p^*} \left(\frac{y}{p}\right) - 1 \right\}, \end{aligned}$$

in other words, $S(u) = -\left(\frac{-1}{p}\right)$. In fact, $1 - ux^{-1}$ takes all values in $\mathbf{F}_p$ except 1, and the sum of the $\left(\frac{y}{p}\right)$ is zero. Therefore,

$$\tau^2 = \left(\frac{-1}{p}\right)(p - 1 - \sum_{u=1}^{p-1} \alpha^u) = \left(\frac{-1}{p}\right) p.$$

For the second formula, since the characteristic is q which is odd, it follows that

$$\tau^q = \sum_{x \in \mathbf{F}_p} \left(\frac{x}{p}\right)^q \alpha^{qx} = \sum_{x \in \mathbf{F}_p} \left(\frac{x}{p}\right) \alpha^{qx} = \left(\frac{q}{p}\right) \sum_{x \in \mathbf{F}_p} \left(\frac{qx}{p}\right) \alpha^{qx} = \left(\frac{q}{p}\right) \tau.$$

By using the fact that $\tau \neq 0$, assertion 2) follows from assertion 1). □

Proof. (of the quadratic reciprocity law) We saw that if q does not divide $a \in \mathbf{Z}$, then $a^{(q-1)/2} \equiv \left(\frac{a}{q}\right) \bmod q$. Therefore, by applying this to $a = p$, we obtain the following equalities in $\mathbf{F}_q(\alpha)$ by successively invoking formulas *1)* and *2)* of the preceding lemma:

$$\begin{aligned}\left(\frac{p}{q}\right) &= p^{(q-1)/2} = \left(\left(\frac{-1}{p}\right)\tau^2\right)^{(q-1)/2} \\ &= (-1)^{(p-1)(q-1)/4}\tau^{q-1} = (-1)^{(p-1)(q-1)/4}\left(\frac{q}{p}\right).\end{aligned}$$

This yields the following equality of signs, first in $\mathbf{F}_q$, then in $\mathbf{Z}$:

$$\left(\frac{p}{q}\right) = (-1)^{(p-1)(q-1)/4}\left(\frac{q}{p}\right),$$

which finishes the proof. □

Other proofs of the quadratic reciprocity law are proposed in Exercises 1-6.13 and 2-7.14.

5. Applications to the Number of Solutions of Equations

We will now explain another application of Gauss sums (and other elementary theorems) to finding the number of solutions of equations in $\mathbf{F}_q$ *or* $\mathbf{Z}/N\mathbf{Z}$.

5.1. Theorem. (Chevalley-Warning) *Let* $k = \mathbf{F}_q$ *be a finite field of characteristic* p. *If* $P \in k[x_1, \ldots, x_n]$ *and* $\deg(P) < n$, *then*

$$\operatorname{card}\{x \in k^n \mid P(x) = 0\} \equiv 0 \bmod p.$$

In particular, if P *is homogeneous of degree* $d < n$, *then* P *has a nontrivial zero (i.e., distinct from* 0*).*

We will start by calculating the sum of values of a monomial.

5.2. Lemma. *Let $x^m := x_1^{m_1} \cdots x_n^{m_n}$ be a monomial. Then $\sum_{x\in k^n} x^m$ is zero except when every m_i is non-zero and divisible by $(q-1)$. In particular, this sum is zero as soon as $m_1 + \cdots + m_n < (q-1)n$.*

Proof. Let us point out that since the polynomial "X^0" is the constant polynomial, it follows naturally that $0^0 = 1$. The calculation

$$\sum_{x\in k^n} x^m = \sum_{(x_1,\dots,x_n)\in k^n} x_1^{m_1} \cdots x_n^{m_n} = \left(\sum_{x_1\in k} x_1^{m_1}\right)\cdots\left(\sum_{x_n\in k} x_n^{m_n}\right)$$

brings us back to the case of one variable. If $m = 0$, then $\sum_{y\in k} y^0 = q\cdot 1_k = 0$. If m is not divisible by $q-1$, take y_0 to be a generator of k^*, so $y_0^m \neq 1$, and therefore,

$$\sum_{y\in k} y^m = \sum_{y\in k} (y_0 y)^m = y_0^m \sum_{y\in k} y^m$$

yields $\sum_{y\in k} y^m = 0$. □

Proof. (of the Chevalley-Warning theorem) We can deduce from the lemma that if $Q \in k[x_1,\dots,x_n]$ and $\deg(Q) < (q-1)n$, then $\sum_{x\in k^n} P(x) = 0$. Now let P be the polynomial in the statement of the Chevalley-Warning theorem. We will apply the previous result to $Q = 1 - P^{q-1}$. Notice that $\deg(Q) = (q-1)\deg(P) < (q-1)n$ and that $Q(x) = 1$ if $P(x) = 0$, while $Q(x) = 0$ if $P(x) \neq 0$ and $x \in k^n$. It follows that in k, we have the equality

$$0 = \sum_{x\in k^n} Q(x) = \sum_{\substack{x\in k^n \\ P(x)=0}} 1 = \mathrm{card}\{x \in k^n \mid P(x) = 0\} 1_k,$$

which completes the proof since k is of characteristic p, and hence $m1_k = 0$ is equivalent to $m \equiv 0 \bmod p$. □

5.3. Definition. If $Q(x) = \sum_{1\leqslant i,j\leqslant n} a_{ij}x_ix_j$ is a quadratic form where $a_{ij} = a_{ji}$, we say that it is *nondegenerate* if $D_Q := \det(a_{ij}) \neq 0$.

5.4. Remark. If we do not impose the symmetry condition $a_{ij} = a_{ji}$, we can (if the characteristic of the field k is not equal to 2) replace Q by $Q'(x) = \sum_{1\leqslant i,j\leqslant n} b_{ij}x_ix_j$, where $b_{ij} := \frac{1}{2}(a_{ij} + a_{ji})$, in such a way that for all x, we have $Q(x) = Q'(x)$. In general, the study of quadratic forms in characteristic 2 is more subtle, and we will therefore avoid it.

We start by showing that if the characteristic of the field k is not equal to 2, then we can replace Q by a diagonal form $Q'(y) = a_1y_1^2 + \cdots + a_ny_n^2$. We write $Q(x) = {}^txAx$ where A is symmetric; if we introduce the symmetric bilinear form $B(x,y) = {}^txAy$, it follows that $Q(x) = B(x,x)$ and $B(x,y) = \frac{1}{2}(Q(x+y) - Q(x) - Q(y))$. Let F be a vector subspace of k^n, and let $F^\perp := \{x \in k^n \mid \forall y \in F,\ B(x,y) = 0\}$. Then we have $\dim F + \dim F^\perp = n$. To see why this is true, we take a basis for F, $e_1, \ldots, e_r$, and let $\Phi(x) = (B(e_1,x), \ldots, B(e_r,x))$. The kernel of the linear map $\Phi : k^n \to k^r$ is $F^\perp$, and its image is all of k^r because if not there would exist $a_1, \ldots, a_r$, all non-zero such that $0 = a_1B(e_1,x) + \cdots + a_rB(e_r,x) = B(a_1e_1 + \cdots + a_re_r, x)$, which contradicts the hypothesis that B (or Q) is nondegenerate. It therefore follows that $n = \dim \operatorname{Ker} \Phi + \dim \operatorname{Im} \Phi = \dim F + \dim F^\perp$.

We now prove by induction on n that there exists an orthogonal basis. Choose e_1 such that $Q(e_1) \neq 0$, so $k^n = \langle e_1 \rangle \oplus \langle e_1 \rangle^\perp$, and we can proceed inductively since $\dim \langle e_1 \rangle^\perp = n-1$ and since the form remains nondegenerate when we restrict it to $\langle e_1 \rangle^\perp$. Now, if $e_1, \ldots, e_n$ is an orthogonal basis such that $Q(e_i) = a_i$, and we denote by $y_1, \ldots, y_n$ the coordinates of the vector $(x_1, \ldots, x_n)$ in the basis $e_1, \ldots, e_n$, we have that

$$Q(x_1, \ldots, x_n) = Q(y_1e_1 + \cdots + y_ne_n) = a_1y_1^2 + \cdots + a_ny_n^2.$$

Let us point out that if we call the quadratic form on the right Q' and the change of basis matrix U, then $D_Q = \det(U)^2 D_{Q'}$. In particular, if we work over $\mathbf{F}_p$, then we have $\left(\frac{D_Q}{p}\right) = \left(\frac{D_{Q'}}{p}\right)$. This remark is used in the proof of the following theorem.

5.5. Theorem. *Let Q be a nondegenerate quadratic form in n variables with coefficients in F_p ($p \neq 2$). Then*

$$\operatorname{card}\{x \in (\mathbf{F}_p)^n \mid Q(x) = 0\} = p^{n-1} + \epsilon(p-1)p^{\frac{n}{2}-1},$$

where

$$\epsilon = \begin{cases} 0 & \text{if } n \text{ is odd,} \\ \left(\dfrac{(-1)^{n/2}D_Q}{p}\right) & \text{if } n \text{ is even.} \end{cases}$$

Proof. From the remarks before the statement of the theorem, we can assume that the form Q is diagonal, in other words, $Q(x) = a_1x_1^2 + \cdots + a_nx_n^2$. Let N be the cardinality that we want to compute. We have

$$\begin{aligned} pN &= \sum_{a=0}^{p-1} \sum_{x \in \mathbf{F}_p^n} \exp\left(\frac{2\pi i a Q(x)}{p}\right) \\ &= p^n + \sum_{a=1}^{p-1} \sum_{x \in \mathbf{F}_p^n} \exp\left(\frac{2\pi i a Q(x)}{p}\right) \\ &= p^n + \sum_{a=1}^{p-1} \sum_{x_1,\dots,x_n \in \mathbf{F}_p} \exp\left(\frac{2\pi i a (a_1 x_1^2 + \cdots + a_n x_n^2)}{p}\right) \\ &= p^n + \sum_{a=1}^{p-1} \prod_{j=1}^{n} \sum_{x_j \in \mathbf{F}_p} \exp\left(\frac{2\pi i a a_j x_j^2}{p}\right) = p^n + \sum_{a=1}^{p-1} \prod_{j=1}^{n} \tau(a a_j) \\ &= p^n + \tau(1)^n \left(\frac{a_1 \cdots a_n}{p}\right) \sum_{a=1}^{p-1} \left(\frac{a}{p}\right)^n. \end{aligned}$$

Now, $a_1 \ldots a_n = D_Q$, and the sum $\sum_{a=1}^{p-1} \left(\frac{a}{p}\right)^n$ is 0 (resp. $p-1$) if n is odd (resp. if n is even). It follows from this that $N_p = p^{n-1}$ if n is odd. If n is even, observe that

$$\tau(1)^n = \left(\tau(1)^2\right)^{n/2} = \left(\frac{-1}{p}\right)^{n/2} p^{n/2},$$

and we have the formula for N_p. □

5.6. Remark. This statement gives us a much more precise formulation of the Chevalley-Warning theorem in the case of quadratic forms. This is obvious if the quadratic form is nondegenerate; we should add that a degenerate form can be written, after a variable change, as $Q(x_1, \ldots, x_n) = a_1 x_1^2 + \cdots + a_r x_r^2$ with $r < n$ and $D'_Q := a_1 \ldots a_r \neq 0$. In this case, it follows that

$$N_p = p^{n-r} \left(p^{r-1} + \epsilon(p-1) p^{\frac{r}{2}-1}\right) = p^{n-1} + \epsilon(p-1) p^{n-\frac{r}{2}-1},$$

where now ϵ is zero if r is odd and is $\left(\frac{(-1)^{r/2} D'_Q}{p}\right)$ if r is even.

We will now consider a quadratic form $Q(x) = \sum_{1 \leqslant i,j \leqslant n} a_{ij} x_i x_j$ with *integer* coefficients. If we want to count the number of solutions modulo N where N is not necessarily prime, we can rely on the two following lemmas (where the first is a variation of the Chinese remainder theorem).

5.7. Lemma. *Let* $\psi_Q(N) := \mathrm{card}\,\{x \bmod N \mid Q(x) \equiv 0 \bmod N\}$. *If* M *and* N *are relatively prime, then* $\psi_Q(MN) = \psi_Q(M)\psi_Q(N)$.

Proof. This is a corollary of the Chinese remainder theorem: $Q(x) \equiv 0 \bmod MN$ if and only if $Q(x) \equiv 0 \bmod N$ and $Q(x) \equiv 0 \bmod M$, and furthermore, each pair of congruence classes $x \equiv a \bmod M$, $x \equiv b \bmod N$ corresponds to a congruence class $\bmod MN$. □

This lemma reduces our case to counting the solutions modulo p^m. This can be done thanks to the following lemma, which is a special case of "Hensel's lemma".

5.8. Lemma. *Let p be an odd prime number which does not divide D_Q. We define the set of "nonsingular" solutions mod p^m by*

$$\mathscr{C}_Q(p^m) := \{x \bmod p^m \mid Q(x) \equiv 0 \bmod p^m \text{ and } x \not\equiv 0 \bmod p\}.$$

Then we have the formula

$$\begin{aligned}\operatorname{card}\mathscr{C}_Q(p^m) &= p^{(m-1)(n-1)} \operatorname{card}\mathscr{C}_Q(p) \\ &= p^{(m-1)(n-1)}\left(p^{n-1} - 1 + \epsilon(p-1)p^{\frac{n}{2}-1}\right).\end{aligned}$$

Proof. The second equality is an immediate corollary of the first equality and of the preceding theorem. We have an obvious map from $\mathscr{C}_Q(p^{m+1})$ to $\mathscr{C}_Q(p^m)$, which sends an n-tuple of integers modulo p^{m+1} to the same n-tuple of integers modulo p^m. It is enough to show that this map is surjective and that each fiber has order p^{n-1} since we would then have $\operatorname{card}\mathscr{C}_Q(p^{m+1}) = p^{n-1}\operatorname{card}\mathscr{C}_Q(p^m)$, and the lemma follows easily from that. So let x_0 be an n-tuple of integers such that $Q(x_0) \equiv 0 \bmod p^m$, or such that $Q(x) = p^m a_0$. We know that

$$\begin{aligned}Q(x_0 + p^m z) &= Q(x_0) + 2p^m B(x_0, z) + p^{2m} Q(z) \\ &\equiv p^m\left(a_0 + 2B(x_0, z)\right) \bmod p^{m+1},\end{aligned}$$

which is zero modulo p^{m+1} if and only if

$$a_0 + 2B(x_0, z) \equiv 0 \bmod p.$$

Since $x_0 \not\equiv 0 \bmod p$ and B is a nondegenerate bilinear form, this last equation is the equation of an (affine) hyperplane in $\mathbf{F}_p^n$; there are therefore exactly p^{n-1} solutions modulo p at z. □

Generalization. The calculation done on the quadrics can now be generalized by considering, on the one hand, the solutions over $\mathbf{F}_q$ and, on the other hand, forms of arbitrary degree (restricting to diagonal forms). We therefore consider solutions $x = (x_1, \dots, x_n) \in \mathbf{F}_q^n$ to the equation

$$a_1 x_1^d + \cdots + a_n x_n^d = 0. \tag{1.4}$$

It will be useful to provisionally introduce the *trace* and the *norm*, but we will give a more general definition in Chap. 3 (Definition 3-4.8).

5.9. Definition. Let $q = p^m$ and $x \in \mathbf{F}_q$. We define the *trace* (resp. the *norm*) of $\mathbf{F}_q$ over $\mathbf{F}_p$ as

$$\mathrm{Tr}_{\mathbf{F}_p}^{\mathbf{F}_q} x := x + x^p + \cdots + x^{p^{m-1}} \qquad \text{and} \qquad \mathrm{N}_{\mathbf{F}_p}^{\mathbf{F}_q} x := x^{1+p+\cdots+p^{m-1}}. \tag{1.5}$$

One can easily check that these maps send $\mathbf{F}_q$ to $\mathbf{F}_p$ and that the trace is $\mathbf{F}_p$-linear (resp. the norm, multiplicative). We first use the trace to construct an additive character: if $q = p^m$ and $a \in \mathbf{F}_q$, we define it by and denote it as

$$\psi(a) := \exp\left(\frac{2\pi i \,\mathrm{Tr}_{\mathbf{F}_p}^{\mathbf{F}_q} a}{p}\right). \tag{1.6}$$

Now we can generalize the calculation over $\mathbf{F}_p$.

5.10. Lemma. *Let $b \in \mathbf{F}_q$. Then we have the formula*

$$\sum_{a \in \mathbf{F}_q} \psi(ab) = \begin{cases} q & \text{if } b = 0, \\ 0 & \text{if } b \neq 0. \end{cases} \tag{1.7}$$

Proof. The formula is obviously true for $b = 0$. If $b \neq 0$, the map $a \mapsto \mathrm{Tr}_{\mathbf{F}_p}^{\mathbf{F}_q}(ab)$ from $\mathbf{F}_q$ to $\mathbf{F}_p$ is $\mathbf{F}_p$-linear and surjective. Therefore, every element of $\mathbf{F}_p$ appears p^{m-1} times in the image of the trace, and hence $\sum_{a \in \mathbf{F}_q} \psi(ab) = p^{m-1} \sum_{x \in \mathbf{F}_p} \exp(2\pi i x/p) = 0$. □

Convention. The *unitary* character χ_0 is defined by $\chi_0(a) = 1$ for every $a \in \mathbf{F}_q$.

If $\chi : \mathbf{F}_q^* \to \mathbf{C}^*$ is a *character* (i.e., a homomorphism), other than the unitary character (over $\mathbf{F}_q^*$), we extend it by $\chi(0) = 0$.

We can therefore define the corresponding Gauss sums for $a \in \mathbf{F}_q^*$.

$$G(\chi, \psi, a) := \sum_{x \in \mathbf{F}_q} \chi(x)\psi(ax) \qquad \text{and} \qquad G(\chi, \psi) := G(\chi, \psi, 1). \tag{1.8}$$

We then have a proposition analogous to Proposition 1-4.4 (and leave the proof as an exercise).

5.11. Proposition. *We have the following formulas.*

i) $G(\chi_0, \psi, a) = 0,$

ii) $G(\chi, \psi, a) := \bar{\chi}(a) G(\chi, \psi),$

iii) $|G(\chi, \psi)| = \sqrt{q}$ *(if* $\chi \neq \chi_0$*).*

Let us now come to the calculation of

$$N := \operatorname{card}\left\{(x_1, \ldots, x_n) \in (\mathbf{F}_q)^n \mid F(x) := a_1 x_1^d + \cdots + a_n x_n^d = 0\right\}.$$

We first point out that

$$\begin{aligned} qN &= \sum_{a \in \mathbf{F}_q} \sum_{x \in (\mathbf{F}_q)^n} \psi(aF(x)) \\ &= q^n + \sum_{a \in \mathbf{F}_q^*} \sum_{x \in (\mathbf{F}_q)^n} \psi(aF(x)) \\ &= q^n + \sum_{a \in \mathbf{F}_q^*} \prod_{j=1}^{n} \sum_{y \in (\mathbf{F}_q} \psi(a a_j y^d) \\ &= q^n + \sum_{a \in \mathbf{F}_q^*} \prod_{j=1}^{n} T(d, a a_j), \end{aligned}$$

where we let $T(d,a) := \sum_{y \in \mathbf{F}_q} \psi(ay^d)$. The key step in the calculation is the following observation.

5.12. Lemma. *If* $d' = \gcd(d, q-1)$, *then* $T(d,a) = T(d',a)$. *Suppose* d *divides* $q-1$. *If* G_d *denotes the set of the* d *characters* χ *which satisfy* $\chi^d = \chi_0$, *and we let* $G'_d := G_d \setminus \{\chi_0\}$, *then we have the equality*

$$T(d,a) = \sum_{\chi \in G'_d} \bar{\chi}(a) G(\chi, \psi). \tag{1.9}$$

Proof. Let us point out that we must first understand the equality $\chi^d = \chi_0$ as saying: $\forall x \in \mathbf{F}_q^*$, $\chi^d(x) = \chi_0(x) = 1$. This is because if $\chi \in G'_d$, then $\chi(0)^d = 0 \neq 1 = \chi_0(0)$. The first assertion is an immediate consequence of the fact that $\mathbf{F}_q^*$ is cyclic of order $q-1$, thus $\mathbf{F}_q^{*d} = \mathbf{F}_q^{*d'}$. We then check that with the hypothesis that d divides $q-1$, we have

$$\sum_{\chi \in G_d} \chi(x) = \begin{cases} d & \text{if } x \in \mathbf{F}_q^{*d}, \\ 1 & \text{if } x = 0, \\ 0 & \text{if not.} \end{cases} \tag{1.10}$$

We can then deduce that

$$T(d,a)=\sum_{y\in\mathbf{F}_q}\psi(ay^d)=\sum_{t\in\mathbf{F}_q}\sum_{\chi\in G_d}\chi(t)\psi(at)=\sum_{\chi\in G'_d}\bar{\chi}(a)G(\chi,\psi)$$

since $G(\chi_0,\psi)=0$. □

5.13. Theorem. *Let d divide $q-1$, and let S_d be the set of n-tuples of characters $(\chi_1,\ldots,\chi_n)$ such that $\chi_j\neq\chi_0$, $\chi_j^d=\chi_0$ and $\chi_1\cdots\chi_n=\chi_0$. Then the number of solutions of the equation $a_1x_1^d+\cdots+a_nx_n^d=0$ is equal to*

$$N=q^{n-1}+\frac{q-1}{q}\sum_{(\chi_1,\ldots,\chi_n)\in S_d}\bar{\chi}_1(a_1)\cdots\bar{\chi}_n(a_n)G(\chi_1,\psi)\cdots G(\chi_n,\psi). \tag{1.11}$$

Proof. Observe that $\sum_{a\in\mathbf{F}_q^*}\chi(a)$ equals $q-1$ if $\chi=\chi_0$ and equals zero if $\chi\neq\chi_0$. It follows from the previous calculations that

$$\begin{aligned}qN&=q^n+\sum_{a\in\mathbf{F}_q^*}\prod_{j=1}^n T(d,aa_j)\\&=q^n+\sum_{a\in\mathbf{F}_q^*}\sum_{\chi_1,\ldots,\chi_n\in G'_d}\prod_{j=1}^n\bar{\chi}_j(aa_j)G(\chi_j,\psi)\\&=q^n+(q-1)\sum_{(\chi_1,\ldots,\chi_n)\in S_d}\prod_{j=1}^n\bar{\chi}_j(a_j)G(\chi_j,\psi).\square\end{aligned}$$

5.14. Example. We can prove by induction or a direct calculation that the cardinality of S_d equals

$$s(n,d)=\frac{1}{d}\left((d-1)^n+(-1)^n(d-1)\right).$$

Therefore, $N=q^{n-1}+(q-1)R$ where R is the sum of the $|S_d|$ terms whose absolute value equals $q^{\frac{n}{2}-1}$. For $d=2$ we find that $s(n,d)$ is zero for n odd and is 1 for n even; if $n=3$, we find that $s(3,d)=(d-1)(d-2)$. For example, for a cubic equation $a_0x_0^3+a_1x_1^3+a_2x_2^3=0$ over $\mathbf{F}_q$ where $q\equiv 1[3]$ and by letting χ be a character of order 3 (the other one being $\chi^2=\bar{\chi}$), we have

$$N=q^2-(q-1)\left(\alpha+\bar{\alpha}\right),$$

where $\alpha:=-\chi(a_0a_1a_2)G(\chi,\psi)^3/q$.

It is interesting to see how this number varies when we choose a tower of finite fields. A key result in this direction is the Davenport-Hasse theorem, which connects the different Gauss sums.

5.15. Theorem. (Davenport-Hasse) *Let* $\mathbf{F}_q$ *be a finite field and* $\mathbf{F}_{q^m}$ *a finite extension. We denote by* $\mathrm{Tr} = \mathrm{Tr}_{\mathbf{F}_q}^{\mathbf{F}_{q^m}}$ *and* $\mathrm{N} = \mathrm{N}_{\mathbf{F}_q}^{\mathbf{F}_{q^m}}$ *the trace and the norm. If* χ *is a character of* $\mathbf{F}_q^*$, *we have the relation*

$$-G(\chi \circ \mathrm{N}, \psi \circ \mathrm{Tr}) = (-G(\chi, \psi))^m . \tag{1.12}$$

Proof. See [5] (Chap. 11, Sect. 4) or Exercise 1-6.25. □

6. Exercises

6.1. Exercise. *Show that in a commutative group, if the order of* x_1 *is* d_1, *the order of* x_2 *is* d_2 *and* d_1 *and* d_2 *are relatively prime, then the order of* x_1x_2 *is* d_1d_2. *Show also that in a cyclic group, if the order of* x_1 *is* d_1 *and the order of* x_2 *is* d_2, *then the order of the subgroup generated by* x_1 *and* x_2 *is equal to the LCM of* d_1 *and* d_2.

6.2. Exercise. *Prove that if the class of* $x \in \mathbf{Z}$ *generates* $(\mathbf{Z}/p^2\mathbf{Z})^*$ *then it also generates* $(\mathbf{Z}/p^\alpha\mathbf{Z})^*$ *(for odd* p*).*

6.3. Exercise. *Prove that if* N *is even and* m *is odd, the last formula of Proposition 1-2.6 is also true. How should you change the formula when both* N *and* m *even?*

6.4. Exercise. *Let* $K := \mathbf{F}_{q^m}$ *and* $k := \mathbf{F}_q$. *Prove that the maps* $\mathrm{N} = \mathrm{N}_k^K : K^* \to k^*$ *and* $\mathrm{Tr} = \mathrm{Tr}_k^K : K \to k$ *are surjective.*

Prove that $\mathrm{Ker}\,\mathrm{N} = \mathbf{F}_{q^m}^{*q-1}$ *and that* $\mathrm{Ker}\,\mathrm{Tr} = \{x^q - x \mid x \in \mathbf{F}_{q^m}\}$.

6.5. Exercise. *If* b *is the base of a numeral system, (i.e., an integer* $\geqslant 2$*), every real number can be written as an expansion in base* b:

$$a_0, a_1a_2 \ldots a_n \ldots = a_0 + a_1b^{-1} + \cdots + a_nb^{-n} + \ldots ,$$

with $a_0 \in \mathbf{Z}$ *and* $0 \leqslant a_i \leqslant b-1$.

1) Prove that this expansion is unique, except for the case where $a_{n_0} < b-1$ *and* $a_n = b-1$ *for every* $n > n_0$, *in which case* $a_0, a_1a_2 \ldots a_n \ldots = a_0, a_1a_2 \ldots (a_{n_0}+1)000 \ldots$.

2) If $a/c \in \mathbf{Q}$, *show that the expansion of* a/c *in base* b *is ultimately periodic (i.e., it is a repeating decimal) and interpret its period.*

6.6. Exercise. *Use* $(n!)^2+1$ *and* $(n!)^2-1$ *to prove that there exist infinitely many prime numbers congruent to* 1 *modulo* 4 *(resp. to* -1 *modulo* 4*).*

Use $5(n!)^2-1$ *to prove that there exist infinitely many prime number congruent to* -1 *modulo* 5*. Use* $2(n!)^2-1$ *to show that there exist infinitely many prime number congruent to* -1 *modulo* 8.

6.7. Exercise. *An integer* N *is said to be a Carmichael number if* N *is not prime and* $a^{N-1}\equiv 1 \bmod N$ *for every* a *relatively prime to* N.

a) Show that N *is a Carmichael number if and only if* N *is square-free and for every prime factor* p *of* N, $p-1$ *divides* $N-1$.

b) Show that if $6m+1$, $12m+1$ *and* $18m+1$ *are primes, then their product is a Carmichael number (for example:* $N := 7{\cdot}13{\cdot}19$*).*

6.8. Exercise. *Let* $M := 21560 = 2^3\cdot 5\cdot 7^2\cdot 11$, $N := 21576 = 2^3\cdot 3\cdot 29\cdot 31$ *and* $G_1 := (\mathbf{Z}/M\mathbf{Z})^*$, $G_2 := (\mathbf{Z}/N\mathbf{Z})^*$.

a) Do the groups G_1 *and* G_2 *have the same order and are they isomorphic?*

b) Calculate the exponent of the group G_1*, in other words the smallest integer* $m \geqslant 1$ *such that if* a *is relatively prime to* M*, then* $a^m \equiv 1 \bmod M$.

c) How many solutions are there to the equation $x^2 = 1$ *for* $x \in G_1$*?*

d) How many solutions are there to the equation $x^2 = -1$ *for* $x \in G_1$*; same question for* $x^2 = 9$*?*

6.9. Exercise. *Let* $L := 11396 = 2^2\cdot 7\cdot 11\cdot 37$, $M := 16200 = 2^3\cdot 3^4\cdot 5^2$ *and* $N := 13176 = 2^3\cdot 3^3\cdot 61$*; and let* $G_1 := (\mathbf{Z}/L\mathbf{Z})^*$, $G_2 := (\mathbf{Z}/M\mathbf{Z})^*$ *and* $G_3 := (\mathbf{Z}/N\mathbf{Z})^*$.

a) Are the orders of the groups G_i *equal and are the groups isomorphic?*

b) Calculate the exponent G_i*, in other words the smallest integer* $m \geqslant 1$ *such that if* a *is relatively prime to* L *(resp.*M, N*), then* $a^m \equiv 1 \bmod L$ *(resp.* $\bmod M$, $\bmod N$*).*

c) How many solutions does the equation $x^2 = 1$ *have in* G_1, G_2, G_3*?*

d) How many solutions does the equation $a^{L-1} = 1$ *have in* G_1*; same question for* $a^{N-1} = 1$ *in* G_3*? (Notice that* $L-1 = 11395 = 5\cdot 43\cdot 53$ *and* $N-1 = 5^2{\cdot}17{\cdot}31$*.)*

6.10. Exercise. *Calculate the number* $N(a,b,p) = N(a,b)$ *of solutions* $(x,y) \in (\mathbf{F}_p)^2$ *of the equation* $ax^2+by^2 = 1$.

Hint.– *You could repeat the steps of Theorem 1-5.5 (or apply the theorem to the conic* $ax^2+by^2-z^2 = 0$*) for the equation* $ax^2+by^2 = 0$ *and then* *finish from there. A generalization, as well as a different approach, is given in the following exercise.*

6.11. Exercise. (Jacobi sums, see [5]) *Let p be odd and let $\chi_1, \dots, \chi_n : \mathbf{F}_p^* \to \mathbf{C}$ be characters. We define the* Jacobi sum *by*

$$J(\chi_1, \dots, \chi_n) := \sum_{x_1+\cdots+x_n=1} \chi_1(x_1) \dots \chi_n(x_n).$$

We also denote the principal (trivial) character by χ_0.

1) Prove that Jacobi sums can be factored with the help of Gauss sums in the following manner. If χ_j are all nontrivial and $\chi_1 \cdots \chi_n \neq \chi_0$, then

$$J(\chi_1, \dots, \chi_n) = \frac{G(\chi_1) \cdots G(\chi_n)}{G(\chi_1 \cdots \chi_n)},$$

and in particular

$$|J(\chi_1, \dots, \chi_n)| = p^{\frac{n-1}{2}}.$$

2) Let $N_d(u) :=$ card $\{x \in \mathbf{F}_p \mid x^d = u\}$. If $d' = \gcd(d, p-1)$, prove that $N_d(u) = N_{d'}(u)$.

3) Suppose that d divides $p-1$. Recall why the following formula holds:

$$N_d(u) = \sum_{\chi \in G_d} \chi(u)$$

(where G_d is the set of characters such that $\chi^d = \chi_0$).

4) Let $N :=$ card $\{x \in (\mathbf{F}_p)^n \mid a_1 x^{d_1} + \cdots + a_n x_n^{d_n} = b\}$. Prove that

$$N = \sum_{a \cdot x = b} N_{d_1}(x_1) \cdots N_{d_n}(x_n),$$

where $a \cdot x = a_1 x_1 + \cdots + a_n x_n$. Deduce from this that the number N does not change if we replace d_i by $\gcd(d_i, p-1)$.

5) Keeping the same notation, if $d_1, \dots, d_n$ divide $p-1$ and $b \neq 0$, prove that

$$N = p^{n-1} + \sum_{(\chi_1, \dots, \chi_n) \in S} \chi_1 \cdots \chi_n(b) \bar{\chi}_1(a_1) \cdots \bar{\chi}_n(a_n) J(\chi_1, \dots, \chi_n),$$

where S denotes the n-tuples of characters $(\chi_1, \dots, \chi_n)$ such that $\chi_j \neq \chi_0$, but $\chi_j^{d_j} = \chi_0$.

6.12. Exercise. *We define a character as a homomorphism $\chi : (\mathbf{Z}/n\mathbf{Z})^* \to \mathbf{C}^*$ that we extend by convention to all of $\mathbf{Z}/n\mathbf{Z}$ by $\chi(x) := 0$ if x is non-invertible. We say that χ is primitive if it does not come from a character* *modulo m, where m is a nontrivial divisor of n, in other words if we cannot factor $\chi : (\mathbf{Z}/n\mathbf{Z})^* \to (\mathbf{Z}/m\mathbf{Z})^* \to \mathbf{C}^*$. Let*

$$G(\chi, a) = \sum_{x \in \mathbf{Z}/n\mathbf{Z}} \chi(x) \exp\left(\frac{2\pi i a x}{n}\right) = \sum_{x \in (\mathbf{Z}/n\mathbf{Z})^*} \chi(x) \exp\left(\frac{2\pi i a x}{n}\right).$$

Prove the following formulas where a is relatively prime to n and χ is primitive modulo n.

i) $G(\chi, a) = \bar{\chi}(a) G(\chi, 1)$.

ii) $|G(\chi, a)|^2 = n$.

iii) $\overline{G(\chi, 1)} = \chi(-1) G(\bar{\chi}, 1)$.

6.13. Exercise. *In this exercise, you are asked study and calculate the sums*

$$G(N) := \sum_{x=0}^{N-1} \exp\left(\frac{2\pi i x^2}{N}\right).$$

a) If $N = 2M$ with M odd, prove that $G(N) = 0$ (divide the sum into the terms from 0 to $M-1$ and the terms from M to $2M-1$).

b) Let p be an odd prime. By decomposing $x = y + p^{r-1}z$ with y modulo p^{r-1} and z modulo p, prove that $G(p^r) = pG(p^{r-2})$; conclude then that $G(p^{2r}) = p^r$ and $G(p^{2r+1}) = p^r G(p)$.

c) We introduce the function $\phi(x) := f(x) + f(x+1) + \cdots + f(x+N-1)$, where $f(x) := \exp\left(\frac{2\pi i x^2}{N}\right)$ on the interval $[0,1]$. Let

$$\hat{\phi}_m := \int_0^1 \phi(t) \exp(-2\pi i m t) dt$$

be the Fourier coefficient of ϕ. Check that

$$G(N) = \frac{\phi(0) + \phi(1)}{2} = \sum_{n \in \mathbf{Z}} \hat{\phi}_m.$$

d) From this, deduce the equality

$$G(N) = (1 + i^{-N}) \int_{-\infty}^{+\infty} \exp\left(\frac{2\pi i y^2}{N}\right) dy = (1 + i^{-N})\sqrt{N} C,$$

and compute the constant C by choosing $N = 1$. Now conclude from this *that*

$$G(N) = \begin{cases} \sqrt{N} & \textit{if } N \equiv 1 \bmod 4, \\ i\sqrt{N} & \textit{if } N \equiv 3 \bmod 4, \\ (1+i)\sqrt{N} & \textit{if } N \equiv 0 \bmod 4, \\ 0 & \textit{if } N \equiv 2 \bmod 4. \end{cases}$$

e) We now introduce $G(a,N) := \sum_{x=0}^{N-1} \exp\left(\frac{2\pi i a x^2}{N}\right)$. *Prove that* $G(a,MN) = G(aM,N)G(aN,M)$ *if* $\gcd(M,N) = \gcd(a,MN) = 1$, *then that if* N *is odd, we have* $G(a,N) = \left(\frac{a}{N}\right) G(N)$. *Conclude from this that for* M, N *relatively prime and odd*

$$G(MN) = \left(\frac{M}{N}\right)\left(\frac{N}{M}\right) G(M)G(N),$$

and deduce the quadratic reciprocity law from this formula and from the previous question.

f) More generally, if $\gcd(2a,N) = 1$, *calculate the sum*

$$G(a,b,c,N) := \sum_{x=0}^{N-1} \exp\left(\frac{2\pi i(ax^2+bx+c)}{N}\right).$$

6.14. Exercise. *1) If* p *is prime and* $a \in \mathbf{Z}$, *we let* $N(a,p) := \operatorname{card}\{(x, y, z) \in \mathbf{F}_p^3 \mid x^2 + y^2 + z^2 \equiv a \bmod p\}$. *If* p *is odd, prove that* $N(a,p) = p^2 + \left(\frac{-a}{p}\right) p$. *What is* $N(a,2)$ *equal to?*

2) Let p *be an odd prime. Assuming that* $N(p,7) = 42$, *calculate* $N(7,p)$.

3) Let p *be a prime number such that* $p \equiv 3 \bmod 4$. *Calculate*

$$M := \operatorname{card}\{(x,y,z) \in \mathbf{F}_p^3 \mid x^4 + y^4 + z^4 \equiv 1 \bmod p\}.$$

6.15. Exercise. *By a similar method, prove the following generalization of the Chevalley-Warning theorem (Theorem 1-5.1). Let* $P_1, \ldots, P_s$ *be polynomials of degree* $d_1, \ldots, d_s$ *with* $d_1 + \cdots + d_s < n$. *Prove that*

$$\operatorname{card}\{x \in k^n \mid P_1(x) = \cdots = P_s(x) = 0\} \equiv 0 \bmod p.$$

In particular, if the polynomials are homogeneous, then they have a common nontrivial zero.

6.16. Exercise. *We consider the quadratic form given by*

$$Q(x,y,z,t) = x^2 - 2xy + 3y^2 + 3z^2 + 7t^2.$$

How many solutions does the equation $Q(x, y, z, t) = 0$ *have modulo 5? Same question modulo 7?*

6.17. Exercise. *We denote by* N_m *the number of solutions* $x, y \in \mathbf{F}_{2^m}$ *of the equation* $y^2 + y = x^3$*. Prove that if* m *is odd,* $N_m = 2^m$*, and that if* m *is even,* $N_m = 2^m - (-1)^{m/2} 2^{1+m/2}$.

Hint.– *The case where* m *is even is more subtle. One way is to introduce the sums* $R(a) = \sum_{y \in \mathbf{F}_{2^m}} \psi(a(y^2 + y))$ *and* $S(a) = \sum_{x \in \mathbf{F}_{2^m}} \psi(ax^3)$ *and conclude that* $N_m = 2^m + 2^{-m} \sum_{a \neq 0} R(a)S(a)$*. The sums* $S(a)$ *can be calculated as in the proof of Lemma 1-5.12, with the help of the Davenport-Hasse relation (Theorem 1-5.15), and then show that* $R(a) = 0$ *except for* $R(1) = 2^m$ *before finishing the proof.*

6.18. Exercise. (Kloosterman sums) *We define the following sum of exponentials:*

$$S(a, b, q) := \sum_{x \in (\mathbf{Z}/q\mathbf{Z})^*} \exp\left(\frac{2\pi i(ax + bx^{-1})}{q}\right),$$

where, by convention, x^{-1} *is an integer (modulo* q*) such that* $x^{-1}x \equiv 1 \bmod q$*. (Notice that, with this convention,* $\exp\left(\frac{2\pi i a n^{-1}}{q}\right) \neq \exp\left(\frac{2\pi i a}{qn}\right)$*.)*

We will use the Weil inequality (see Chap. 6, Formula 6.11):

$$|S(a, b, p)| \leqslant 2\sqrt{p},$$

which is valid for any odd prime number p *which does not divide* ab*. We denote by* $e(z) = \exp(2\pi i z)$ *and* $e_q(z) = \exp(2\pi i z/q)$ *and also*

$$\sum_{x \bmod q} = \sum_{x \in \mathbf{Z}/q\mathbf{Z}} \quad \text{and} \quad \sum_{x \bmod^* q} = \sum_{x \in (\mathbf{Z}/q\mathbf{Z})^*}$$

so that we can also write $S(u, v, q) = \sum_{x \bmod^* q} e_q(ux + vx^{-1})$.

1) Prove that the absolute value of the sums $S(a, b, q)$ *with respect to the "root mean square" is approximately* $\sqrt{q}$*, or to be more precise, that*

$$\sum_{a, b \bmod q} |S(a, b, q)|^2 = \phi(q) q^2.$$

Therefore, on average, the size of $|S(a, b, q)|$ *is* $\sqrt{\phi(q)}$*, or approximately* $\sqrt{q}$.

The point of this exercise is to fix an upper bound on the individual sums by using the result due to Weil cited above.

2) Prove that these sums can be factored and reduced to the case $q = p^m$*: if* $q = q_1 q_2$ *where* $\gcd(q_1, q_2) = 1$*,* $a = q_2 a_1 + q_1 a_2$ *and* $b = q_2 b_1 + q_1 b_2$*, then*

$$S(a, b, q) = S(a_1, b_1, q_1) S(a_2, b_2, q_2).$$

3) Prove that if $a = p^h a_0$ *and* $b = p^h b_0$*, then*

$$S(a, b, p^m) = \sum_{x \bmod^* p^m} e\left(\frac{a_0 x + b_0 x^{-1}}{p^{m-h}}\right) = p^h S(a_0, b_0, p^{m-h}).$$

(Which implies that we can reduce to the case $p \nmid \gcd(a, b)$*.)*

4) Show that if $m/2 \leqslant n < m$ *and* p *does not divide* y*, then* $(y + p^n z)^{-1} \equiv y^{-1} - p^n y^{-2} z \bmod p^m$*. Now suppose that* $m = 2n + 1$ *(or more generally* $m \leqslant 3n$*) and that* p *never divides* y*. Show that* $(y + p^n z)^{-1} \equiv y^{-1} - p^n y^{-2} z + p^{2n} y^{-3} z^2 \bmod p^m$*.*

5) Prove that if $m/2 \leqslant n < m$ *and* p *does not divide* $\gcd(a, b)$*, then*

$$|S(a, b, p^m)| \leqslant A p^n,$$

with $A := \begin{cases} 4 & \text{if } p = 2 \text{ and } m - n \geqslant 3, \\ 2 & \text{if not.} \end{cases}$

Hint.– *Decompose the sum over* $x \bmod^* p^m$ *into* $x = y + p^n z$*, with* $y \bmod^* p^n$ *and* $z \bmod p^{m-n}$*, and take* A *to be an upper bound for the number of solutions to the congruence* $a - by^{-2} \equiv 0 \bmod p^{m-n}$*.*

6) Deduce from the previous calculation that $S(a, b, p^m) = 0$ *if* $m \geqslant 2$ *and* p *divides* a *but not* b *or vice versa.*

7) If p *is odd and* m *even, prove that if* $\gcd(p, a, b) = 1$*, then*

$$|S(a, b, p^m)| \leqslant 2p^{m/2}.$$

Hint.– *Using question 6), reduce to the case where* a *and* b *are invertible modulo* p*, and apply the result from question 5) with* $n := m/2$*.*

8) Let p *be odd. Prove that*

$$\left|\sum_{t \bmod p^{n+1}} e\left(\frac{at^2}{p} + \frac{ht}{p^{n+1}}\right)\right| = \begin{cases} p^{n+\frac{1}{2}} & \text{if } p^n \text{ divides } h \text{ but not } a, \\ 0 & \text{if } p^n \text{ does not divide } h. \end{cases}$$

Hint.– *If* p^n *divides* h *(and* $a \not\equiv 0 \bmod p$*), we would bring in a Gauss sum, if not, we would decompose the sum over* $t = r + ps$ *with* $r \bmod p$ *and* $s \bmod p^n$*.*

9) Let $p \neq 2$ *be a prime which does not divide* $\gcd(a, b)$ *and let* m *be an odd number. Prove that*

$$|S(a, b, p^m)| \leqslant 2p^{m/2}.$$

Hint.– *If* $m = 2n+1$, *write* $x = y + p^n z$ *with* $y \bmod^* p^n$ *and* $z \bmod p^{n+1}$ *in the sum and use the preceding question.*

10) From this, deduce the following theorem.

Theorem. (Weil, Estermann) The following estimates hold, where we denote by $\omega(q)$ the number of distinct primes which divide q and by $d(q)$ the number of divisors of q.

i) If $\gcd(2ab, q) = 1$ then

$$|S(a,b,q)| \leqslant 2^{\omega(q)}\sqrt{q}.$$

ii) In the general case, we have

$$|S(a,b,q)| \leqslant d(q)\gcd(a,b,q)^{1/2}q^{1/2}.$$

6.19. Exercise. *Let* p *be a prime number and* $F \in \mathbf{Z}[X_1, \ldots, X_n]$ *be a homogeneous polynomial; we denote by* $\nabla F(x) = \left(\frac{\partial F}{\partial X_1}(x), \ldots, \frac{\partial F}{\partial X_n}(x)\right)$ *and assume moreover that* $\nabla F(x) \equiv 0 \bmod p$ *only if* $x \equiv 0 \bmod p$ *(we say that* F *is "smooth modulo* p*"). We define the following sum of exponentials*

$$S(a,q) := \sum_{x \bmod q} \exp\left(\frac{2\pi i a F(x)}{q}\right),$$

where the sum is over $x \in (\mathbf{Z}/q\mathbf{Z})^n$ *and* $\gcd(a,q) = 1$.

1) Whenever $q = q_1 q_2$ *with* $\gcd(q_1, q_2) = 1$, *find* a_1 *and* a_2 *such that*

$$S(a,q) = S(a_1,q_1)S(a_2,q_2).$$

2) Check that $F(y + p^{m-1}z) \equiv F(y) + p^{m-1}\nabla F(y) \cdot z \bmod p^m$.

3) Let $m \geqslant 2$. *By transforming the sum* $S(a,p^m)$ *over* $x = y + p^{m-1}z$ *into a sum over* $y \bmod p^{m-1}$ *and* $z \bmod p$, *prove that*

$$S(a,p^m) = \begin{cases} p^{n(d-1)}S(a,p^{m-d}) & \text{if } m > d, \\ p^{n(m-1)} & \text{if } m \leqslant d. \end{cases}$$

4) By using Deligne's upper bound, which says that whenever F *is smooth and of degree* d *where* d *is relatively prime to* p, *we have* $|S(a,p)| \leqslant B_{n,d}p^{n/2}$ *(see Chap. 6, (6.10)), prove the following upper bound*

$$|S(a,q)| \leqslant C^{\omega(q)} q^{n\left(1-\frac{1}{d}\right)},$$

where $\omega(q)$ *is the number of primes which divide* q *and where* C *is a constant which only depends on* F.

6.20. Exercise. *Let* $N_p := \left|\{(x,y,z,t) \in (\mathbf{F}_p)^4 \mid ax^4 + by^4 + z^2 + t^2 = 0\}\right|$. *Assume also that* $ab \neq 0$.

1) Prove that if $p \equiv 3 \bmod 4$, *we have*

$$N_p = \begin{cases} p^3 + p^2 - p & \text{if } ab \in \mathbf{F}_p^{*2}, \\ p^3 - p^2 + p & \text{if } ab \in \mathbf{F}_p^* \setminus \mathbf{F}_p^{*2}. \end{cases}$$

2) Prove that if $p \equiv 1 \bmod 4$, *we have*

$$N_p = \begin{cases} p^3 + 3p^2 - 3p & \text{if } -a/b \in \mathbf{F}_p^{*4}, \\ p^3 - p^2 + p & \text{if } -a/b \in \mathbf{F}_p^* \setminus \mathbf{F}_p^{*4}. \end{cases}$$

Hint. – *By following the procedure in the proof of Theorem 1-5.13, show that*

$$pN_p = p^4 + (p-1)\left\{\left(\frac{ab}{p}\right)\tau^4 + \tau^2 G(\chi)G(\bar{\chi})(\chi(b/a) + \chi(a/b))\right\},$$

where τ *is the Gauss sum associated to the Legendre character (of order 2) and* χ *is one of the characters of order 4 (the other one being* $\bar{\chi}$*).*

3) Finish by finding N_p *if* $p = 2$ *or if* $ab = 0$.

6.21. Exercise. *We will now try to find integer solutions* (x, y) *of the equation* $x^2 + 15y^2 = m$, *denoted* $(\mathscr{E}_m)$.

1) Let p *be a prime number* $\neq 2, 3, 5$. *If* p *divides* m, *prove that either* p *divides* x *and* y *and then* p^2 *divides* m, *or* $\left(\frac{-15}{p}\right) = 1$.

2) Let p *be a prime number* $\neq 2, 3, 5$. *Deduce from this that a necessary condition for the equation* $(\mathscr{E}_p)$ *to have a solution is that* p *must belong to certain congruence classes modulo 15, and specify these classes.*

3) Does the equation $x^2 + 15y^2 = 77077$ *have an integer valued solution (notice that* $77077 = 7^2 \cdot 11^2 \cdot 13$*)?*

6.22. Exercise. *In this exercise, you are asked to calculate, for each prime number* $p \neq 2, 17$, *the number* $N_p := \operatorname{card}\{(x,y,) \in (\mathbf{F}_p)^2 \mid 2y^2 = x^4 - 17\}$.

1) Calculate the numbers $L_p := \operatorname{card}\{(x,y,z) \mid 2y^2 = x^2 - 17z^2\}$ *and use this to calculate* $M_p := \operatorname{card}\{(x,y,z) \mid 2y^2 = x^2 - 17\}$.

2) Whenever $p \equiv 3 \bmod 4$, *prove that* $N_p = M_p$ *and, as a consequence, that*

$$N_p = \begin{cases} p+1 & \text{if } p \equiv 3 \bmod 8, \\ p-1 & \text{if } p \equiv 7 \bmod 8. \end{cases}$$

3) As usual, let $\mathrm{e}(z) := \exp(2\pi iz)$ *and let*

$$\tau(a) := \sum_{x\in\mathbf{F}_p} \mathrm{e}\left(ax^2/p\right) \quad \text{and} \quad \rho(a) := \sum_{x\in\mathbf{F}_p} \mathrm{e}\left(ax^4/p\right).$$

Prove that

$$N_p = p + p^{-1}\sum_{a=1}^{p-1} \mathrm{e}(17a/p)\tau(2a)\rho(-a).$$

From now on, we assume that $p \equiv 1 \bmod 4$. *We introduce* $G = \{\chi_0, \chi_1, \chi_2, \chi_3\}$, *the set of characters of* $\mathbf{F}_p^*$ *such that* $\chi_0(x) = 1$ *and* $\chi^4(x) = 1$ *for* $x \in \mathbf{F}_p^*$. *We extend them to* $\mathbf{F}_p$ *by the convention* $\chi_0(0) = 1$ *and* $\chi_j(0) = 0$ *for* $j = 1, 2, 3$. *Suppose that* χ_1 *is the Dirichlet character* $\chi_1(x) := \left(\frac{x}{p}\right)$. *We also introduce the associated Gauss sums:*

$$G(\chi, a) := \sum_{x\in\mathbf{F}_p} \chi(x)\,\mathrm{e}(ax/p) \quad \text{and} \quad G(\chi) := G(\chi, a).$$

4) Recall why $G(\chi_0, a) = 0$, $G(\chi, a) = \bar{\chi}(a)G(\chi)$ *and also that if* $\chi \neq \chi_0$, *then* $|G(\chi)| = \sqrt{p}$.

5) Prove the formula

$$\rho(a) = \bar{\chi}_1(a)G(\chi_1) + \bar{\chi}_2(a)G(\chi_2) + \bar{\chi}_3(a)G(\chi_3).$$

6) Using this, find a formula for N_p *in terms of Gauss sums of the form*

$$N_p = p - \epsilon_0 + \frac{\tau(1)}{p}\left(\epsilon_1 G(\chi_2)^2 + \epsilon_2 G(\chi_3)^2\right),$$

where $|\epsilon_i| = 1$.

7) Conclude that $N_p \geqslant 1$ *for every* $p \neq 2, 17$.

6.23. Exercise. *In this exercise, we ask you to prove that the equation*

$$2y^2 = x^4 - 17$$

has solutions modulo N *for every* N, *but does not have any rational solutions over* $\mathbf{Q}$.

1) Assume that there exist $x = a/b$ *and* $y = c/d$, *which are a solution to the equation, with* $a, c \in \mathbf{Z}$, $b, d \in \mathbf{N}^*$ *and* $\gcd(a, b) = \gcd(c, d) = 1$. *Prove that* b^4 *divides* d^2 *and that* d^2 *divides* $2b^4$ *and deduce from this that* $d = b^2$ *and* $2c^2 = a^4 - 17b^4$.

2) Let $p \neq 2$ *which divides* c. *Prove that* p *is a square modulo 17, and deduce from this that* c *itself is a square modulo 17. Conclude then that 2 would be a fourth power, which is a contradiction.*

3) Let $p \neq 2, 17$. It was proven in the previous exercise (Exercise 1-6.22) that there exist $u, v \in \mathbf{F}_p^*$ where $2u^2 = v^4 - 17$. By using Lemma 1-5.8, prove that the equation in question has solutions modulo p^n for every n.

4) Prove that the equation also has solutions modulo 2^n and modulo 17^n, by refining the previous argument. (You might want to use that $2 \cdot 5^2 \equiv 2^4 \bmod 17$ and $3^4 - 17 \equiv 0 \bmod 2^6$.)

5) Using the Chinese remainder theorem, conclude that the equation $2y^2 = x^4 - 17$ has solutions modulo N for every N.

6.24. Exercise. *In this exercise, let $\mathrm{e}(x) := \exp(2\pi i x)$ and notice that for $x \in \mathbf{F}_p$, then the expression $\mathrm{e}(x/p)$ is well-defined. Let p be odd and let $Q_1(x) = a_1x_1^2 + \cdots + a_nx_n^2$ and $Q_2(x) = b_1x_1^2 + \cdots + b_nx_n^2$ be two quadratic forms with coefficients in $\mathbf{F}_p$. Assume that n is odd and that the following condition is fulfilled.*

$$\text{For } 1 \leqslant i < j \leqslant n, \text{ we have } a_ib_j - a_jb_i \neq 0. \qquad (*)$$

We will calculate $N := \operatorname{card}\{x \in \mathbf{F}_p^n \mid Q_1(x) = Q_2(x) = 0\}$.

a) Prove that $\sum_{a,b\in\mathbf{F}_p} \sum_{x\in\mathbf{F}_p^n} \mathrm{e}\left(\frac{aQ_1(x) + bQ_2(x)}{p}\right) = p^2N$, and deduce from this the following formula:

$$N = p^{n-2} + p^{-2} \sum_{(a,b)\neq(0,0)} \sum_{x\in\mathbf{F}_p^n} \mathrm{e}\left(\frac{aQ_1(x) + bQ_2(x)}{p}\right),$$

where the sum is over nonzero pairs $(a, b) \in \mathbf{F}_p^2$.

b) Let $\tau := \sum_{x\in\mathbf{F}_p} \mathrm{e}(x^2/p)$ and let $Q(x) = c_1x_1^2 + \cdots + c_nx_n^2$. Recall the formula which gives $\sum_{x\in\mathbf{F}_p^n} \mathrm{e}(Q(x)/p)$ in terms of the c_i and of the Gauss sum τ, whenever $c_1 \cdots c_n \neq 0$. Deduce that if $c_1 \cdots c_{n-1} \neq 0$ but $c_n = 0$, then

$$\sum_{x\in\mathbf{F}_p^n} \mathrm{e}(Q(x)/p) = \left(\frac{c_1 \cdots c_{n-1}}{p}\right) \tau^{n-1} p,$$

where $\left(\frac{\cdot}{p}\right)$ designates the Legendre symbol. Also, recall what the value of τ^2 is.

c) To lighten the notation, we let $T(a, b) := \sum_{x\in\mathbf{F}_p^n} \mathrm{e}\left(\frac{aQ_1(x) + bQ_2(x)}{p}\right)$. Show that if (a, b) is not proportional to one of the $(b_i, -a_i)$, then

$$\sum_{\lambda\in\mathbf{F}_p^*} T(\lambda a, \lambda b) = 0.$$

Calculate this last sum for $(a, b) = (b_i, -a_i)$.

d) Let

$$D_i = \prod_{1\leqslant j\leqslant n, j\neq i} (b_i a_j - a_i b_j) \qquad \text{and} \qquad \epsilon_i = \left(\frac{D_i}{p}\right).$$

From the preceding arguments, deduce the formula

$$N = p^{n-2} + (p-1)\left(\frac{-1}{p}\right)^{(n-1)/2} \left(\sum_{i=1}^{n} \epsilon_i\right) p^{(n-3)/2}.$$

6.25. Exercise. (Proof of the Davenport-Hasse formula, Theorem 1-5.15) *Let χ be a (nontrivial) character of $\mathbf{F}_q^*$ and let $f \in \mathbf{F}_q[X]$ be monic of degree n, i.e., $f(X) = X^n - a_1X^{n-1} + \cdots + (-1)^n a_n$. We set $\lambda(f) = \psi(a_1)\chi(a_n)$. Show that λ is multiplicative, i.e., that $\lambda(fg) = \lambda(f)\lambda(g)$.*

Prove that if N, Tr *are the norm and trace of $\mathbf{F}_{q^m}$ to $\mathbf{F}_q$, then*

$$G(\chi \circ \mathrm{N}, \psi \circ \mathrm{Tr}) = \sum_{f, \deg(f)\,|\,m} \deg(f)\lambda(f)^{m/\deg(f)},$$

where the sum is over the monic irreducible polynomials f in $\mathbf{F}_q[X]$ whose degree divides m.

Prove the identity

$$1 + G(\chi, \psi)T = \sum_f \lambda(f)T^{\deg(f)} = \prod_g \left(1 - \lambda(g)T^{\deg(g)}\right)^{-1},$$

where the sum is over the monic polynomials and the product over the irreducible monic polynomials in $\mathbf{F}_q[X]$.

By taking the logarithmic derivative, deduce the Davenport-Hasse relation.

6.26. Exercise. *Prove that for every N, the equation*

$$3x^3 + 4x^3 + 5z^3 = 0$$

has primitive solutions modulo N (i.e., such that $\gcd(x, y, z, N) = 1$*). Same question for $5x^3 + 22y^3 + 2z^3 = 0$.*

Chapter 2

Applications: Algorithms, Primality and Factorization, Codes

"Elle est retrouvée.
Quoi ? - L'Éternité.
C'est la mer allée
Avec le soleil."

ARTHUR RIMBAUD

This chapter describes some industrial applications of number theory, via computer science. We succinctly describe the main algorithms as well as their theoretical complexity or computation time. We use the notation $O(f(n))$ to denote a function $\leqslant Cf(n)$; furthermore, the unimportant—at least from a theoretical point of view—constants which appear will be ignored. In the following sections, we introduce the basics of cryptography and of the "RSA" system, which motivates the study of primality tests and factorization methods. We finish the chapter with an introduction to error-correcting codes, which will lead us into the study of cyclotomic polynomials.

1. Basic Algorithms

Let n be an integer. Once we have chosen a base $b \geqslant 2$, we write n in base b, in other words, with the *digits* $a_i \in [0, b-1]$:

$$n = a_0 + a_1 b + \cdots + a_r b^r = \overline{a_r a_{r-1} \ldots a_1 a_0}^{\,b}, \text{ where } a_r \neq 0$$

(the two most standard base choices are $b = 10$ for usual decimal notation and $b = 2$ for binary notation, which is especially well-adapted to computer

M. Hindry, *Arithmetics*, Universitext,
DOI 10.1007/978-1-4471-2131-2_2,

programming). We will consider an operation on the digits to be a single operation (or an operation which needs $O(1)$ computation time). It is natural to refer to the number of digits necessary in order to describe n, in other words $r+1$, as its *complexity*. Since we can see that $b^r \leqslant a_r b^r < n \leqslant b^{r+1}$, we know that

$$r \leqslant \frac{\log n}{\log b} < r+1$$

and can therefore describe the complexity as proportional to $\log n$. It is clear that the manipulation of random numbers of size n requires at least $\log n$ elementary operations. We consider, as much from a practical point of view as from a theoretical one, an algorithm to be "good" if it is a *polynomial* algorithm; that is to say, it uses $O\left((\log n)^{\kappa}\right)$ elementary operations. Conversely, we consider an *exponential* algorithm, meaning that its execution time or required number of operations is greater than $\exp(\kappa \log n) = n^{\kappa}$, to be infeasible (for large n, of course).

Addition. In order to add two numbers m and n with at most r digits, we must perform at most r additions of two digits and (possibly) carry a digit. The cost is therefore $O(\log \max(n,m)) = O(r)$. The number of operations used in subtraction is similar.

Multiplication. In order to calculate $n \times m$, where n and m are two numbers with at most r digits (with the usual elementary school algorithm), we must perform at most r^2 elementary multiplications and r additions, and possibly carry a digit, and therefore, the cost is $O\left((\log \max(n,m))^2\right) = O\left(r^2\right)$.

Remark. The addition algorithm is (up to constants) optimal, but some more sophisticated methods (notably the "fast Fourier transform") lets us perform multiplications at a much better cost, for example in $O\left(r(\log r)^2\right)$. See Exercises 2-7.3 and 2-7.4.

Division algorithm. Given a and $b \geqslant 1$, if we compute (q,r) such that $a = qb + r$ and $0 \leqslant r \leqslant b-1$ with (a variation of) the algorithm learned in elementary school, we perform a number of elementary operations similar to that of multiplication, i.e., $O(\log \max(a,b)^2)$. In order to give an example of a *turtle algorithm* (do not use!), we could perform the following procedure. We start by setting $q_0 = 0$ and $r_0 = a$. Then we have $a = q_0 b + r_0$; if $r_0 < b$, we stop, and if not, we compute $q_1 = q_0 + 1$ and $r_1 = r_0 - b$ in such a way that $a = q_1 b + r_1$, and we get the result by iteration and by stopping when $r_n < b$ and $a = q_n b + r_n$. If $a > b$, we must perform approximately a/b subtractions, therefore the cost is $O((\log a) \times (a/b))$ (which is exponential).

Euclidean algorithm. Given two integers, a and b, the goal is to compute $d := \gcd(a,b)$ and $(u,v) \in \mathbf{Z}^2$ such $au + bv = d$ (Bézout's lemma). The principle is the following: we divide a by b, $a = bq_1 + r_1$; then divide b by r_1, $b = r_1q_2 + r_2$, and in subsequent steps divide r_n by r_{n+1}, $r_n = r_{n+1}q_{n+2} + r_{n+2}$. Keep in mind that the sequence r_n is strictly decreasing and stops when $r_{n+1} = 0$, and therefore $\gcd(a,b) = r_n$. In fact,

$$\gcd(a,b) = \gcd(b,r_1) = \gcd(r_1,r_2) = \cdots = \gcd(r_n,r_{n+1}) = r_n.$$

In order to compute (u,v), we could proceed as follows: we set $u_0 = 1$, $u_1 = 0$, $v_0 = 0$ and $v_1 = 1$ and then recursively define $u_n = u_{n-2} - q_nu_{n-1}$ and $v_n = v_{n-2} - q_nv_{n-1}$. One can immediately check by induction that $au_n + bv_n = r_n$. We will now estimate the maximal number of times we need to use the division algorithm. We can assume that $r_0 = a \geqslant r_1 = b$ and see that $r_n = r_{n+1}q_{n+2} + r_{n+2} \geqslant r_{n+1} + r_{n+2}$. If $r_0 > r_1 > \cdots > r_n = d$ is the sequence which gives the gcd, set $d_i = r_{n-i}$. We then have $d_{i+2} \geqslant d_{i+1} + d_i$. Let $\alpha := (1+\sqrt{5})/2$ be the positive root of $X^2 = X + 1$; it follows that $d_i \geqslant \alpha^i$. This is true because $d_0 = d \geqslant 1 = \alpha^0$, $d_1 \geqslant d_0 + 1 \geqslant 2 \geqslant \alpha^1$ and if the inequality is true until $i+1$, we have $d_{i+2} \geqslant d_{i+1} + d_i \geqslant \alpha^{i+1} + \alpha^i = \alpha^{i+2}$. From this we conclude that $a = d_n \geqslant \alpha^n$, and the number of steps is bounded above by $\log(a)/\log(\alpha) = O(\log a)$. We should point out that this argument implies that the longest computation happens when a and b are terms in Fibonacci sequence (see Exercise 2-7.5). The total cost is therefore $O\left(\log\max\{|a|,|b|\}^3\right)$.

Computations in $\mathbf{Z}/N\mathbf{Z}$. The goal is to perform addition and multiplication of two integers smaller than N, then to take the remainder gotten from dividing by N in the division algorithm. In order to calculate the inverse of a modulo N, we proceed as follows: if a is an integer, the Euclidean algorithm tells us that either $\gcd(a,N) > 1$—in which case a is not invertible modulo N—or there exist u, v (gotten from the algorithm) such that $au + Nv = 1$ and therefore the inverse of a is the class of u modulo N. The cost is therefore the same as that of the Euclidean algorithm.

Exponentiation. In order to calculate a^m, we could of course calculate $a \times a \times \cdots \times a$, but this will force us to perform $m-1$ multiplications; we could do a lot better by performing the computation in $O(\log m)$ multiplications. For example, if $m = 2^r$ we would carry out r multiplications. In the general case, we write m in binary notation $m = \epsilon_0 + \epsilon_1 2 + \cdots + \epsilon_r 2^r$ and we would calculate

$$a^m = \left(\left(\left(a^{\epsilon_r}\right)^2 a^{\epsilon_{r-1}}\right)^2 a^{\epsilon_{r-2}} \cdots\right)^2 a^{\epsilon_0}.$$

Or we could do the calculation in the other direction; the algorithm can be defined iteratively. In order to do this, we start with the initial data chosen to be $(u,v,n) := (1,a,m)$ and we iterate as follows: if n is even, we replace (u,v,n) by $(u,v^2,n/2)$ and if is n odd, we replace (u,v,n) by $(uv, v^2, (n-$

1)/2); we stop when $n = 0$, and we therefore have $u = a^m$. Since n is at least divisible by 2 in each step, the number of steps r satisfies $2^r \leqslant m$, and hence we must perform $O(\log m)$ multiplications. If we calculate $\bmod N$, we reduce each result $\bmod N$, and so in each step we multiply integers $\leqslant N$. The total cost to compute $a^m \bmod N$ is therefore $O\left(\log m(\log N)^2\right)$.

Computations in $\mathbf{F}_q$ and $\mathbf{F}_q^*$. We will assume that the finite field $\mathbf{F}_q = \mathbf{F}_{p^f}$ is defined by an irreducible monic polynomial $S(X) = X^f + s_{f-1}X^{f-1} + \cdots + s_0 \in \mathbf{F}_p[X]$ of degree f. We therefore identify $\mathbf{F}_q$ with $\mathbf{F}_p[X]/S\mathbf{F}_p[X]$, which can be seen as the vector space over $\mathbf{F}_p$ with basis $1, x, x^2, \ldots, x^{f-1}$ with addition on the individual coordinates and multiplication defined by $x^i \cdot x^j = x^{i+j}$ and $x^f = -s_{f-1}x^{f-1} - \cdots - s_0$. An element of $\mathbf{F}_q$ is therefore seen as an f-tuple of integers modulo p or as a polynomial of degree $\leqslant f-1$. To perform an addition, we must perform f additions in $\mathbf{F}_p$, so at a cost of $O(f \log p) = O(\log q)$. To carry out a multiplication, we take the product of two polynomials, or essentially f^2 multiplications in $\mathbf{F}_p$, then divide the result by $S(X)$ using the division algorithm, or essentially $O(f)$ divisions and $O(f^2)$ multiplications in $\mathbf{F}_p$. The cost of a multiplication in $\mathbf{F}_q$ is therefore $O(f^2(\log p)^2) + O(f(\log p)^3)$. Let us point out that this cost is still $O((\log q)^3)$, but that if we choose $q = 2^f$ for example, it is $O(f^2) = O((\log q)^2)$.

2. Cryptography, RSA

We are only interested here in one aspect of cryptography and in one system of "public keys", known as RSA from the name of its three inventors, Rivest, Shamir and Adleman [61], and which is one of the most widely used.

Cryptography is the art (or science) of secret messages: we want to send information so that only one other person, the recipient, can see it. A related problem is to be able to identify with certainty the sender of the message. We generally think that the only method is to use a "secret code"; in fact the originality of "public key" cryptography comes precisely from the fact that the code is not secret, but is known (for the most part) by everybody! This is not only a mathematical curiosity, it is also the principle governing credit cards, internet transactions, etc.

The general principle is the following. We call $\mathscr{M}$ the set of messages (in practice we take $\mathscr{M} = [0, N-1]$ or $\mathbf{Z}/N\mathbf{Z}$). Two people, A and B, who wish to exchange messages in such a way that a third person, C, cannot decipher them each choose bijections $f_{\mathrm{A}}, f_{\mathrm{B}} : \mathscr{M} \to \mathscr{M}$. The set $\mathscr{M}$ (say the integer N) is known to everybody, as well as f_{A} and f_{B}, however—and this is the key idea—the inverse function f_{A}^{-1} (resp. f_{B}^{-1}) is only known by A (resp. by B). This does not mean of course that, knowing f_{A}, it is theoretically impossible to compute f_A^{-1}, but this calculation would be so long, that it would be out of the question to carry out in a reasonable time frame. We will later see how to construct such functions.

When A wants to send B a message $m \in \mathcal{M}$ (say an integer modulo N), he or she simply sends $m' = f_B \circ f_A^{-1}(m)$; remember that A knows f_B (which is public) and f_A^{-1} (which only he or she knows). In order to decode this message, B computes $f_A \circ f_B^{-1}(m')$, which will give m; remember that B knows f_A (which is public) and f_B^{-1} (which only he or she knows). The system has two advantages: not only can C not decipher the message without computing f_B^{-1} (which we assume to be out of the question), but B can be sure that it is A who sent the message since it must have been encoded using f_A^{-1}, which only A knows!

This procedure is a simplified form of the known methods under the name of the Diffie-Hellman protocol (1976); its security relies on the choice of the "one-way" functions f, in other words such that f is quick and easy to compute, but f^{-1} is in practice impossible to determine. Many constructions of functions have been suggested, but one of the most hardy and most widely used, relies on the fact that if p and q are very large prime numbers (say 100 or more digits), then their product $N := pq$ can be calculated very quickly (say 10,000 elementary operations), whereas if you only know N, it is an extremely long calculation to factor it, impossible in practice.

We now construct the functions f_A of the RSA system. We choose two very large prime numbers, p and q, compute $N := pq$ and also choose a medium-sized integer d which is relatively prime to $\phi(N) = (p-1)(q-1)$. The public key is therefore (N, d); however, p and q are secret and we set, for a any integer smaller than N,

$$f(a) := a^d \bmod N.$$

To decode a message, we calculate the inverse e of d modulo $\phi(N)$ and we observe that

$$f^{-1}(b) = b^e \bmod N,$$

since $\left(a^d\right)^e = a^{ed} \equiv a \bmod N$, because $a^{\phi(N)} \equiv 1 \bmod N$.

2.1. Remarks. 1) There is one little constraint on the "message" a: it should be relatively prime to N[1]. Nonetheless, observe that the proportion of integers which are relatively prime to N is $\phi(N)/N = (1-1/p)(1-1/q)$; so if p, q are for example $\geqslant 10^{50}$, the proportion of integers which are not relatively prime to N is $\leqslant 2 \cdot 10^{-50}$.

[1] If by mistake, a message $a = pa'$ was sent, we could certainly still decode it by $f(a)^e = p^{de}a'^{ed} = p^{ed}a' = a$, but C, or whoever else, would only have to compute $\gcd(a, N)$ to discover p and crack the code!

2) Once p, q and d have been chosen, the computation of N, $\phi(N)$ and e is performed in polynomial time (fast); likewise, the operation $a \mapsto f(a)$ is just as fast as $a \mapsto f^{-1}(a)$ *if we know* e.

3) We can see, at least heuristically, that knowing the number e allows us to factor N: if we write $de - 1 = 2^r M$ (with M odd), by computing $\gcd(a^{2^j M} \pm 1, N)$ for $j = 1, 2, \ldots$ and some values of a, we have a good chance of quickly factoring N.

4) Therefore, if someone knows only the public key (N, d), they should *a priori* factor N in order to compute $\phi(N)$ then e. In fact, the knowledge of $\phi(N)$ is equivalent to that of p and q, because $\phi(N) = N - (p + q) + 1$ (the knowledge of the product and the sum of two integers lets you easily determine the integer pair).

This system gives rise to many problems, the solutions to which are more or less satisfactory.

i) How do you construct (very) large prime numbers?
ii) What methods do we have for factoring an integer?
iii) How should you choose p and q in RSA that resist factorization methods?

Since it is clear from question *iii*) that the prime numbers should not be too "special", question *i*) is essentially equivalent to the following problem.

- (I) (Primality Test) Give a fast algorithm which determines whether a number N is prime.

If we had access to such an algorithm $\mathscr{P}$, we could in fact decide on the size of the integer (for example $N \sim 10^{50}$), randomly choose an odd integer N_1 of this size, and test $\mathscr{P}(N_1)$ then $\mathscr{P}(N_1 + 2)$, $\mathscr{P}(N_1 + 4)$ until we find a prime number. By the theorems on the distribution of prime numbers, the number of primes in an interval $[N_1, N_1 + H]$ is approximately $H/\log(N_1)$; so we expect to find a prime number in $O(\log(N_1))$ tries.

We will see that satisfactory answers to problem *i*) are available, but we only know partial answers to the other questions.

3. Primality Test (I)

We consider an *odd* integer N and the problem of determining whether N is prime. We denote by (M, N) the gcd of M and N. The letter p is reserved for a number which we already know is prime. The first of all of the primality tests, and in some sense the "grandfather", is the following lemma.

3.1. Lemma. (Fermat) *If N is prime and $(a, N) = 1$, then $a^{N-1} \equiv 1 \bmod N$.*

Proof. The group $\mathbf{Z}/N\mathbf{Z}^*$ has order $N-1$ and the lemma follows from the Lagrange's theorem.[2] □

This is a "good" test, in the sense that computing $a^{N-1} \bmod N$ requires $O(\log N)$ multiplications (under the condition of course that you use the binary notation for $N-1$). However, it is also a "bad" test, because there are numbers, called *Carmichael numbers*, which satisfy the test without being prime. We even know that there are infinitely many of them [11], the smallest being $561 = 3 \cdot 11 \cdot 17$. We can easily see that a number N is a Carmichael number if and only if N is square-free and $p-1$ divides $N-1$ for every p which divides N. In general, we could introduce $\lambda(N)$, the exponent of the group $(\mathbf{Z}/N\mathbf{Z})^*$, sometimes called the *Carmichael function*: it is the smallest positive integer (in the sense of divisibility or the usual order) such that for all a relatively prime to N, $a^{\lambda(N)} \equiv 1 \bmod N$. By what we have seen, we know that if $N = p_1^{m_1} \cdots p_k^{m_k}$ is odd, we have

$$\lambda(N) = \operatorname{lcm}\left(p_1^{m_1-1}(p_1-1), \ldots, p_k^{m_k-1}(p_k-1)\right). \tag{2.1}$$

It is always true that $\lambda(N)$ divides $\phi(N)$ and the equality holds if and only if $(\mathbf{Z}/N\mathbf{Z})^*$ is cyclic, i.e., if $N = p^\alpha$ or $2p^\alpha$ or 4.

3.2. Lemma. (Euler[3]) *If N is prime and $(a, N) = 1$, then*

$$a^{\frac{N-1}{2}} \equiv \left(\frac{a}{N}\right) \bmod N.$$

Proof. This is simply a restatement of assertion *ii)* from Theorem 1-3.3. □

The *Solovay-Strassen test* is an algorithm which checks the congruences given below for a randomly chosen a. This test is always polynomial (for any value of a, we can always quickly calculate the Jacobi symbol thanks to the quadratic reciprocity law, see Exercise 2-7.7) and is better than Fermat's test.

3.3. Lemma. *Let $H := \left\{a \in (\mathbf{Z}/n\mathbf{Z})^* \mid a^{\frac{N-1}{2}} \equiv \left(\frac{a}{N}\right) \bmod N\right\}$, then $H = (\mathbf{Z}/n\mathbf{Z})^*$ if and only if N is a prime number.*

[2] To prove Fermat's little theorem by using Lagrange's theorem is obviously an anachronism.

[3] Calling a statement which uses the Legendre or Jacobi symbol "Euler's criterion" is also an anachronism.

Proof. We have seen that if N is prime, then $H = (\mathbf{Z}/n\mathbf{Z})^*$. If p^2 divides N, there exists a of order $p(p-1)$, and p does not divide $N-1$. Therefore, $a^{N-1} \neq 1$. If $N = pp_2 \cdots p_r$ with $r \geqslant 2$, choose (by the Chinese remainder theorem) $a \equiv 1$ modulo $p_2, \ldots, p_r$ and which is not a square modulo p; hence $\left(\frac{a}{N}\right) = -1$, but $a^{(N-1)/2} \equiv 1 \bmod p_2 \cdots p_r$ and thus $a^{(N-1)/2} \not\equiv -1 \bmod N$. □

Applications.

i) Probabilistic polynomial test. If N is composite then $(\mathbf{Z}/N\mathbf{Z}^* : H) \geqslant 2$ and hence by randomly choosing a, we have at least a one in two chance that $a \notin H$. Hence if N successively passes k tests, we can say that it is prime with a probability greater than $1 - 2^{-k}$.
ii) Deterministic polynomial test (assuming GRH). Analytic theory has provided a proof that if the Dirichlet $L(\chi, s)$ functions do not vanish on $\mathrm{Re}(s) > 1/2$ (generalized Riemann hypothesis, GRH), then for every nontrivial character $\chi : (\mathbf{Z}/N\mathbf{Z})^* \to \mathbf{C}^*$, there exists an $a \leqslant 2(\log N)^2$ such that $\chi(a) \neq 0, 1$. We can deduce from this that if N were composite, there would exist $a \leqslant 2(\log N)^2$ which would not pass the Solovay-Strassen test. If $N = p_1^{m_1} \cdots p_k^{m_k}$, we introduce $f(a) := a^{\frac{N-1}{2}} \left(\frac{a}{N}\right)$ and
$$\chi_i : (\mathbf{Z}/N\mathbf{Z})^* \xrightarrow{f} (\mathbf{Z}/N\mathbf{Z})^* \to (\mathbf{Z}/p_i^{m_i}\mathbf{Z})^* \hookrightarrow \mathbf{C}^*.$$
We see that H is the intersection of the kernels of χ_i. By trying all of the $a \in [2, 2(\log N)^2]$, we therefore get a primality certificate (i.e., a proof of primality), under the condition that the Riemann hypothesis is true.

We could improve the Solovay-Strassen test and algorithm.

3.4. Lemma. *(Rabin-Miller) Let N be odd. Set $N - 1 = 2^s M$, with M odd. If N is prime and $(a, N) = 1$, then either $a^M \equiv 1 \bmod N$ or there exists $0 \leqslant r \leqslant s-1$ such that $a^{2^r M} \equiv -1 \bmod N$.*

Proof. The order of a modulo N is $2^t M'$, where $0 \leqslant t \leqslant s$ and M' is an odd integer which divides M. If $t = 0$, then $a^{M'} = 1$ hence $a^M = 1$. If $t \geqslant 1$, then, since N is prime, $a^{2^{t-1}M'} = -1$, and therefore $a^{2^{t-1}M} = -1$. □

This test is better than Euler's test, because, for one thing, if the pair a, N passes the Rabin-Miller test, then it also must pass Euler's test. Furthermore, if N is composite, the proportion of a which pass the refined test is $\leqslant 1/4$ and often smaller than that. Of course there exists a probabilistic polynomial version of the refined test and a deterministic polynomial version, assuming that the Riemann hypothesis is true.

3.5. Remark. If $N \equiv 3 \bmod 4$, then "Rabin-Miller" is identical to "Solovay-Strassen", and even equivalent to $a^{(N-1)/2} \equiv \pm 1 \bmod N$. We know that $(N-1)/2$ is odd, and we can observe that if $\epsilon = \pm 1$, then $\left(\frac{\epsilon}{N}\right) = \epsilon$, and if $a^{(N-1)/2} \equiv \pm 1 \bmod N$, then

$$\left(\frac{a}{N}\right) = \left(\frac{a \cdot (a^2)^{(N-3)/4}}{N}\right) = \left(\frac{a^{(N-1)/2}}{N}\right) = a^{(N-1)/2} \bmod N.$$

Proof. ("Rabin-Miller" > "Solovay-Strassen", in the general case) Now, we know that $a^{(N-1)/2} = a^{2^{s-1}M}$ equals $-1 \bmod N$ if $r = s-1$ and equals $1 \bmod N$ in all of the other cases. Therefore, we need to compute $\left(\frac{a}{N}\right)$. If $a^M \equiv 1 \bmod N$, then $\left(\frac{a}{N}\right) = \left(\frac{a}{N}\right)^M = \left(\frac{a^M}{N}\right) = 1$, hence $a^{\frac{N-1}{2}} \equiv \left(\frac{a}{N}\right) \bmod N$. Now assume that $a^{2^r M} \equiv -1 \bmod N$. Let p_i divide N and write $p_i - 1 = 2^{s_i} M_i$. Then, since $a^{2^r M} \equiv -1 \bmod p_i$, the order of a modulo p_i is of the form $2^{r+1} L_i$ (with L_i odd). Therefore, modulo p_i, we get

$$\left(\frac{a}{p_i}\right) \equiv a^{(p_i-1)/2} \equiv a^{2^{s_i-1}M_i} \equiv \begin{cases} 1 & \text{if } s_i > r+1, \\ -1 & \text{if } s_i = r+1. \end{cases}$$

Now notice that $r+1 \leqslant s_i$. Let h be the number of indices i such that $s_i = r+1$. Therefore, we have $\left(\frac{a}{N}\right) = (-1)^h$. Modulo 2^{r+2}, we have $N = 1 + 2^s M = \prod_i p_i = \prod_i (1 + 2^{s_i}) \equiv 1 + h 2^{r+1} \bmod 2^{r+2}$. In the case where $r < s-1$, h must be even, so that $\left(\frac{a}{N}\right) = 1$, and we get $a^{(N-1)/2} \equiv 1 \bmod N$. In the case where $r = s-1$, then h is odd and $\left(\frac{a}{N}\right) = -1 \equiv a^{(N-1)/2} \bmod N$. □

We can summarize the previous discussion by introducing the following sets:

$G_0 := (\mathbf{Z}/N\mathbf{Z})^*$,

$G_1 := \{a \in (\mathbf{Z}/N\mathbf{Z})^* \mid a^{N-1} \equiv 1 \bmod N\}$,

$G_2 := \{a \in (\mathbf{Z}/N\mathbf{Z})^* \mid a^{(N-1)/2} \equiv \pm 1 \bmod N\}$,

$G_3 := \left\{a \in (\mathbf{Z}/N\mathbf{Z})^* \mid a^{(N-1)/2} \equiv \left(\frac{a}{N}\right) \bmod N\right\}$,

$$S := \left\{a \in (\mathbf{Z}/N\mathbf{Z})^* \mid a^M \equiv 1 \bmod N \text{ or } \exists r \in [0, s-1] \text{ such that } a^{2^r M} \equiv -1 \bmod N\right\}.$$

We always have the inclusions $S \subset G_3 \subset G_2 \subset G_1 \subset G_0$, and these are equalities if and only if N is prime, or also if and only if $G_3 = G_0$. Furthermore, G_1, G_2 and G_3 are subgroups, but in general S is not, even though in the case $N \equiv 3 \bmod 4$ we have seen that $G_2 = G_3 = S$. In fact, S is stable under inversion, and if $a, b \in S$ do not satisfy the same congruence or both $a^M = b^M = 1$, then $ab \in S$. But if $a^{2^r M} = b^{2^r M} = -1$, it could happen that $ab \notin S$. For example, if $\epsilon^2 = 1$ but $\epsilon \neq \pm 1$ and if $a^{2M} = -1$ (which would force $N \equiv 1 \bmod 4$), then $a \in S$ and $a\epsilon \in S$, because $(a\epsilon)^{2M} = -1$. However, $(\epsilon a^2)^M = \epsilon^M a^{2M} = -\epsilon \neq \pm 1$ and $(\epsilon a^2)^{2M} = 1$, hence $\epsilon a^2 \notin S$. By considering $a \mapsto \left(\frac{a}{N}\right) a^{(N-1)/2}$ from G_2 to $\{\pm 1\}$, we see that $(G_2 : G_3) = 1$ or 2. We are now going to compute the cardinality of the set S and, in particular, verify the following statement.

3.6. Proposition. *Let N be an odd, composite number. If $N \neq 9$, then*

$$\frac{|S|}{|G_0|} \leqslant \frac{1}{4}.$$

3.7. Definition. Let A, B be integers. We define

$$\phi(A; B) = \operatorname{card}\left\{a \in (\mathbf{Z}/A\mathbf{Z})^* \mid a^B \equiv 1 \bmod A\right\}.$$

3.8. Lemma. *Let $t \geqslant 0$ and $N = 1 + 2^s M = p_1^{\alpha_1} \cdots p_k^{\alpha_k}$ (with M odd). We set $p_i - 1 = 2^{s_i} M_i$, $s_i' = \min(t, s_i)$ and $t_i := \gcd(M, M_i)$. Then*

$$\phi(N, 2^t M) = 2^{s_1' + \cdots + s_k'} t_1 \cdots t_k.$$

Moreover, the cardinality of the set

$$\left\{a \in (\mathbf{Z}/N\mathbf{Z})^* \mid a^{2^t M} \equiv -1 \bmod N\right\}$$

is 0 if $t \geqslant \min_i s_i$, and equal to $\phi(N, 2^t M) = 2^{tk} t_1 \cdots t_k$ if $t < \min_i s_i$.

Proof. We know that $a^{2^t M} \equiv 1 \bmod N$ if and only if $a^{2^t M} \equiv 1 \bmod p_j^{\alpha_j}$ for $j = 1, \ldots, k$. Now, the group $(\mathbf{Z}/p_j^{\alpha_j}\mathbf{Z})^*$ is cyclic of order $(p_j - 1)p_j^{\alpha_j - 1}$, so the number of solutions is

$$\gcd(2^t M, (p_j - 1)p_j^{\alpha_j - 1}) = \gcd(2^t M, 2^{s_j} M_j) = 2^{\min(t, s_j)} t_j.$$

By the Chinese remainder theorem, the number of solutions modulo N is therefore the product of these numbers, and hence we have proven the first claim. For the second claim, we see right away that either there does not exist any solution, or there does exist a solution and therefore the set of solutions is in bijection with the solutions of the previous congruence. The congruence $a^{2^tM} \equiv -1 \bmod p_j^{\alpha_j}$ is solvable if and only if 2^{t+1} divides $(p_j-1)p_j^{\alpha_j-1}$, in other words if and only if $t+1 \leqslant s_j$, hence we have the desired result. □

Proof. (of Proposition 2-3.6) Assume that $s_1 \leqslant s_2 \leqslant \ldots \leqslant s_k$. By decomposing the set S into $S_0 := \{a \in (\mathbf{Z}/N\mathbf{Z})^* \mid a^M \equiv 1 \bmod N\}$ and $T_j := \left\{a \in (\mathbf{Z}/N\mathbf{Z})^* \mid a^{2^jM} \equiv -1 \bmod N\right\}$ for $0 \leqslant j \leqslant s_1 - 1$ and by applying Lemma 2-3.8 to each one of these sets, we have

$$\operatorname{card}(S) = t_1 \cdots t_k \left(1 + 1 + 2^k + \cdots + 2^{k(s_1-1)}\right) = t_1 \cdots t_k \left(\frac{2^{ks_1} + 2^k - 2}{2^k - 1}\right).$$

The ratio of $a \in G_0$ which pass the Rabin-Miller test is therefore

$$\frac{\operatorname{card}(S)}{\operatorname{card}(G_0)} = \frac{t_1 \cdots t_k}{M_1 \cdots M_k} \frac{2^{-(s_1+\cdots+s_k)}}{p_1^{\alpha_1-1} \cdots p_k^{\alpha_k-1}} \left(\frac{2^{ks_1} + 2^k - 2}{2^k - 1}\right). \tag{2.2}$$

If $k = 1$, the ratio is equal to $\dfrac{t_1}{M_1 p_1^{\alpha_1-1}} \leqslant \dfrac{1}{p_1^{\alpha_1-1}}$, and is therefore $\leqslant \dfrac{1}{5}$, except when $N = 3^2$ in which case we have $|S|/|G_0| = 1/3$. If $k \geqslant 2$, we can assume that $\alpha_1 = \cdots = \alpha_k = 1$, if not, the ratio is $\leqslant 1/p_i$, which in practice we can assume to be arbitrarily small. If one of the M_i is different from t_i, then $t_1 \ldots t_k / M_1 \ldots M_k \leqslant 1/3$. Furthermore,

$$2^{-s_1-\cdots-s_k}\left(\frac{2^{ks_1} + 2^k - 2}{2^k - 1}\right) \leqslant 2^{-ks_1}\frac{2^k - 2}{2^k - 1} + \frac{1}{2^k - 1} \leqslant 2^{1-k},$$

so the ratio is $\leqslant 1/8$ if $k \geqslant 4$ and $\leqslant 1/4$ if $k = 3$.

If $k = 2$ and if one of the M_i is distinct from all of the t_i, then the ratio is $\leqslant 1/6$. If $k = 2$ and $M_1 = t_1$ (i.e., M_1 divides M) and $M_2 = t_2$ (i.e., M_2 divides M), we see that $M_1 = M_2$, hence $s_1 < s_2$ (if not $p_1 = p_2$). We then have that the ratio is $\leqslant 2^{s_1-s_2}\dfrac{1 + 2^{1-2s_1}}{3} \leqslant \dfrac{1 + 2^{1-2s_1}}{6} \leqslant \dfrac{1}{4}$. □

3.9. Remark. By looking at the upper bounds above, we can prove that the two "worst" cases are the following.

i) The number N is equal to pq with $q = 2p - 1$ and $p \equiv 3 \bmod 4$. For example, $N = 3 \cdot 5$, $N = 7 \cdot 13$, etc. It follows that $p = 1 + 2M_1$ and $q = 1 + 4M_1$ and $N = (1 + 2M_1)(1 + 4M_1) = 1 + 2M_1(3 + 4M_1)$, hence $t_1 = t_2 = M_1 = M_2$ and so

$$\frac{\operatorname{card}(S)}{\operatorname{card}(G_0)} = \frac{1}{4}.$$

ii) The number N is equal to $pqr = 1 + 2M$, where $p = 1 + 2M_1$, $q = 1 + 2M_2$, $r = 1 + 2M_3$ and M_i divides M. It follows from this that the ratio is also 1/4. Take for example: $N = 8911 = 7 \cdot 19 \cdot 67$ (where $M_1 = 3$, $M_2 = 9$, $M_3 = 33$ and $M = 4455 = 3^4 \cdot 5 \cdot 11$).

4. Primality Test (II)

In this section we present the Agrawal-Kayal-Saxena algorithm [10], which dates back to July 2002, and was introduced in their article "PRIMES is in P". *It gives a primality test in polynomial time.*

The original idea was to perform tests in $\mathbf{Z}[X]$. For example, we easily see that if N is prime, then $(X-a)^N \equiv X^N - a \bmod N$, but this test has the major default of requiring the computation of N coefficients. That will just not do!

4.1. Lemma. *Let N be prime and $h(X) \in \mathbf{Z}[X]$ a polynomial of degree r. Then*

$$(X-a)^N \equiv X^N - a \ \operatorname{mod}(N, h(X)).$$

Recall that in a ring, the notation $a \equiv b \bmod I$ means that $a-b$ belongs to the ideal I and that $(a_1, \dots, a_m)$ is the notation used for the ideal generated by $a_1, \dots, a_m$. Thus the congruence in the lemma can be restated as: there exists $P, Q \in \mathbf{Z}[X]$ such that $(X-a)^N - (X^N - a) = NP(X) + h(X)Q(X)$.

It should be noted that if r is $O((\log N)^k)$, then this test remains polynomial. The problem is to choose pairs a, $h(X)$ in such a way that they detect non-primality. The solution proposed by Agrawal, Kayal and Saxena is to choose $h(X) = X^r - 1$ with r being a "very well-chosen" prime, in particular $r = O((\log N)^k)$, and to prove that it is then sufficient to test the $a \in [1, L]$ with $L = O(\sqrt{r} \log N)$ in order to ensure that N is prime, or possibly a prime power, which is not so bad.

The argument is essentially algebraic and combinatorial, but nevertheless uses a result on the distribution of prime numbers, in fact a weak form of the prime number theorem (see Chap. IV, (4.10)), which says that the sum of the $\log p$ for p prime and smaller than x is $\geqslant c_1 x$ for some constant $c_1 > 0$. We summarize what we are going to use in a lemma.

4.2. Lemma. *Let $Y > 1$ and let $N \geqslant 2$ be an integer. There exists a prime number r which satisfies the following two properties.*

i) The order of N modulo r is at least Y.
ii) Furthermore, $r = O\left(Y^2 \log N\right)$.

Proof. Set $A := \prod_{1 \leqslant y \leqslant Y}(N^y - 1)$. Let r be the smallest prime number which does not divide A. Then for $y \leqslant Y$, we have $N^y \not\equiv 1 \bmod r$, and hence condition *i*). Moreover, every $p < r$ divides A, whereas $A \leqslant N^{Y(Y+1)/2}$ and consequently

$$c_1 r \leqslant \sum_{p<r} \log p \leqslant \log A \leqslant \frac{Y(Y+1)}{2} \log N.$$

From this we have that $r = O\left(Y^2 \log N\right)$. □

Remark. We could add that, since the order of N modulo r divides $r-1$, we necessarily have $r > Y$.

We will also use the following elementary combinatorial lemma.

4.3. Lemma. *The cardinality of the set of monomials in L variables of degree $\leqslant k$ is*

$$\operatorname{card}\{(m_1, \ldots, m_L) \mid m_i \geqslant 0 \text{ and } m_1 + \cdots + m_L \leqslant k\} = \binom{L+k}{k}.$$

Furthermore, we have the estimate

$$\binom{L+k}{k} \geqslant 2^{\min(L,k)}.$$

Proof. The first formula is classical and can be proven, for example, by induction (call the cardinality in question $f(L,k)$, check that $f(L,0) = 1$ and $f(1,k) = k+1$, and then prove that $f(L,k) = f(L,k-1) + f(L-1,k)$). For the lower bound, observe that if $k \leqslant L$, then

$$\binom{L+k}{k} = \frac{(L+k)(L+k-1)\cdots(L+2)(L+1)}{k(k-1)\cdots 2\cdot 1} = \prod_{i=0}^{k-1}\left(\frac{L+k-i}{k-i}\right) \geqslant 2^k,$$

and if $L \geqslant k$, reverse the roles of L and k. □

Remark. We can often improve this inequality; for example, if $1 \leqslant k \leqslant L$, then $\binom{L+k}{k} \geqslant 2^k(L+1)/2$, and thus if $L \geqslant 5$, we have $\binom{L+k}{k} \geqslant 2^{k+1}$.

We will now state a version of the main theorem of Agrawal-Kayal-Saxena.

4.4. Theorem. *Let $N \geqslant 2$ and let r be a prime number satisfying:*

i) no prime number $\leqslant r$ divides N;
ii) we have $\operatorname{ord}(N \bmod r) \geqslant (2\log N/\log 2)^2 + 1$;

iii) for $1 \leqslant a \leqslant r-1$, *we have*

$$(X-a)^N \equiv X^N - a \mod (N, X^r - 1).$$

Then N *is a prime power.*

Remarks. In order to prove this theorem, we only assume that hypothesis *iii)* is satisfied for $1 \leqslant a \leqslant L$ and we will see that we can take L smaller than $r-1$. By Lemma 2-4.2, we can choose $r = O\left((\log N)^5\right)$ such that *ii)* is satisfied, and it would necessarily follow that $r \geqslant (2\log N/\log 2)^2 + 1$. Thus it is clear that the theorem implies that the following algorithm is correct and polynomial.

ALGORITHM. [10] We put in N and the algorithm returns "Prime" or "Composite".

1) We check to see if $N = a^b$ where $b \geqslant 2$; if so, then N is "Composite".
2) We try the prime numbers $r = 2, 3 \ldots$. If r divides N, N is "Composite". If not, we check whether r is relatively prime $N^y - 1$ for $y = 1, 2, \ldots, Y$, where $Y = \lfloor (2\log N/\log 2)^2 \rfloor + 1$; if so we keep r and go to the next step, if not we look for a larger r.
3) For $a = 1, 2, 3, 4, \ldots$ (stop at $r-1$), we check whether $(X-a)^N \not\equiv X^N - a \mod (N, X^r - 1)$. If so, then N is "Composite", if not, we proceed to $a+1$.
4) If the algorithm keeps going until $a = r-1$, then N is "Prime".

Let us briefly discuss its complexity (without trying to optimize it). We easily see that the longest step is step (3), which requires $O(r \log N)$ multiplications in the ring $\mathbf{Z}[X]/(N, X^r - 1)$, where each one uses at most $O((r\log N)^2)$ elementary operations. We thus have $O((r\log N)^3)$ in all. If we add that $r = O\left((\log N)^5\right)$, we obtain a complexity of at most $O\left((\log N)^{18}\right)$.

We now proceed to the proof of the theorem. Let p be a prime divisor of N. We denote by $d_1 := \mathrm{ord}(N \bmod r)$, $d_2 = \mathrm{ord}(p \bmod r)$ and $d := \mathrm{lcm}(d_1, d_2)$. It should be noted that d_1 (resp. d_2) is the order of the subgroup generated by N (resp. by p) in $(\mathbf{Z}/r\mathbf{Z})^*$ and that d is therefore the order of the subgroup generated by N and p in $(\mathbf{Z}/r\mathbf{Z})^*$. We then choose $h(X)$ to be an irreducible factor of $\Phi_r(X) := (X^r - 1)/(X - 1)$ in $\mathbf{F}_p[X]$. Let us point out, even if we do not need it, that $\deg(h) = d_2$ (see Theorem 2-6.2.8). We will work in the *field* $K := \mathbf{F}_p[X]/(h(X))$, which is a finite field (isomorphic to $\mathbf{F}_{p^{d_2}}$) and which we obtain by adding a primitive rth root of unity to $\mathbf{F}_p$. By construction, $x := X \bmod h(X)$ is of order r in K^*. It is natural to look at the subgroup G of K^* generated by the classes of $(X-a)$ for $1 \leqslant a \leqslant L$. The heart of the proof consists of finding an upper and lower bound for the order of G.

4.5. Lemma. *We have the lower bound*

$$\operatorname{card}(G) \geqslant \binom{L+d-1}{d-1} \geqslant 2^{\min(L,d-1)}.$$

From the remark immediately following the combinatorial lemma (Lemma 2-4.3), we have for example that if $1 \leqslant d-1 \leqslant L$, then $\operatorname{card}(G) \geqslant 2^d$, and if $L \leqslant d$, then $\operatorname{card}(G) \geqslant 2^{L+1}$.

Proof. In light of the combinatorial lemma mentioned above, it suffices to show that the classes of elements,

$$\prod_{1 \leqslant a \leqslant L} (X-a)^{m_a}, \quad \text{for } m_a \geqslant 0 \quad \text{and} \quad \sum_{a=1}^{L} m_a \leqslant d-1,$$

are all distinct in K. First of all, the a are distinct modulo p, because if not, then $p \leqslant L < r$ and we assumed that N was not divisible by any prime number smaller than r, so $p > r$. Thus our polynomials are all distinct in $\mathbf{F}_p[X]$. Now we bring in the key point that if $P = \prod_{1 \leqslant a \leqslant L}(X-a)^{m_a}$, then we have, on one hand, $P(X)^N \equiv P(X^N) \bmod (N, X^r-1)$, but also $P(X)^p \equiv P(X^p) \bmod p$, so the two congruences are valid $\bmod(p, X^r-1)$. For $m = N^i p^j$, it therefore follows that

$$P(X)^m \equiv P(X^m) \mod(p, X^r-1) \text{ or even } \mod(p, h(X)).$$

In fact, the set of m such that $P(X)^m \equiv P(X^m) \bmod(p, X^r-1)$ is multiplicative (the fairly simple proof is given in detail in part *ii*) of 2-4.7 below). Now let P and Q be two polynomials of the form given above (considered in $\mathbf{F}_p[X]$), and suppose that they are in the same class in K, i.e., suppose $P \equiv Q \bmod(p, h(X))$. Let x be the class of X, which is an rth primitive root of unity in K, and therefore

$$(P-Q)(x^m) = 0, \qquad \text{for} \quad m \in \langle N, p\rangle \subset (\mathbf{Z}/r\mathbf{Z})^*.$$

But we know that N and p generate a subgroup of order d in $(\mathbf{Z}/r\mathbf{Z})^*$, thus the polynomial $P-Q$ has at least d roots, and since $\deg(P-Q) \leqslant d-1$, we see that $P = Q$ (first in $\mathbf{F}_p[X]$, then, if we want, in $\mathbf{Z}[X]$). □

In order to find an upper bound for $|G|$, we choose a generator of G (it is a subgroup of K^* and is thus cyclic) and define the following set.

4.6. Definition. Let g be a generator of G. We define

$$\mathscr{I} = \mathscr{I}_g := \{m \in \mathbf{N} \mid g(X)^m \equiv g(X^m) \bmod(X^r-1, p)\}.$$

The main properties of $\mathscr{I}$ are summarized in the following lemma.

4.7. Lemma. *The set $\mathscr{I}$ satisfies the following properties.*

i) N and p are in $\mathscr{I}$.

ii) $\mathscr{I}$ is multiplicative, i.e., if m_1 and $m_2 \in \mathscr{I}$, then $m_1 m_2 \in \mathscr{I}$.

iii) If m_1 and $m_2 \in \mathscr{I}$ satisfy $m_1 \equiv m_2 \mod r$, then $m_1 \equiv m_2 \mod \operatorname{card}(G)$.

Proof. The first property has already been established. For *ii*), write

$$g(X)^{m_1 m_2} = (g(X)^{m_1})^{m_2} \equiv (g(X^{m_1}))^{m_2} \operatorname{mod}(p, X^r - 1),$$

and notice that since $m_2 \in \mathscr{I}$, we have $g(Y)^{m_2} \equiv g(Y^{m_2}) \operatorname{mod}(p, Y^r - 1)$. Therefore, by substituting $Y = X^{m_1}$, we obtain

$$\begin{aligned}(g(X^{m_1}))^{m_2} &= g(X^{m_1 m_2}) + pQ_1(X^{m_1}) + (X^{m_1 r} - 1)Q_2(X^{m_1}) \\ &\equiv g(X^{m_1 m_2}) \operatorname{mod}(p, X^r - 1).\end{aligned}$$

In order to prove *iii*), suppose that m_1 and $m_2 \in \mathscr{I}_g$ and that $m_2 = m_1 + kr$, where $k \geqslant 0$. It follows from this that

$$g(X)^{m_2} \equiv g(X^{m_2}) \operatorname{mod}(X^r - 1, p) \qquad \text{and thus} \qquad \operatorname{mod}(h(X), p);$$

hence $g(X)^{m_1+kr} = g(X^{m_1+kr})$ in K. But $X^{m_1+kr} \equiv X^{m_1} \operatorname{mod}(X^r - 1)$ and therefore $\operatorname{mod}(h(X))$. Thus we obtain the equality in K^*

$$g(X)^{m_1} g(X)^{kr} = g(X^{m_1}) = g(X)^{m_1},$$

where the last equality comes from the hypothesis that $m_1 \in \mathscr{I}$. From this, we of course have that $g(X)^{kr} = 1 \in K^*$ and hence $\operatorname{card}(G)$ divides $kr = m_2 - m_1$. □

Proof. (end of the proof of Theorem 2-4.4) In order to apply the lemma, we use that N, p and hence all of the products of powers $N^i p^j$ are in $\mathscr{I}$. Recall that these elements generate a subgroup of order d in $(\mathbf{Z}/r\mathbf{Z})^*$. If we set

$$E := \{(i,j) \in \mathbf{N} \times \mathbf{N} \mid 0 \leqslant i, j \leqslant \sqrt{d}\},$$

then the cardinality of E is $(\lfloor\sqrt{d}\rfloor + 1)^2 > d$. By the pigeonhole principle[4], there are two elements $N^{i_1}p^{j_1}$ and $N^{i_2}p^{j_2}$, which are congruent modulo r, and such that (i_1, j_1) and (i_2, j_2) are distinct in E. These two elements $N^{i_1}p^{j_1}$ and $N^{i_2}p^{j_2}$ are therefore congruent modulo $\operatorname{card}(G)$. First suppose that $N^{i_1}p^{j_1} \neq N^{i_2}p^{j_2}$, which implies that

$$\operatorname{card}(G) \leqslant |N^{i_1}p^{j_1} - N^{i_2}p^{j_2}| \leqslant N^{2\sqrt{d}}.$$

If we combine this upper bound with the lower bound gotten above, we see that

$$\min(L + 1, d) \log 2 \leqslant (2\sqrt{d}) \log N.$$

[4]The pigeonhole principle says that if we put $n+1$ pigeons into n boxes, at least one of the boxes will contain at least two pigeons.

We will prove that this inequality is impossible.

1) If we had $L \geqslant d$, we could deduce that $\sqrt{d} \leqslant 2\log N/\log 2$ or moreover that $d \leqslant (2\log N/\log 2)^2$. But this inequality is a contradiction since, by construction, $d \geqslant d_1$, and we assumed that $d_1 > (2\log N/\log 2)^2$.
2) Now if $L < d$, we deduce that $(L+1)\log 2 \leqslant (2\sqrt{d})\log N$, and since $d \leqslant r-1$, this would give us $(L+1)\log 2 \leqslant 2\sqrt{r-1}\log N$.

It is therefore a sufficient condition that $L \geqslant 2\sqrt{r-1}\log N/\log 2$ is large enough in order to conclude that $N^{i_1}p^{j_1} = N^{i_2}p^{j_2}$. The choice $L = r-1$ is suitable[5] since then the desired equality would be equivalent to the inequality $\sqrt{r-1} \geqslant 2\log N/\log 2$, which is where the hypothesis $r \geqslant (2\log N/\log 2)^2 + 1$ comes from. We finish the proof by pointing out that the inequality $N^{i_1}p^{j_1} = N^{i_2}p^{j_2}$ immediately implies that $N = p^\alpha$. □

4.8. Remark. One variation of this proof consists of abandoning the constraint that r is a prime number; we choose a factor, $h(X)$, of $\Phi_r \in \mathbf{F}_p[X]$ where Φ_r is the rth cyclotomic polynomial (cf. Sect. 6 of this chapter), and we could then omit every analytic estimate of the distribution of prime numbers (see [33] for this version, as well as a finer estimate of the complexity).

5. Factorization

We briefly consider, and necessarily very unsatisfactorily, the problem of factorization: having established, by a primality test, that an integer N is not prime, how could we go about factoring it? We start by pointing out that the (complete) factorization problem is essentially equivalent to the problem of finding one factor, because of course, by iterating this procedure, we would achieve a complete factorization.

The naive factorization method consists of checking if 2 divides N, then if 3 divides N, etc. If $N = pq$ where p and q are roughly of the same size, i.e., $p \sim q \sim \sqrt{N}$, we see that we would need to perform $O(\sqrt{N})$ divisions before arriving at a factorization of N. The naive algorithm is thus exponential.

There do exist more efficient algorithms. In fact, one of the best algorithms known [49] (using elliptic curves) has a number of operations estimated by $\exp(C\sqrt{\log p \log\log p})$, where p is the smallest prime factor of N. In the case where $N = pq$ where $p \sim q \sim \sqrt{N}$, we therefore get an algorithm with an order of complexity $\exp(C'(\log N)^\kappa)$ (where $\kappa < 1$), which grows

[5] We point out however that we could take $L = O(\sqrt{r}\log N)$, which would allow us to slightly improve the estimate of the complexity.

less quickly than N^κ but more quickly than $(\log N)^\kappa$. We say that such an algorithm is *subexponential.* Another algorithm [19] ("number field sieve") has a complexity on the order of $\exp\left(C(\log N)^{1/3}(\log\log N)^{2/3}\right)$. In 2006, it was known in practice how to factor an integer with 100 digits in a couple of hours, and by using many computers over many months, how to factor an integer with 150 digits. But we still cannot factor, over the course of a human lifetime, an RSA number, with say 300 digits. A surprising fact is that the complexity of various algorithms (proven probabilistically or heuristically) tends to take the form of a function (see [48]):

$$L(b, N) := \exp\left(C(\log N)^b(\log\log N)^{1-b}\right).$$

The case $b = 0$, in other words $(\log N)^C$, corresponds to polynomial algorithms, the case $b = 1$, in other words N^C, corresponds to exponential algorithms and the cases $0 < b < 1$ correspond to subexponential algorithms; the two algorithms cited above have a complexity estimated at $L(1/2, N)$ and $L(1/3, N)$.

We are not going to present the most powerful algorithms right away, since they use tools which surpass the level of this chapter; the algorithms which use elliptic curves and the number field sieve are presented in Appendix A, which is about factorization. For the moment, we will settle for describing an algorithm which improves on the naive algorithm by providing an even more efficient one.

From now on, we use the convention that the letter p is reserved for a factor of N.

Pollard's ρ algorithm. We proceed as follows. We choose a_0 between 1 and N and we compute the sequence given by $a_{i+1} = f(a_i)$, where $f(a) := a^2 + 1 \bmod N$. We then choose k "big enough, but not too big" and we calculate $\gcd(a_{2k} - a_k, N)$, hoping that it is nontrivial; if that is the case, we have found a factorization, if not, we try again with larger k. We will explain below why, at least statistically, there exists k of size $O(\sqrt{p})$, where p divides $\gcd(a_{2k} - a_k, N)$. Assuming that, we see that the average complexity of the algorithm is $O(\sqrt{p})$, thus $O(\sqrt[4]{N})$.

The analysis of the complexity is based on the hypothesis that the sequence a_i modulo p is sufficiently "random", which has been satisfactorily confirmed in practice. Now, the probability that r numbers modulo p chosen "at random" are all distinct is[6]

$$P_r = \left(1 - \frac{1}{p}\right)\left(1 - \frac{2}{p}\right)\cdots\left(1 - \frac{r-1}{p}\right) \leqslant \exp\left(-\frac{r(r-1)}{2p}\right).$$

If we take r on the order of $\sqrt{p}$, say $r \geqslant 2\sqrt{p}$, the probability that two

[6]Example. If $n \geqslant 23$, the probability that, among n people, two have the same birthday is greater than $1/2$.

of the numbers are equal (modulo p) will be $> 1/2$, thus we have a good chance to have two indices $i < j < r$ such that $a_i \equiv a_j \bmod p$. Considering the construction that follows, we would have $a_{i+m} \equiv a_{j+m} \bmod p$ for every $m \geqslant 0$, and in particular, by taking $m = j - 2i$ and $k = j - i$, we would have $a_k \equiv a_{2k} \bmod p$ (see [22] for more details).

"Difference of squares" algorithm. The second algorithm, that we will only sketch, is based on the fact that the number of elements $a \in (\mathbf{Z}/N\mathbf{Z})^*$ such that $a^2 = 1$ is at least equal to 4 if N has at least two distinct prime factors. If we knew how to compute a square root in $(\mathbf{Z}/N\mathbf{Z})^*$, say $\mathscr{A}(x)$, with a fast algorithm $\mathscr{A}$, then we could factor N like this: take a at random and calculate $b = \mathscr{A}(a^2)$. Then we of course have that $a^2 \equiv b^2 \bmod N$, or even that N divides $(a + b)(a - b)$. Now, there is (at least) a one in two chance that $\pm a \bmod N$ is not the square root calculated by $\mathscr{A}$ and, in this case, the calculation of $\gcd(N, a + b)$ or of $\gcd(N, a - b)$ would give us a factorization. Unfortunately, or luckily, we do not know of any fast algorithm $\mathscr{A}$ (it is even possible that one does not exist). One extension of this idea is the following: instead of directly looking for an equality $a^2 \equiv b^2 \bmod N$, we try to construct one. In order to do this, we randomly take a close to $\sqrt{N}$, we reduce a^2 modulo N (taking care to take the representative in $[-N/2, N/2]$) and we try to factor it with small prime numbers. In this way, we get a family of congruences $a_j^2 \equiv \prod_{p \in S} p^{n_{p,j}}$. We therefore look for a combination of these numbers which provides an equality of the type $\prod_i a_i^2 \equiv \prod_j b_j^2 \bmod N$ (this is a linear algebra problem over $\mathbf{F}_2$). This idea, presented very vaguely here, is expanded on in more detail in Appendix A, when we describe the number field sieve algorithm. Property quantified, this algorithm has an average (heuristic) complexity on the order of $L(1/2, N)$—which is already remarkable, even if it is insufficient for very large numbers.

Examples of precautions to take when choosing p and q for the RSA method. We will only give some very elementary indications, since the question is fairly complex, and in fact largely open.

1) The absolute value, $|p - q|$, must be large. We can see why by writing $q = p + \delta$ where δ is much smaller than p. Since $N = pq$, then $\sqrt{N} = p\sqrt{1 + \delta/p} \sim p + \delta/2$ and we could find p with the "naive" algorithm in $O(\delta)$ steps!

2) It must be that $p - 1$ (resp. $q - 1$) are not too *smooth*, in other words, cannot be factored too quickly, for example the product of small prime numbers. To see why this is true, choose $C > 0$, and let $p_1, \ldots, p_k$ be the prime numbers smaller than C; the set $S := \{s = p_1^{m_1} \cdots p_k^{m_k} \mid s \leqslant N\}$ has cardinality $O((\log N)^k)$, and we can therefore calculate $\gcd(a^s - 1, N)$ for some values of a and $s \in S$ in polynomial time. If $p - 1 \in S$ (in other words

if $p-1$ only has prime factors $\leqslant C$), then we have a very good chance of being able to factor N.

3) A less obvious constraint is that it must be that the "secret" exponent e is not too small. It is clear that if $e = O(\log N)$ for example, then by trying $O(\log N)$ times, we will find e, but in fact it can be shown that you must avoid having $e \ll N^{1/4}$ (see Exercise 3-6.12 of Chap. IV).

These relatively trivial remarks could cast doubt the security of the RSA system (see [17] for a more precise description of the catalogued attacks on the RSA system). However, theoretical support for it is provided by the following considerations. Let us call P the class of problems for which there exists a polynomial algorithm (for example the problem of deciding whether a number is prime is in P, by Agrawal-Kayal-Saxena). We can define a class NP, a priori much larger than P, which is the class of problems for which there exists a polynomial verification (for example, the problem of factorization of a number is clearly in NP, since if we are given a factorization, we can verify it in polynomial time). However, the factorization problem has a subexponential solution. The security of the RSA system rests, from a theoretical point of view, on the hypothesis that the factorization problem is not in P. In fact, it is a special case of a large problem in complexity theory[7]:

Is it true that $\mathrm{P} \neq \mathrm{NP}$?

6. Error-Correcting Codes

We give a glimpse of another industrial application of algebra and arithmetic: the construction of "error-correcting codes", which can, to a certain degree, reconstruct a message if its transmission was slightly defective. This technique is for example needed to produce CD readers, to transmit images by space probes, etc. If this introduction leaves you hungry to learn more, I recommend Demazure's book, Cours d'algèbre *[3].*

6.1. Generalities about Error-Correcting Codes

In order to transmit information, we assume that we are using a finite alphabet $\mathcal{Q}$, containing q symbols or letters and that we are sending words of a fixed length n; a word is therefore and element of $\mathcal{Q}^n$. We can think of binary language, i.e., $\mathcal{Q} := \{0,1\}$, or of genetic codes, for example $\mathcal{Q} := \{A, C, G, U\}$ (the bases found in RNA are A for adenine, C for cytosine, G for guanine and U for uracil). We will most often take the example of

[7]This problem $\mathrm{P} \neq \mathrm{NP}$ is one of the seven problems, for the solution of which a million dollars is offered by the Clay Mathematics Institute.

$\mathscr{Q} := \mathbf{F}_q$, which has the disadvantage of limiting the possible values of q but the advantage of providing a richer structure.

The set of words $\mathscr{Q}^n$ can be endowed with a *Hamming distance*, defined as follows. If $x = (x_1, \ldots, x_n) \in \mathscr{Q}^n$ and $x' = (x'_1, \ldots, x'_n) \in \mathscr{Q}^n$, then

$$d(x, x') := \operatorname{card}\{i \in [1, n] \mid x_i \neq x'_i\}.$$

It can easily be checked that is in fact a distance.

A *code* is a subset $\mathscr{C} \subset \mathscr{Q}^n$ containing at least two distinct elements in $\mathscr{Q}^n$; we define the *distance* of a code as

$$d(\mathscr{C}) := \min_{x \neq x' \in \mathscr{C}} d(x, x').$$

Once we have chosen a code $\mathscr{C}$, the principle consists of only sending those messages which belong to $\mathscr{C}$. If we know that at most $d(\mathscr{C}) - 1$ transmission errors have been committed, then using the error-correcting code will enable us to establish the existence of one or more errors. Furthermore, if t errors have been committed during the transition of a word and if $2t + 1 \leqslant d(\mathscr{C})$, we see that there exists one single word in $\mathscr{C}$ located at a distance $\leqslant t$ from the received word. In conclusion, the code allows us to correct t errors and we say that it is *t-correcting*. If we denote by $d = d(\mathscr{C})$ the distance of the code and $t = t(\mathscr{C})$ the number of errors that are systematically corrected by the code, we easily see that relationship between the two is given by $t = \left\lfloor \frac{d-1}{2} \right\rfloor$ and conversely $d = 2t + 1$ or $2t + 2$. Except for some examples, we leave aside the question of *decoding*, which is essentially the study of algorithms which allow you to find the word of the code located at a minimal distance from a given word (it should be noted that you cannot in general guarantee the uniqueness of this word except under certain conditions). One of the properties required of a code is obviously that it corrects or finds the most possible errors (we could also insist that the decoding be the simplest possible). An intuitively obvious requirement is that it uses the least amount of space; we could formalize this idea by introducing the *code rate* t/n, and the *information rate* that we define as the ratio $\log \operatorname{card}(\mathscr{C}) / n \log q$. Information theory, developed by Shannon (see the founding article [67]), says that if we are willing to send longer and longer messages (i.e., to let n be very large), then there exist codes as safe we want them to be, with an information rate close to 1. Shannon's theorem is however an existence theorem, it does not specify how to construct such codes.

We are actually going to exclusively concentrate on *linear codes*, where the alphabet is (in bijection with) $\mathbf{F}_q$, the space of words is (in bijection with) the vector space $(\mathbf{F}_q)^n$ and $\mathscr{C}$ is a subspace. In the case of $q = 2$, we are talking about binary codes, in the case $q = 3$, we are talking about ternary codes, etc.

The most important parameters of a linear code are the cardinality of the alphabet $q = \operatorname{card} \mathscr{Q}$, its *length* say n, its *dimension* say $k := \dim \mathscr{C}$, its distance $d(\mathscr{C})$, its code rate and its information rate k/n.

Remark. Let $\mathscr{C} \subset \mathbf{F}_q^n$ be a linear code. We define the *weight* of an element $w(x)$ as the number of non-zero components of x. We can easily see that

$$d(\mathscr{C}) = \min_{0 \neq x \in \mathscr{C}} d(0, x) = \min_{0 \neq x \in \mathscr{C}} w(x).$$

6.1.1. Examples. 1) The most basic example of a code is the use of a *parity bit*: in order to transmit a word $x = (x_1, \dots, x_{n-1}) \in (\mathbf{F}_2)^{n-1}$, we send $\bar{x} = (x_1, \dots, x_{n-1}, x_1 + \dots + x_{n-1}) \in (\mathbf{F}_2)^n$. To see if the received message $x' = (x_1, \dots, x_n)$ is correct, we check whether $x_n = x_1 + \dots + x_{n-1}$. This code has length n and dimension $n - 1$. It allows us to find an error but not to correct it. Its distance is 2.

2) Hamming code. Take the set of words with seven binary digits, $q = 2$, $n = 7$, and let $\mathscr{C}$ be the code with basis

$$e_0 = \begin{pmatrix} 1 \\ 1 \\ 0 \\ 1 \\ 0 \\ 0 \\ 0 \end{pmatrix}, \quad e_1 = \begin{pmatrix} 0 \\ 1 \\ 1 \\ 0 \\ 1 \\ 0 \\ 0 \end{pmatrix}, \quad e_2 = \begin{pmatrix} 0 \\ 0 \\ 1 \\ 1 \\ 0 \\ 1 \\ 0 \end{pmatrix}, \quad e_3 = \begin{pmatrix} 0 \\ 0 \\ 0 \\ 1 \\ 1 \\ 0 \\ 1 \end{pmatrix}.$$

The coding principle is simple: in order to transmit a message $m = (m_0, m_1, m_2, m_3)$, we transmit $x = m_0 e_0 + m_1 e_1 + m_2 e_2 + m_3 e_3$. For this simple example, we will explain the decoding *under the hypothesis that at most one error was committed.* Equations of the vector subspace $\mathscr{C}$ are given by

$$L(x) = (x_0 + x_3 + x_5 + x_6, x_1 + x_3 + x_4 + x_6, x_2 + x_4 + x_5 + x_6) = 0.$$

For each vector e of weight 1, we then calculate the triplet $L(e)$. From this, we obtain the following algorithm of correction and decoding. After having received the message $x = (x_0, \dots, x_6)$, we check whether $L(x) = 0$. If $L(x) = 0$, the message is correct, if $L(x) = (1, 0, 0)$, then x_0 must be corrected, if $L(x) = (0, 1, 0)$, then x_1 must be corrected and if $L(x) = (1, 0, 1)$, then x_5 must be corrected. Finally, if $L(x) = (1, 1, 1)$, then x_6 must be corrected. Thus we have $m = (x_0, x_0 + x_1, x_5, x_6)$.

We denote by $T(x_1, \dots, x_7) := (x_7, x_1, \dots, x_6)$ the "shift", so we have that $T(e_0) = e_1$, $T(e_1) = e_2$, $T(e_2) = e_3$ and $T(e_3) = e_0 + e_1 + e_2$. Thus

$T(\mathscr{C}) = \mathscr{C}$ ($\mathscr{C}$ is then called cyclic). It is easy to see that each non-zero vector in $\mathscr{C}$ has at least three non-zero coordinates, and therefore $d(\mathscr{C}) = 3$. Therefore, this code is 1-correcting and allows us to identify two errors but not to correct them.

An amusing example. The previous code suggests that it is possible to recover an element of $\mathbf{F}_2^4$ (or say an integer between 0 and 15) starting with an element of $\mathbf{F}_2^7$ (or say seven yes/no pieces of information) if at most one error has been committed (granted at most one of the bits of information is false). One version of this is the seven following questions which allow us to determine an integer N between 0 and 15.

1) Is the integer $N \geqslant 8$?
2) Is the integer N in the set $\{4, 5, 6, 7, 12, 13, 14, 15\}$?
3) Is the integer N in the set $\{2, 3, 6, 7, 10, 11, 14, 15\}$?
4) Is the integer N odd?
5) Is the integer N in the set $\{1, 2, 4, 7, 9, 10, 12, 15\}$?
6) Is the integer N in the set $\{1, 2, 5, 6, 8, 11, 12, 15\}$?
7) Is the integer N in the set $\{1, 3, 4, 6, 8, 10, 13, 15\}$?

We leave as an exercise the justification of the following algorithm. We denote the answers to the above questions by $m = (m_1, \ldots, m_7)$ ($m_i = 1$ if the ith answer is yes, $m_i = 0$ if not), and we compute $a_1 = m_4 + m_5 + m_6 + m_7$, $a_2 = m_2 + m_3 + m_6 + m_7$ and $a_3 = m_1 + m_3 + m_5 + m_7$. If $a_1 = a_2 = a_3 = 0$, we conclude that there is not an error, if not we change the rth answer m_r into $r = \overline{a_1 a_2 a_3}$ (binary numeral notation), and the number we are looking for is therefore written

$N = \overline{m_1 m_2 m_3 m_4}$.

We will now show how to characterize and construct codes and how to deduce new codes from the given ones by using elementary linear algebra. We denote by n the length of the codes and by k their dimension, unless specified otherwise.

6.1.2. Definition. A *generator* matrix of a code $\mathscr{C}$ is a matrix whose rows form a basis of $\mathscr{C}$. (It is therefore a matrix of rank k having k rows and n columns.) A *parity-check* matrix of a code $\mathscr{C}$ is a matrix whose rows form a basis for the linear forms which are zero over $\mathscr{C}$. (It is therefore a matrix of rank $n-k$ having $n-k$ rows and n columns.)

6.1.3. Remarks. Being given a generator matrix is of course equivalent to being given a basis of the vector space $\mathscr{C}$, and given a parity-check matrix is of course equivalent to being given a basis of linear equations which define $\mathscr{C}$ in $\mathbf{F}_q^n$. If A is a generator matrix and B a parity-check matrix, we easily see that $A\,{}^tB = 0$, or also $B\,{}^tA = 0$. Moreover, we can recognize the distance of the code as the smallest number d such that there exist d dependent column vectors in B.

Assume that we are given a code $\mathscr{C}$ with parity-check matrix B and assume that the code is 1-correcting. We show you how to decode a received message, x', which is different in at least one coordinate from the sent message, x. First of all, if we denote the error committed by $\epsilon = x' - x$, we see that $B(x') = B(\epsilon)$. We will therefore compute $B(x')$; if this is non-zero, then no error has been committed, if not, we compute the images of the vectors e_i in the canonical basis $f_i = B(e_i)$. If only one error has been committed, we find a unique i such that $B(x')$ is proportional to f_i, say $B(x') = a_i f_i$, and therefore $\epsilon = a_i e_i$ and $x = x' - a_i e_i$.

If $\mathscr{C}$ is a code of length n over the field $\mathbf{F} = \mathbf{F}_q$, we can associate to it the following codes.

i) *Shortened* code. Let $d(\mathscr{C}) \leqslant \ell \leqslant n$. We set $\mathscr{C}^{(\ell)} := \{x \in \mathbf{F}_q^\ell \mid (x; 0, \ldots, 0) \in \mathscr{C}\}$. It is a code of length ℓ, and we easily see that $d\left(\mathscr{C}^{(\ell)}\right) \geqslant d\left(\mathscr{C}\right)$.
ii) *Extended* code. We can create the analogue of the "parity bit" by constructing $\bar{\mathscr{C}} := \{(x_1, \ldots, x_{n+1}) \in \mathbf{F}_q^{n+1} \mid (x_1, \ldots, x_n) \in \mathscr{C}$ and $x_1 + \cdots + x_n + x_{n+1} = 0\}$. We can easily see that $d\left(\mathscr{C}\right) \leqslant d\left(\bar{\mathscr{C}}\right) \leqslant d\left(\mathscr{C}\right) + 1$. One variation is the *even subcode* defined as $\mathscr{C}' = \{x \in \mathscr{C} \mid x_1 + \cdots + x_n = 0\}$. We have $d\left(\mathscr{C}\right) \leqslant d\left(\mathscr{C}'\right)$.
iii) *Dual* code. We define the scalar product $\langle x, y\rangle := x_1 y_1 + \cdots + x_n y_n$, and we set $\mathscr{C}^* := \{x' \in \mathbf{F}_q^n \mid \forall x \in \mathscr{C},\ \ \langle x, x'\rangle = 0\}$. We have that $\dim \mathscr{C}^* = n - \dim \mathscr{C}$. An interesting category of binary codes is that of *self-dual* codes, i.e., such that $\mathscr{C}^* = \mathscr{C}$; such codes have dimension $n/2$, and the weight of an element is even since $w(x) \equiv \langle x, x\rangle \bmod 2$.

As an exercise, you could try to figure out how to construct a parity-check (or generator) matrix of each of these codes, starting with the parity-check (or generator) matrix of the original code.

6.1.4. Lemma. *Let $\mathscr{C}$ be a code of dimension k and of length n over $\mathbf{F}_q$. The following inequalities hold:*

i) $d(\mathscr{C}) \leqslant n + 1 - k$;
ii) if $\mathscr{C}$ is t-correcting $1 + \binom{n}{1}(q-1) + \binom{n}{2}(q-1)^2 + \cdots + \binom{n}{t}(q-1)^t \leqslant q^{n-k}$.

Proof. *i)* The vectors of the form $(x_1, \ldots, x_{n+1-k}, 0, \ldots, 0)$ form a vector subspace $\mathscr{D}$ of $(\mathbf{F}_q)^n$. Since $\dim \mathscr{D} + \dim \mathscr{C} = n+1$, we see that $\mathscr{D} \cap \mathscr{C} \neq \{0\}$, hence the existence of a non-zero vector of $\mathscr{C}$ of weight $\leqslant n + 1 - k$. For *ii*), we can observe that for every $x \in \mathbf{F}_q^n$ and $0 \leqslant t \leqslant n$,

$$\operatorname{card}(B(x,t)) = 1 + \binom{n}{1}(q-1) + \binom{n}{2}(q-1)^2 + \cdots + \binom{n}{t}(q-1)^t.$$

If the code t-correcting, the balls $B(x,t)$ with center $x \in \mathscr{C}$ are disjoint and thus

$$\operatorname{card}(\cup_{x\in\mathscr{C}} B(x,t)) = q^k \operatorname{card}(B(0,t)) \leqslant q^n. \qquad \square$$

6.1.5. Definition. A code such that $d(\mathscr{C}) = n+1-k$ is called MDS *maximal distance separable*. A t-correcting code such that $\mathscr{C} = \cup_{x\in\mathscr{C}} B(x,t)$ (forcibly a disjoint union) is called *perfect* t-correcting.

The Hamming code of length 7 studied in the examples is perfect 1-correcting since, in this case, we can show that $\operatorname{card} B(x,1) = 1+7 = 8$ and $8 \operatorname{card} \mathscr{C} = 2^7$. We could also notice that this code is not MDS, because $d(\mathscr{C}) = 3 < 4 = n-k+1$.

6.2. Linear Cyclic Codes

We will explicitly describe an interesting class of codes which in particular contains some of the classical codes, such as that of Hamming, Reed-Solomon and Golay and which will lead us into the study of cyclotomic polynomials.

6.2.1. Definition. A linear cyclic code is a linear code, $\mathscr{C}$, of length n, which is stable under the transformation $T(a_0, a_1, \ldots, a_{n-1}) = (a_{n-1}, a_0, \ldots, a_{n-2})$.

We can give a nice algebraic characterization of cyclic codes by introducing the natural isomorphism of vector spaces $\mathbf{F}_q^n \cong \mathbf{F}_q[X]_n \cong \mathbf{F}_q[X]/Q\mathbf{F}_q[X]$, where $\mathbf{F}_q[X]_n$ represents the polynomials of degree $< n$ and where Q is a polynomial of degree n. Since the characteristic (or minimal) polynomial of the endomorphism T is $Q = X^n - 1$, we therefore choose this value. Hence we denote by $\psi : \mathbf{F}_q^n \to \mathbf{F}_q[X]_n \cong \mathbf{F}_q[X]/(X^n-1)$ defined as $\psi(a_0, a_1, \ldots, a_{n-1}) \mapsto a_0 + a_1 X + \cdots + a_{n-1}X^{n-1} \bmod (X^n - 1)$. We immediately see that

$$\psi \circ T(a_0, a_1, \ldots, a_{n-1}) = X(a_0 + a_1 X + \cdots + a_{n-1}X^{n-1}) \bmod (X^n - 1).$$

Thus a vector subspace $\mathscr{C} \subset \mathbf{F}_q^n$ is stable under T if and only if its image under ψ is stable under multiplication by X. We should point out that an $\mathbf{F}_q$ vector subspace of $\mathbf{F}_q[X]/(X^n-1)$ which is stable under multiplication by X is nothing other than an *ideal* of $\mathbf{F}_q[X]/(X^n-1)$. Finally, the ideals of $\mathbf{F}_q[X]/(X^n-1)$ correspond to the ideals of $\mathbf{F}_q[X]$ which contain the polynomial $X^n - 1$ and therefore are of the form $P\mathbf{F}_q[X]$ where P divides $X^n - 1$. We summarize this discussion in the following theorem.

6.2.2. Theorem. *Let $K := \mathbf{F}_q$ and let $\mathscr{C}$ be a cyclic code of length n. We identify K^n with $K[X]/(X^n - 1)$ via $(a_0, a_1, \dots, a_{n-1}) \mapsto a_0 + a_1X + \cdots + a_{n_1}X^{n-1}$. There exist natural bijections between the following objects:*

i) a cyclic code of length n;
ii) an ideal $K[X]/(X^n - 1)$;
iii) a monic polynomial which divides $X^n - 1$ in $K[X]$.

One of the bijections associates P, which divides $X^n - 1$, to the ideal $\mathscr{C}$ of $K[X]/(X^n-1)$ generated by its class modulo X^n-1, and another associates an ideal of $K[X]/(X^n - 1)$ to the vector subspace corresponding to $\mathscr{C}$ of K^n. Furthermore, $\dim \mathscr{C} = n - \deg(P)$.

This leads to the following problem: how to decompose the polynomial $X^n - 1$ in $\mathbf{F}_q[X]$?

It is of course better to start with a decomposition in $\mathbf{Z}[X]$ (or $\mathbf{Q}[X]$), which is provided by *cyclotomic polynomials*. In order to define these, we denote by $\mu_n = \{\zeta \in \mathbf{C} \mid \zeta^n = 1\}$ the group of nth roots of unity and μ_n^* the subset of nth primitive roots of unity, and hence $\operatorname{card} \mu_n = n$ and $\operatorname{card} \mu_n^* = \phi(n)$.

We will need Gauss's lemma.

6.2.3. Lemma. *If $P = p_0 + p_1X + \cdots + p_dX^d \in \mathbf{Z}[X]$ is a non-zero polynomial, we define its content as $c(P) := \gcd(p_0, \dots, p_d)$. We therefore have that*

$$c(PQ) = c(P)c(Q).$$

Proof. By factoring $P = c(P)P^*$ and $Q = c(Q)Q^*$, we see that $c(PQ) = c(P)c(Q)c(P^*Q^*)$. So we have reduced the proof to showing that if P and Q are primitive (i.e., $c(P) = c(Q) = 1$), then $c(PQ) = 1$. If p is a prime number, we denote by $\bar{P}$ the image of P in $\mathbf{F}_p[X]$. We have that $\bar{P} \neq 0$ and $\bar{Q} \neq 0$, thus $\bar{P} \cdot \bar{Q} = \overline{PQ} \neq 0$ because $\mathbf{F}_p[X]$ is integral. So no p divides $c(PQ)$, which implies that it is invertible. □

6.2.4. Corollary. *Let $P \in \mathbf{Z}[X]$. Suppose that there exist $Q, R \in \mathbf{Q}[X]$ such that $P = QR$. Then there exists $\lambda \in \mathbf{Q}^*$ such that λQ and $\lambda^{-1}R$ have integer coefficients.*

Proof. We can write $Q = \dfrac{a}{b}Q_1$ (resp. $R = \dfrac{c}{d}R_1$), where a, b, c, d are integers and where Q_1 and R_1 are primitive polynomials with integer coefficients. We can deduce from this that $bd\,P = ac\,Q_1R_1$ and, since the equality is in $\mathbf{Z}[X]$, we can deduce, using Gauss's lemma, that $bd\,c(P) = ac$ and, in particular, that bd divides ac. Thus $P = c(P)Q_1R_1$. □

6.2.5. Corollary. *If $\alpha \in \mathbf{C}$ is a root of a monic polynomial with integer coefficients, then the minimal (monic) polynomial of α has integer coefficients.*

Proof. Let P, a priori in $\mathbf{Q}[X]$, be the minimal polynomial of α and let Q be monic with integer coefficients such that $Q(\alpha) = 0$. Then $Q = PR$, where R is in $\mathbf{Q}[X]$. Gauss's lemma says that there exists $\lambda \in \mathbf{Q}^*$ such that $R_0 = \lambda R$ and $P_0 = \lambda^{-1}P$ have integer coefficients. By observing that $Q = P_0R_0$, it follows that the leading coefficient of P_0 is invertible, and hence $P = \pm P_0$ has integer coefficients. □

6.2.6. Definition. The nth cyclotomic polynomial, denoted Φ_n, is defined as

$$\Phi_n(X) := \prod_{\zeta \in \mu_n^*} (X - \zeta).$$

These polynomials, a priori with complex coefficients, in fact have integer coefficients and moreover provide a decomposition of $X^n - 1$ into irreducible factors, as shown in the following theorem.

6.2.7. Theorem. *The polynomials Φ_n have the following properties.*

i) $\Phi_n \in \mathbf{Z}[X]$ *and* $\deg \Phi_n = \phi(n)$.

ii) $X^n - 1 = \prod_{d \mid n} \Phi_n(X)$.

iii) The polynomials Φ_n are irreducible in $\mathbf{Z}[X]$ and in $\mathbf{Q}[X]$.

Proof. With the given definition, $\Phi_n \in \mathbf{C}[X]$. Formula *ii)* is clear, as well as the fact that $\deg(\Phi_n) = \phi(n)$; however it is less clear that in fact $\Phi_n \in \mathbf{Z}[X]$ and that Φ_n is irreducible in $\mathbf{Q}[X]$ (or $\mathbf{Z}[X]$). We shall start by showing that the coefficients of Φ_n are integers. It is clear that $\Phi_1(X) = X - 1 \in \mathbf{Z}[X]$, and formula *ii)* leads us to try induction on n. The polynomial $B := \prod_{d \mid n,\, d \neq n} \Phi_d(X)$ is monic and, by applying induction, has integer coefficients. We can therefore carry out the division algorithm in $\mathbf{Z}[X]$, and obtain $X^n - 1 = BQ + R$. Formula *ii)* then guarantees that B divides R (in $\mathbf{Q}[X]$), so $R = 0$ and $Q = \Phi_n$. We will now show that Φ_n is irreducible in $\mathbf{Z}[X]$. Let ζ be a primitive nth root of unity and P its minimal polynomial over $\mathbf{Q}$. We therefore need to show that $P = \Phi_n$. First, observe that $P \in \mathbf{Z}[X]$. Then choose a prime number p which does not divide n, so ζ^p is still an nth primitive root of unity. Let Q be its minimal polynomial, which is also in $\mathbf{Z}[X]$. If P and Q were distinct, the product PQ would divide Φ_n. But since $Q(\zeta^p) = 0$, we see that ζ is a root of $Q(X^p)$ and thus $Q(X^p) = P(X)R(X)$, for some $R \in \mathbf{Z}[X]$. By reducing the coefficients modulo p, we have

$$\bar{Q}(X^p) = \bar{Q}(X)^p = \bar{P}(X)\bar{R}(X),$$

and so $\bar{P}(X)$ divides $\bar{Q}(X)^p$ in $(\mathbf{Z}/p\mathbf{Z})[X]$. Moreover, the factors of $X^n - 1$, and hence of $\bar{P}(X)$, are simple in $(\mathbf{Z}/p\mathbf{Z})[X]$ (the derivative of $X^n - 1$ is nX^{n-1}, and we made a point of choosing p so that it does not divide n): the polynomial $\bar{P}(X)$ in fact divides $\bar{Q}(X)$. But then, $\bar{P}(X)^2$ divides $\bar{\Phi}_n(X)$ in $(\mathbf{Z}/p\mathbf{Z})[X]$, which contradicts the fact that the factors of $\bar{\Phi}_n(X)$ are simple. To summarize, we have established that if p is a prime number which does not divide n, the minimal polynomial of ζ kills ζ^p. We easily deduce from this that if m is relatively prime to n, then $P(\zeta^m) = 0$. Thus $\deg(P) \geqslant \phi(n)$ and since P divides Φ_n, we have that $P = \Phi_n$, and it is therefore irreducible. □

Since Φ_n has integer coefficients, we can reduce its coefficients modulo p and consider it as a polynomial in $\mathbf{F}_p[X]$ (or in $\mathbf{F}_q[X]$ with $q = p^f$).

6.2.8. Theorem. *The decomposition into irreducible factors of the polynomial $\Phi_n \in \mathbf{F}_q[X]$ (with $q = p^f$) depends on whether n modulo p is zero or not.*

i) If $n = p^s m$ where $p \nmid m$, we have $\Phi_n(X) = \Phi_m(X)^{p^s - p^{s-1}}$.
ii) If $\gcd(n, q) = 1$ and if r is the order of $q \bmod n$ in $(\mathbf{Z}/n\mathbf{Z})^*$, then Φ_n can be decomposed into the product of $\phi(n)/r$ distinct irreducible factors of degree r.

Proof. Assume first that $n = p^r m$. By Fermat's little theorem and the formulas from Exercise 2-7.12, it follows that $\Phi_m(X)^p \equiv \Phi_m(X^p) = \Phi_{mp}(X)\Phi_m(X)$, hence $\Phi_{mp}(X) \equiv \Phi_m(X)^{p-1}$, and subsequently that

$$\Phi_{mp^r}(X) = \Phi_{mp}\left(X^{p^{r-1}}\right) \equiv \Phi_{mp}(X)^{p^{r-1}} \equiv \Phi_m(X)^{p^{r-1}(p-1)},$$

which proves the first assertion. From now on, suppose that p is relatively prime to n. Let β be an nth primitive root in an extension of $\mathbf{F}_q$. Every factor of Φ_n can be written as $Q = \prod_{i \in I}(X - \beta^i)$, with $I \subset (\mathbf{Z}/n\mathbf{Z})^*$. The polynomial Q has coefficients in $\mathbf{F}_q$ if and only if

$$Q(X)^q = Q(X^q). \tag{$*$}$$

In fact, $\left(\sum_j a_j X^j\right)^q = \sum_j (a_j)^q X^{qj}$ and $a \in \mathbf{F}_q$ if and only if $a^q = a$. Thus the polynomial Q has coefficients in $\mathbf{F}_q$ if and only if

$$\prod_{i \in I}(X^q - \beta^{iq}) = \prod_{i \in I}(X - \beta^i)^q = \prod_{i \in I}(X^q - \beta^i),$$

or even if and only if I is stable under multiplication by q (in $(\mathbf{Z}/n\mathbf{Z})^*$). The smallest stable subset is clearly of the form $I := \{i, iq, iq^2, \ldots, iq^{r-1}\}$. Also, the irreducible factors of $\Phi_n(X)$ in $\mathbf{F}_q[X]$ are of the form

$$Q = \prod_{s=0}^{r-1}(X - \beta^{iq^s}),$$

and, in particular, all have degree r. □

6.2.9. Examples. 1) Take $n = 11$ and $q = 3$; we see that the order of $3 \bmod 11$ is equal to 5. Thus $X^{11} - 1 = (X-1)\Phi_{11}(X)$ in $\mathbf{Z}[X]$ and $\Phi_{11} = P_1P_2 \in \mathbf{F}_3[X]$, where $\deg(P_i) = 5$. We can check that, in $\mathbf{F}_3[X]$,

$$X^{11} - 1 = (X-1)(X^5 - X^3 + X^2 - X - 1)(X^5 + X^4 - X^3 + X^2 - 1).$$

2) Take $n = 23$ and $q = 2$; we see that the order of $2 \bmod 23$ is equal to 11. Thus $X^{23} - 1 = (X-1)\Phi_{23}(X)$ in $\mathbf{Z}[X]$ and $\Phi_{23} = P_1P_2 \in \mathbf{F}_2[X]$, with $\deg(P_i) = 11$. We can check that, in $\mathbf{F}_2[X]$,

$$\begin{aligned} X^{23} - 1 = (X-1)(X^{11} + X^{10} + X^6 + X^5 + X^4 + X^2 + 1) \\ \times (X^{11} + X^9 + X^7 + X^6 + X^5 + X + 1). \end{aligned}$$

3) Take $n = 15$ and $q = 2$; thus $X^{15} - 1 = (X-1)\Phi_3(X)\Phi_5(X)\Phi_{15}(X)$ in $\mathbf{Z}[X]$, with $\Phi_{15} = X^8 - X^7 + X^5 - X^4 + X^3 - X + 1$. The order of $2 \bmod 3$ is equal to 2, the order of $2 \bmod 5$ is equal to 4 and the order of $2 \bmod 15$ is equal to 4. The polynomials $\Phi_3 = X^2 + X + 1$ and $\Phi_5 = X^4 + X^3 + X^2 + X + 1$ are therefore irreducible in $\mathbf{F}_2[X]$, and $\Phi_{15} = P_1P_2 \in \mathbf{F}_2[X]$, where $\deg(P_i) = 4$. We can check that, in $\mathbf{F}_2[X]$,

$$X^{15}-1 = (X-1)(X^2+X+1)(X^4+X^3+X^2+X+1)(X^4+X^3+1)(X^4+X+1).$$

4) More generally, if $\gcd(q, n) = 1$, a cyclic code of length n corresponds, by Theorem 2-6.2.2, to a subset $I \subset \mathbf{Z}/n\mathbf{Z}$, which is stable under multiplication by q. More explicitly, the associated code is the ideal of $\mathbf{F}_q[X]/(X^n - 1)$ generated by the polynomial $Q = \prod_{i \in I}(X - \beta^i)$, where β is an nth primitive root of unity. To estimate the distance of such a code, we can use the following result.

6.2.10. Theorem. *Let $\mathscr{C}$ by a linear cyclic code of length n over $\mathbf{F}_q$ associated to $I \subset (\mathbf{Z}/n\mathbf{Z})$. If there exist i and s such that $\{i+1, i+2, \ldots, i+s\} \subset I$, then $d(\mathscr{C}) \geqslant s+1$.*

Proof. Let β be an nth primitive root in an extension of $\mathbf{F}_q$ and let Q be a polynomial modulo $X^n - 1$ which belongs to $\mathscr{C}$. We know that $Q(\beta^{i+j}) = 0$ for $j = 1, \ldots, s$. Assume that the weight w of Q (viewed as an element of $\mathbf{F}_q^n$) is $\leqslant s$, which means that $Q = a_1X^{i_1} + \cdots + a_wX^{i_w}$ with $0 \leqslant i_1 <$

$i_2 < \cdots < i_w < n$. We need to show that Q is in fact zero. Now, we have the equations $a_1\beta^{i_1(i+j)} + \cdots + a_w\beta^{i_w(i+j)} = 0$ for $j = 1, \ldots, s$. Let $a'_1 := a_1\beta^{i_1 i}, \ldots, a'_w := a_w\beta^{i_w i}$. The equations can be rewritten as

$$\beta^{i_1 j}a'_1 + \cdots + \beta^{i_w j}a'_w = 0, \qquad \text{for } j = 1, \ldots, s.$$

The matrix of the $\beta^{i_r j}$ can be extracted from a Vandermonde matrix with $\beta^{i_r} \neq \beta^{i_{r'}}$, because β has order n, and its rank therefore equals $w = \min\{w, s\}$. This means that $a'_1 = \cdots = a'_w = 0$, and hence $a_1 = \cdots = a_w = 0$. □

6.2.11. Remark. The bound given in the theorem is generally not optimal. We can see this below in the example of Golay codes.

6.2.12. Examples. (Linear cyclic codes.)

We will now describe in detail some examples gotten from choosing q, n and a subset $I \subset \mathbf{Z}/n\mathbf{Z}$ which is stable under multiplication by q. To be rigorous, we should clarify that the code that we construct also depends on the nth primitive root β that we choose. However, it is not difficult to see that the various codes gotten from the choices of β are all isomorphic. We will therefore omit β.

Hamming codes. One first interesting choice of parameters is $n = (q^r - 1)/(q-1)$, and we can easily check that the order of $q \bmod n$ is r. We set $I := \{1, q, q^2, \ldots, q^{r-1}\}$, which defines a code $\mathscr{C}$ of dimension $n - r$ (once β, a primitive nth root of unity, is chosen). We will now directly verify that $d(\mathscr{C}) \geqslant 3$. A polynomial of weight 2 can be written $f = aX^i + bX^j$ with say $0 \leqslant i < j \leqslant n-1$, and the condition that it is killed by β^{q^ℓ} for $0 \leqslant \ell \leqslant r-1$ is therefore written as $a + b\beta^{(j-i)q^\ell} = 0$. Since β is of order n, we see that this is impossible except when $a = b = 0$. Thus the code $\mathscr{C}$ is 1-correcting, and since $\operatorname{card} B(x,1) = 1 + n(q-1) = q^r$, we see that $\mathscr{C}$ is perfect 1-correcting, and thus $d(\mathscr{C}) = 3$ or 4 (we show below that the distance is 3 and that the code is therefore MDS if and only if $r = 2$). Binary Hamming codes are obtained by taking $q = 2$ and by choosing $I := \{1, 2, 4, \ldots, 2^{r-1}\}$ and hence $k = n - r = 2^r - r - 1$. Since $\{1, 2\} \subset I$, we see that $d(\mathscr{C}) \geqslant 3$. For $r = 3$, $q = 2$, $n = 7$, we get the code studied in the first example (2-6.1.1).

In order to see that the distance of a Hamming code is equal to $d(\mathscr{C}) = 3$, we write a parity-check matrix A for the code (a matrix with r rows and n columns). The columns $e_1, \ldots, e_n$ of A are vectors in $(\mathbf{F}_q)^r$, and we have just shown that any pair of them is linearly independent. Now, there are $n = (q^r-1)/(q-1)$ of them, and they therefore represent exactly one vector from each line in $(\mathbf{F}_q)^r$. Since two of the vectors e_i are never dependent, but of course there exists triples of linearly dependent vectors, we see that $d(\mathscr{C}) = 3$.

Reed-Solomon codes. These codes correspond to the choice $n = q - 1$, most often with $q = 2^f$. Let α be a generator of $\mathbf{F}_q^*$. Once we have chosen k, we set

$$g(X) := \prod_{i=1}^{q-1-k} (X - \alpha^i) .$$

It follows of course that $k = \dim \mathscr{C}$ and, since $I = \{1, 2, 3, \ldots, q-1-k\}$, we have $d(\mathscr{C}) \geqslant q-k$. But we know that for every linear code, $d(\mathscr{C}) \leqslant n+1-k$, hence $d(\mathscr{C}) = q-k$, and the code constructed in this way is therefore MDS. Now suppose that $q = 2^f$. We can consider $\mathscr{C}$ as a *binary* code $\mathscr{C}'$, with the parameters $n' = (2^f - 1)f$, $k' = kf$ and distance $d(\mathscr{C}') \geqslant 2^f - k$. One special feature of this code is that it can correct large numbers of errors: if t satisfies $2t + 1 \leqslant d(\mathscr{C}) = q - k$, the code can correct t elements of $\mathbf{F}_{2^f}$, hence tf binary errors if these errors are distributed in bunches! This feature explains why this type of code is used in the technology of compact discs.

Ternary Golay code. We know that $3^5 - 1 = 11 \cdot 23$. We choose $q = 3$, $n = 11$ and the subset of $(\mathbf{Z}/11\mathbf{Z})^*$ generated by 3, in other words $I := \{1, 3, 4, 5, 9\}$; this code, denoted by $\mathscr{G}_{11}$, is therefore of dimension 6. We point out (but do not use) that $I = \mathbf{F}_{11}^{*2}$. By Theorem 2-6.2.10 on the distance of a cyclic code, we see that $d(\mathscr{G}_{11}) \geqslant 4$ and, by considering the factorization of Φ_{11} in $\mathbf{F}_3[X]$ (cf. Examples 2-6.2.9), we see that $\mathscr{G}_{11}$ contains a polynomial of weight 5, hence $d(\mathscr{G}_{11}) \leqslant 5$. An extensive calculation (which is postponed to Exercise 2-7.22 below) allows us to establish that actually $d(\mathscr{G}_{11}) = 5$. Thus $\mathscr{G}_{11}$ is 2-correcting, and since $\operatorname{card} B(x, 2) = 1 + 2\binom{11}{1} + 2^2\binom{11}{2} = 3^5$, it is clear that the code $\mathscr{G}_{11}$ is perfect 2-correcting (but notice that it is not MDS).

Binary Golay code. We know that $2^{11} - 1 = 23 \cdot 89$ (it is actually the smallest number of the form $2^p - 1$ which is not prime). We therefore choose $q = 2$, $n = 23$ and I as the subset of $(\mathbf{Z}/23\mathbf{Z})^*$ generated by 2, in other words $I := \{1, 2, 3, 4, 6, 8, 9, 12, 13, 16, 18\}$, and we denote by $\mathscr{G}_{23}$ the associated code. Observe also that $I = \mathbf{F}_{23}^{*2}$. By Theorem 2-6.2.10 on the distance of a cyclic code, we see that $d(\mathscr{G}_{23}) \geqslant 5$ and, by considering the factorization of Φ_{23} in $\mathbf{F}_2[X]$ (cf. Examples 2-6.2.9), we see that $\mathscr{G}_{23}$ contains a polynomial of weight 7, hence $d(\mathscr{G}_{23}) \leqslant 7$. An extensive calculation (which is postponed to Exercise 2-7.22, suggested below) allows us to determine that actually $d(\mathscr{G}_{23}) = 7$. Thus $\mathscr{G}_{23}$ is 3-correcting, and since $\operatorname{card} B(x, 3) = 1 + \binom{23}{1} + \binom{23}{2} + \binom{23}{3} = 2^{11}$, it follows that the code $\mathscr{G}_{23}$ is perfect 3-correcting (but notice that it is not MDS).

6.2.13. Remark. We can show that if we exclude trivial codes (i.e., of dimension 1, $n-1$ or n), the only perfect t-correcting codes are those that we have already constructed: the Hamming 1-correcting codes and the two Golay binary and ternary codes [73].

7. Exercises

7.1. Exercise. (Newton's method) *Recall that Newton's iterative method (for approximating the zeros of a function) is applicable to differentiable functions. Let f be a function with a unique zero at α; the iteration is given by*

$$x_{n+1} = x_n - \frac{f(x_n)}{f'(x_n)}.$$

The rate of convergence of this approximation is quadratic, i.e., $|x_{n+1}-\alpha| \leqslant C|x_n-\alpha|^2$. Clarify and prove this assertion for the function $f(x) := x^m - a$, and deduce a fast calculation algorithm for approximating $\sqrt[m]{a}$ from this.

7.2. Exercise. *1) Give a fast algorithm which checks if a given integer N is a power a^m, where $m \geqslant 2$.*

2) If we now want to test whether $N = p^m$ where p is prime and $m \geqslant 2$, we take $a \in [2, N-1]$ and we test if $\gcd(a, N) = 1$. If that is the case, we compute $d = \gcd(a^{N-1}-1, N)$. Prove in this case that p divides d and that, with a high probability, $d \neq N$ and also that $d = p$. Deduce from this an algorithm to check whether $N = p^m$.

7.3. Exercise. (Multiplication algorithm—see [42]) *Suppose that the integers m and n are written in at most $2t$ binary digits, $n = n_1 2^t + n_0$ and $m = m_1 2^t + m_0$. Observe that*

$$mn = m_1 n_1 (2^{2t} - 2^t) + 2^t (m_1 + m_0)(n_1 + n_0) + m_0 n_0 (1 - 2^t)$$

and can therefore be calculated with three multiplications of numbers of size t and some additions and shifts (multiplication by 2 consists of one shift of digits). Deduce from this an algorithm, where the cost $T(r)$ of the multiplication of two numbers with r digits satisfies

$$T(2r) \leqslant 3T(r) + cr,$$

for some appropriate constant c. Deduce from this that $T(r) = O(r^\alpha)$, where $\alpha > \log 3/\log 2$. (Notice that, asymptotically, this algorithm is better that the usual algorithm, whose complexity is $O(r^2)$.)

7.4. Exercise. (Multiplication by fast Fourier transform) *In this exercise, we will give a theoretical presentation of the finite Fourier transform, which will allow us to multiply very large numbers faster than the usual algorithm. The hints are fairly brief, so you could also use a specialized reference, [42] Sect. 4.3.3., to help you finish this exercise.*

Let $N \geqslant 2$ *be an integer and let* A *be a ring. We identify the set* E *of functions from* $\mathbf{Z}/N\mathbf{Z}$ *to* A *with the set of polynomials with coefficients in* A *of degree* $< N$*, in other words, to polynomials associated to the ring* $A[X]/(X^N - 1)$*. If* $a = (a_i)_{0 \leqslant i \leqslant N-1}$ *is a sequence indexed by* $\mathbf{Z}/N\mathbf{Z}$*, we denote by* P_a *the corresponding polynomial. We define a "convolution" by* $(a * b)_i = \sum_{j+h=i} a_j b_h$*, and we can easily check that* $P_{a*b} = P_a P_b$*.*

If ζ *is an* N*th primitive root of unity in* A*, we define the "Fourier transform",* $\mathscr{F} : E \to E$ *and its conjugate* $\bar{\mathscr{F}} : E \to E$ *by the formulas*

$$(\mathscr{F}a)_j = \sum_{i \in \mathbf{Z}/N\mathbf{Z}} \zeta^{ij} a_i = P_a(\zeta^j) \quad \text{and} \quad (\bar{\mathscr{F}}a)_j = \sum_{i \in \mathbf{Z}/N\mathbf{Z}} \zeta^{-ij} a_i = P_a(\zeta^{-j}).$$

1) Prove that the following formulas hold: $\mathscr{F}(a*b) = \mathscr{F}(a) \cdot \mathscr{F}(b)$, $\mathscr{F}\left(\bar{\mathscr{F}}a\right) = Na$ *and* $\bar{\mathscr{F}}\left(\mathscr{F}a\right) = Na$*.*

2) Whenever $N = 2N'$*, we set* $\zeta' := \zeta^2$ *and* $E' := A[X]/(X^{N'} - 1)$*, and we define* $\mathscr{F}', \bar{\mathscr{F}}' : E' \to E'$ *with the help of* ζ'*. For* $a \in E$*, we define* $a^0, a^1 \in E'$ *by setting* $a_i^0 = a_{2i}$ *and* $a_i^1 = a_{2i+1}$*. Check that, for* $0 \leqslant j \leqslant N' - 1$*, the following formulas hold:*

$$(\mathscr{F}a)_j = \left(\mathscr{F}'a^0\right)_j + \zeta^j \left(\mathscr{F}'a^1\right)_j \quad \text{and} \quad (\mathscr{F}a)_{N'+j} = \left(\mathscr{F}'a^0\right)_j - \zeta^j \left(\mathscr{F}'a^1\right)_j.$$

3) Now suppose that $N = 2^r$*. Use the previous arguments to derive a recursive procedure for calculating a Fourier transform. If we denote by* $M(r)$ *the number of multiplications and* $A(r)$ *the number of additions necessary to carry out this procedure, show that* $A(r) + M(r) = O(r2^r) = O(N \log N)$*.*

4) By using the first formula (convolution transformation and ordinary product) and the preceding results, derive a multiplication algorithm for polynomials with coefficients in A*.*

5) The choice of a numeral basis b *lets us write integers in the form* $P_a(b) = a_0 + a_1 b + \cdots + a_d b^d$*. Using the polynomial multiplication algorithm, derive an algorithm for multiplying integers.*

7.5. Exercise. *A Fibonacci sequence of integers is defined by* $u_0 = a$, $u_1 = b$ *and* $u_n = u_{n-1} + u_{n-2}$ *for* $n \geqslant 2$*, where* $1 \leqslant a \leqslant b$ *are integers (the classical Fibonacci sequence corresponds to* $a = b = 1$*).*

1) Prove that $\log |u_n| \sim n \log \left(\dfrac{1+\sqrt{5}}{2}\right)$*.*

2) Prove that $\gcd(u_{n+1}, u_n) = \gcd(b, a)$ *and that the Euclidean algorithm gives this result in n steps. Deduce from this that the complexity estimation given at the beginning of this chapter is generally optimal.*

7.6. Exercise. *Prove that following algorithm allows us to calculate the gcd of two integers, and estimate its complexity. If* n *and* m *are even, factor out 2; if* n *is even and* m *is odd (or conversely), replace* n *by* $n/2$*; if* m *and* n *are odd, replace* n *by* $(n-m)/2$.

7.7. Exercise. *Let* $M \in \mathbf{Z}$ *and let* N *be an odd positive integer. Prove that the Euclidean algorithm, together with the quadratic reciprocity law, gives a fast algorithm (and estimate its complexity) for calculating the Jacobi symbol* $\left(\frac{M}{N}\right)$.

7.8. Exercise. *Let* $M := 85$*; we define the sets* $G_0 := (\mathbf{Z}/M\mathbf{Z})^*$, $G_1 := \{a \in G_0 \mid a^{M-1} = 1\}$, $G_2 := \{a \in G_0 \mid a^{(M-1)/2} = \pm 1\}$, $G_3 := \{a \in G_0 \mid a^{(M-1)/2} = \left(\frac{a}{M}\right)_J\}$ *and finally* $S := \{a \in G_0 \mid a^{21} = 1$ *or* $a^{21} = -1$ *or* $a^{42} = -1\}$.

3.a) Prove that if $a \in S$*, then* $-a \in S$*, and use this to deduce that the cardinality of* S *is even.*

3.b) Calculate the cardinality of G_0, G_1, G_2 *and* S.

3.c) Use this to find the cardinality of G_3.

3.d) Is the set S *a subgroup of* G_0*?*

7.9. Exercise. *For* $n \geqslant 2$*, we denote by* Φ_n *the nth cyclotomic polynomial.*

1) Recall how to decompose Φ_n *in* $\mathbf{F}_p[X]$.

2) Let $a \in \mathbf{Z}$ *and let* p *be a prime number which does not divide* n *but which divides* $\Phi_n(a)$*. Prove that* $p \equiv 1 \bmod n$ *(you could start by observing that the class of* a *modulo* p *is a root of* Φ_n*).*

3) Prove that $\Phi_n(0) = 1$ *and deduce from this that for all* $m \geqslant 2$, $\Phi_n(m)$ *is relatively prime to* m*. Also prove that there are only finitely many* $a \in \mathbf{Z}$ *such that* $\Phi_n(a) = \pm 1$.

4) Deduce from this (without using Dirichlet's theorem on arithmetic progressions) that there exist infinitely many prime numbers, p *such that* $p \equiv 1 \bmod n$ *(resp. infinitely many prime numbers* p *such that* $p \not\equiv 1 \bmod n$*).*

7.10. Exercise. *Let* G *be a finite abelian group.*

1) Prove that there exists an integer N such that G is isomorphic to a subgroup (resp. a quotient) of $(\mathbf{Z}/N\mathbf{Z})^*$.

Hint.– *We can reduce to the case where $G = \mathbf{Z}/n_1\mathbf{Z}\times\cdots\times\mathbf{Z}/n_s\mathbf{Z}$. By using the result proven in the previous exercise, we can choose prime numbers $p_i \equiv 1 \bmod n_i$, and show that $N := p_1 \cdots p_s$ works.*

2) (This question requires some knowledge of Galois theory, see for example Appendix C, in particular Examples C-1.1.) Prove that there exists a finite Galois extension, $K/\mathbf{Q}$, such that $\mathrm{Gal}(K/\mathbf{Q}) \cong G$.

7.11. Exercise. *Let $P = X^4 + 1$. We will study its factorization over various fields.*

1) Prove that P is irreducible in $\mathbf{Q}[X]$ and calculate its factorization over the fields $\mathbf{Q}(i)$, $\mathbf{Q}(\sqrt{2})$ and $\mathbf{Q}(i\sqrt{2})$.

2) Show that for every prime number p, P is not irreducible over $\mathbf{F}_p$.

Hint.– *Construct a factorization by using the fact that -1, 2 or -2 is a square. Variation: observe that $P = \Phi_8$ and invoke Theorem 2-6.2.8.*

7.12. Exercise. *1) Prove that the following relations hold (you could compare the degrees and the roots of both sides):*

$$\Phi_n(X^p) = \begin{cases} \Phi_{np}(X) & \text{if } p \text{ divides } n, \\ \Phi_{np}(X)\Phi_n(X) & \text{if } p \text{ does not divide } n. \end{cases}$$

2) Prove that $\Phi_{p^r} = X^{p^{r-1}(p-1)} + X^{p^{r-1}(p-2)} + \cdots + X^{p^{r-1}} + 1$ (for $r \geqslant 1$).

7.13. Exercise. *For $n \geqslant 3$, we denote by $\Phi_n^+(X)$ the monic polynomial with the property that $(\Phi_n^+(X))^2 = \prod_{\zeta\in\mu_n^*}(X - \zeta - \zeta^{-1})$.*

1) Compute Φ_3^+, Φ_5^+ and Φ_7^+.

2) Prove that $\deg \Phi_n^+ = \phi(n)/2$ and $\Phi_n(X) = X^{\phi(n)/2}\Phi_n^+(X + X^{-1})$. Deduce from this that $\Phi_p^+(2) = \Phi_p(1) = p$.

3) Prove that Φ_n^+ is in $\mathbf{Z}[X]$ and is irreducible (in particular, it is the minimal polynomial of $2\cos(2\pi/n)$).

7.14. Exercise. *Let $P = \prod_{i=1}^r (X - \alpha_i)$ and $Q = \prod_{j=1}^s (X - \beta_j)$ be two polynomials in $K[X]$. We define their resultant by the formula*

$$\mathrm{res}(P,Q) := \prod_{i=1}^{r} Q(\alpha_i) = \prod_{i=1}^{r}\prod_{j=1}^{s}(\alpha_i - \beta_j).$$

We refer you to a classical algebra text (cf. for example [43]) to see how

$\operatorname{res}(P,Q)$ *can be expressed as a determinant in the coefficients of* P *and* Q, *which shows in particular that* $\operatorname{res}(P,Q) \in K$ *and, more generally, that if* $P, Q \in A[X]$, *then* $\operatorname{res}(P,Q) \in A$.

1) Prove that $\operatorname{res}(Q,P) = (-1)^{rs}\operatorname{res}(P,Q)$.

2) We will assume from now on that $P, Q \in \mathbf{Z}[X]$, *and we choose* q *to be an odd prime. We denote by* $\tilde{P}$ *(resp.* $\tilde{Q}$*) the reduction modulo* q *of* P *(resp. of* Q*). Prove that the class of* $\operatorname{res}(P,Q)$ *modulo* q *is equal to* $\operatorname{res}(\tilde{P},\tilde{Q})$.

3) Prove that $\tilde{\Phi}_q^+ = (X-2)^{(q-1)/2}$ *in* $\mathbf{F}_q[X]$ *(*Φ_n^+ *is defined in Exercise 2-7.13).*

4) Use the previous questions and question 2) *of Exercise 2-7.13 to show that if* p *and* q *are distinct odd primes, then*

$$\operatorname{res}(\tilde{\Phi}_q^+, \tilde{\Phi}_p^+) \equiv p^{(q-1)/2} \equiv \left(\frac{p}{q}\right) \operatorname{mod} q.$$

5) Prove that $\operatorname{res}(\Phi_q^+, \Phi_p^+) = \prod_{\eta \in \mu_q^*} \eta^{-(p-1)/2}\Phi_p(\eta)$ *and deduce from this that* $\operatorname{res}(\Phi_q^+, \Phi_p^+) \in \{+1, -1\}$.

6) Prove that the following formula holds,

$$\operatorname{res}(\Phi_q^+, \Phi_p^+) = \left(\frac{p}{q}\right),$$

and use this to give a proof of the quadratic reciprocity law.

7.15. Exercise. *Let* N *be an odd integer.*

1) If its factorization can be written as $N = p_1^{m_1} \cdots p_k^{m_k}$, *where* $p_i - 1 = 2^{s_i}L_i$ *and* L_i *are odd, prove that*

$$\frac{\operatorname{card}\{a \in (\mathbf{Z}/N\mathbf{Z})^* \mid \operatorname{ord}(a \operatorname{mod} N) \textit{ is odd}\}}{\operatorname{card}\{a \in (\mathbf{Z}/N\mathbf{Z})^*\}} = 2^{-s_1 \cdots - s_k}.$$

2) Deduce from this that if we had a fast algorithm, $\mathscr{P}$, *which calculates the period (the order of* $a \operatorname{mod} N$*), then we have a fast probabilistic factorization algorithm.*

Hint.– *Randomly choose* a, *test to see whether* $\gcd(a,N) = 1$, *then whether the period* $\mathscr{P}(a)$ *is even; in this case compute* $\gcd(a^{\mathscr{P}(a)/2} \pm 1, N)$.

7.16. Exercise. *Prove that* $2^m + 1$ *can only be prime if* $m = 2^n$. *Set* $F_n := 2^{2^n} + 1$ (known as a Fermat number). *Prove that* F_n *is prime if and only if* F_n *divides* $3^{\frac{F_n - 1}{2}} + 1$. *Check that* F_0, F_1, F_2, F_3 *and* F_4 *are prime, but not* F_5 *(which is divisible by 641).*

7.17. Exercise. (Lucas test and Mersenne numbers) *Start by proving that* $M_n := 2^n - 1$ *can only be prime if* n *is itself prime. Check that* M_2, M_3, M_5, M_7 *are prime, but that* M_{11} *is not prime. The numbers* $M_p = 2^p - 1$ *are called* Mersenne numbers. *In this exercise, we ask you to prove the Lucas primality test for these numbers.*

a) We define a sequence with values in a ring A *by* $V_0 = 2$, $V_1 = a$ *and* $V_{n+1} - aV_n + V_{n-1} = 0$. *Verify the following formulas:* $V_{2n-1} = V_n V_{n-1} - a$, $V_{2n} = V_n^2 - 2$, *and also* $V_n V_m = V_{n+m} - V_{n-m}$.

b) Let M *be odd,* a *an integer such that* $\gcd(a^2 - 4, M) = 1$ *and* V_n *the sequence defined above. If* $V_{M+1} \equiv 2 \bmod M$ *and if for every prime number* q *which divides* $M + 1$ *we have* $\gcd(V_{\frac{M+1}{q}} - 2, M) = 1$, *prove that* M *is prime.*

c) We define the following sequence by $L_1 := 4$ *and* $L_{i+1} := L_i^2 - 2$. *Prove that the Mersenne number* M_p *is prime if and only if* $L_{p-1} \equiv 0 \bmod M_p$.

7.18. Exercise. (Perfect numbers) *This nice problem has been handed down to us from Euclid: we say that an integer is* perfect *if it is equal to the sum of its proper divisors, symbolically:*

$$n = \sum_{\substack{d \mid n \\ d \neq n}} d \qquad \text{or} \qquad 2n = \sigma(n) := \sum_{d \mid n} d.$$

a) Show that if $M_p = 2^p - 1$ *is a prime Mersenne number (cf. previous exercise), then* $P_p := 2^{p-1} M_p$ *is a perfect number (this fact as well as the examples* $P_2 = 6$, $P_3 = 28$, $P_5 = 496$ *were known to Euclid).*

b) Prove the following result due to Euler: an even perfect number n *is of the form* P_p.

Hint.– *Write* $n = 2^m M$ *with* M *odd and* $m \geqslant 1$; *prove that* $2n = \sigma(2^m)\sigma(M)$ *and deduce from this that* M *must be prime, then finish the exercise.*

Remark. *Nobody knows whether there exists an odd perfect number; it is generally conjectured that there do not exist any and that the perfect numbers are in bijection with the prime Mersenne numbers.*

7.19. Exercise. (Pocklington-Lehmer test or certificate) *Let* $N \geqslant 2$. *Suppose that* $N - 1$ *is (partially) factored as* $N - 1 = p_1^{e_1} \cdots p_k^{e_k} M$, *with* $M < \sqrt{N}$, *and moreover that for each* p_i, *we have an* a_i *such that*

$$\begin{cases} a_i^{N-1} \equiv 1 \bmod N, \\ \gcd\left(a_i^{\frac{N-1}{p_i}} - 1, N\right) = 1. \end{cases}$$

Use this to show that if q divides N, then $q \equiv 1 \bmod p_i^{e_i}$, and also that N is prime.

7.20. Exercise. *Let β be a 17th primitive root of unity in an extension of $\mathbf{F}_2$. We let $I := \mathbf{F}_{17}^{*2}$ and set*

$$f(X) = \prod_{i \in I} (X - \beta^i) .$$

Prove that the polynomial $f(X)$ defines a cyclic code $\mathscr{C}$ of length 17, and calculate its dimension and bounds on its distance $d(\mathscr{C})$, for example $3 \leqslant d(\mathscr{C}) \leqslant 6$. Then give the exact value of $d(\mathscr{C})$.

7.21. Exercise. *1.a) Describe the degrees of the decomposition into irreducible factors of $X^{85} - 1$ in $\mathbf{Q}[X]$.*

1.b) Give the number of irreducible factors, as well as their degrees, of the decomposition of $X^{85} - 1$ in $\mathbf{F}_2[X]$.

1.c) Explain how to construct a binary cyclic code of length 85 and dimension 64. It is possible to construct such a code with dimension 63?

7.22. Exercise. *(Where we show that $d(\mathscr{G}_{11}) = 5$ and $d(\mathscr{G}_{23}) = 7$ and use the notion of a self-dual code.)*

A) Let $\mathscr{C}$ be a cyclic code of length n generated by the polynomial $g = g(X)$ of degree d. Let $\mathscr{C}'$ be its even subcode $\mathscr{C}^*$ its dual code.

1) Prove that $\mathscr{C}' = \mathscr{C}$ if and only if $g(1) = 0$. If $g(1) \neq 0$, check that $\mathscr{C}'$ is cyclic and generated by the polynomial $(X-1)g(X)$.

2) Prove that $\mathscr{C}^*$ is cyclic and generated by the polynomial $h^*(X) = X^{n-d}$ $h(1/X)$ where $g(X)h(X) = X^n - 1$.

Hint.– *You can show that if $\deg(f) \leqslant n-d-1$ and $\deg(e) \leqslant d-1$, then $\langle fg, eh^* \rangle$ is equal to the coefficient of X^{n-1} in the product $f(X)g(X)e^*(X)$ $h(X) = f(X)e^*(X)(X^n - 1)$, and is therefore zero.*

B) Suppose that $\mathscr{C} \subset \mathscr{C}^*$ (i.e., for all $x, y \in \mathscr{C}$, we have $\langle x, y \rangle = 0$).

1) If $q = 2$, prove that for all $x, y \in \mathscr{C}$, we have $w(x+y) \equiv w(x) + w(y) \bmod 4$.

2) If $q = 3$, prove that for all $x, y \in \mathscr{C}$, we have $w(x+y) \equiv w(x) + w(y) \bmod 3$.

C) We introduce the subcode $\mathscr{D}$ of $\mathscr{G}_{11}$, composed of vectors whose sum of the coordinates equals zero (the "even" subcode).

1) Prove that if $g(X)$ is the generating polynomial of $\mathscr{G}_{11}$, the code $\mathscr{D}$ is cyclic and its generator is $(X-1)g(X)$.

2) Prove that $\mathscr{D} \subset \mathscr{D}^*$ *(i.e., for every* $x, y \in \mathscr{D}$ *we have* $\langle x, y\rangle = 0$*). Deduce from this that for every* $x \in \mathscr{D}$*, we have* $w(x) \equiv 0 \bmod 3$.

3) We denote by $\bar{\mathscr{D}}$ *and* $\bar{\mathscr{G}}_{11}$ *the extended codes. Set* $e_{11} = (1, \ldots, 1) \in \mathbf{F}_3^{11}$ *and* $e_{12} = (1, \ldots, 1) \in \mathbf{F}_3^{12}$*. Prove that* $e_{11} \in \mathscr{G}_{11}$*,* $e_{12} \in \bar{\mathscr{G}}_{11}$ *and hence* $\bar{\mathscr{G}}_{11} = \bar{\mathscr{D}} \oplus \mathbf{F}_3 e_{12}$.

4) Prove that $\bar{\mathscr{G}}_{11}$ *is self-dual. Deduce from this that for every* $x, y \in \bar{\mathscr{G}}_{11}$*, we have* $w(x+y) \equiv w(x) + w(y) \bmod 3$*, and hence that* $d(\bar{\mathscr{G}}_{11}) \equiv 0 \bmod 3$.

5) Knowing that $4 \leqslant d(\mathscr{G}_{11}) \leqslant 5$ *and* $d(C) \leqslant d(\bar{C}) \leqslant d(C)+1$*, conclude that* $d(\mathscr{G}_{11}) = 5$ *and* $d(\bar{\mathscr{G}}_{11}) = 6$.

D) Let p be an odd prime such that $\left(\frac{2}{p}\right) = 1$*,* $S := \mathbf{F}_p^{*2}$ *and* $\mathscr{C}$ *a binary code of length p which corresponds to the set S (which, by hypothesis, is stable under multiplication by 2). We denote by* $\bar{\mathscr{C}}$ *the extended code of length* $p+1$.

1) If $g = g(X)$ *is a generator of* $\mathscr{C}$ *and if* $g^*(X) = X^{(p-1)/2} g(1/X)$ *is its reciprocal polynomial, show that* $g(X) = g^*(X)$ *if* $p \equiv 1 \bmod 8$*, and that* $\Phi_p(X) = g(X) g^*(X)$ *if* $p \equiv -1 \bmod 8$.

2) We suppose from now on that $p \equiv -1 \bmod 8$*. Prove that* $\bar{\mathscr{C}}$ *is self-dual (i.e.,* $\bar{\mathscr{C}} = \bar{\mathscr{C}}^*$*, or for all* $\bar{x}, \bar{y} \in \bar{\mathscr{C}}$*, we have* $\langle \bar{x}, \bar{y}\rangle = 0$*).*

3) Let $x = \sum_{i \in I} X^i$ *and* $y = \sum_{i \in J} X^i$*. Show that* $\langle x, y\rangle = |I \cap J| \bmod 2$ *and that* $w(x+y) = |I| + |J| - 2|I \cap J|$*. Conclude from this that if* $\langle x, y\rangle = 0$*, then* $w(x+y) \equiv w(x) + w(y) \bmod 4$.

4) Use the previous question to show that if $\mathscr{D}$ *is a self-dual code generated by the elements whose weight is a multiple of 4, then every element of* $\mathscr{D}$ *has weight which is a multiple of 4, and in particular,* $d(\mathscr{D}) \equiv 0 \bmod 4$.

5) Apply the preceding questions to the case $p = 23$*. Observe that if g is the generator of* $\mathscr{C} = \mathscr{G}_{23}$*, we have* $w(g) = 7$*, so* $w(\bar{g}) = 8$*. Conclude from this that* $d(\bar{C}) \equiv 0 \bmod 4$*. Knowing that* $5 \leqslant d(\mathscr{G}_{23}) \leqslant 7$ *and* $d(C) \leqslant d(\bar{C}) \leqslant d(C)+1$*, deduce that* $d(\mathscr{G}_{23}) = 7$ *and* $d(\bar{\mathscr{G}}_{23}) = 8$.

Chapter 3

Algebra and Diophantine Equations

"... it is a thing of beauty and of joy for ever..."

JAMES JOYCE

In this chapter, we address some classical problems in number theory, such as finding integer solutions to polynomial equations. The examples that we will look at cover three large topics.

1) The decomposition of an integer n into the sum of two, three or four squares, in other words, the search for solutions of the equation $n = x_1^2 + x_2^2 + \cdots + x_k^2$.

2) "Fermat's last theorem" (proven by Andrew Wiles in 1995): the only solutions to the equation $x^n + y^n = z^n$ for $n \geqslant 3$ are the trivial ones (i.e., $xyz = 0$).

3) Solutions to the Pell's equation $x^2 - dy^2 = 1$ (or more generally $x^2 - dy^2 = n$). The study of congruences—the theme of Chap. 1—gives us necessary conditions for the existence of solutions to such an equation. The methods introduced in this chapter are the use of rings more general than $\mathbf{Z}$ *and also results about rational approximations.*

To be more precise, we will study rings such as $\mathbf{Z}[i]$, $\mathbf{Z}[\exp(2\pi i/n)]$, $\mathbf{Z}[\sqrt{d}]$ and even the noncommutative ring of Hurwitz quaternions, a subring of the division ring of quaternions defined by Hamilton. On the other hand, we will have a look at how fast a sequence of rational numbers can converge to a real number.

We will finish with an outline of the main properties of these rings by introducing some supplementary notions from algebra: algebraic integers and

M. Hindry, *Arithmetics*, Universitext,
DOI 10.1007/978-1-4471-2131-2_3,

Dedekind rings and from the geometry of numbers: lattices and Minkowski's theorem.

1. Sums of Squares

We want to find out under which conditions an integer $n \in \mathbf{N}$ can be written as the sum of squares. Let us first have a look at some constraints that we can write in terms of congruences.

We know that $x^2 \equiv 0$ or 1 modulo 4, thus a number $n = 4n' + 3$ cannot be the sum of two squares. To be more precise, notice that if $p \equiv 3 \bmod 4$ and p divides $n = x^2 + y^2$, then p must divide y. This is true because otherwise, we could write $(xy^{-1})^2 \equiv -1 \bmod p$ and then deduce that -1 is a square modulo p, which cannot be true. Since p divides y, it also divides x, and we can conclude that $x = px'$, $y = py'$ and $n = p^2 n'$. By repeatedly applying this argument, we see that *if $p \equiv 3 \bmod 4$ and if $n = p^{2a+1}m$, where m and p are relatively prime, then n is not the sum of two squares.*

Notice that if x is even, then $x^2 \equiv 0$ or 4 modulo 8, whereas if x is odd, then $x^2 \equiv 1$ modulo 8. We then see that $x^2 + y^2 + z^2$ is never congruent to 7 modulo 8. We can slightly refine this argument: if $n = 4n'$ and if $n = x^2 + y^2 + z^2$, then we see that x, y and z must be even, hence $x = 2x'$, $y = 2y'$ and $z = 2z'$, with $n' = x'^2 + y'^2 + z'^2$. By repeatedly applying this reasoning, we see that *if n is of the form $n = 4^a(8m + 7)$, then n is not the sum of three squares.*

It is a remarkable fact that the obstructions given by these congruences are, in the case of the sums of squares, the only ones.

1.1. Theorem. (Two-square theorem) *An integer $n \in \mathbf{N}$ is the sum of two squares if and only if every prime number p congruent to 3 modulo 4 appears with an even exponent in the decomposition of n into prime factors.*

1.2. Theorem. (Three-square theorem) *An integer $n \in \mathbf{N}$ is the sum of three squares of integers if and only if it is not of the form $n = 4^a(8m + 7)$.*

1.3. Theorem. (Four-square theorem) *Let $n \in \mathbf{N}$, then there exist integers, x, y, z, t such that $n = x^2 + y^2 + z^2 + t^2$.*

We are going to postpone the proof the second theorem until later (see, for example, Serre's book [8] or Exercise 3-6.8 together with the Hasse-Minkowski theorem 6-3.18 and its Corollary 6-3.19, or also Exercises 3-6.9 and 3-6.10). To prove the first theorem, we introduce the ring $\mathbf{Z}[i]$, and to prove the third, we introduce the ring of Hurwitz quaternions.

We can immediately see from the statements of these theorems that the set of sums of two squares (resp. of four squares) is stable under multiplication, but not the set of sums of three squares. For example, $18 = 2 \cdot 3^2 = 4^2 + 1^2 + 1^2$ and $14 = 2 \cdot 7 = 3^2 + 2^2 + 1^2$, but $18 \cdot 14 = 4 \cdot 9 \cdot 7$ is not the sum of three squares. The multiplicativity of the set of sums of two (resp. four) squares can be explained by the following formulas:

$$(x^2+y^2)(a^2+b^2) = (ax-by)^2 + (ay+bx)^2$$

and $$(x^2+y^2+z^2+t^2)(a^2+b^2+c^2+d^2) =$$

$$(ax-by-cz-dt)^2+(ay+bx-ct+dz)^2+(az+bt+cx-dy)^2+(at-bz+cy+dx)^2.$$

The origin of these formulas will be clear once we give an interpretation of them in $\mathbf{Z}[i]$ or in the quaternions.

If we set $\mathscr{C}_2 := \{n \in \mathbf{N} \mid \exists x, y \in \mathbf{N},\ n = x^2 + y^2\}$ and

$\mathscr{C}_4 := \{n \in \mathbf{N} \mid \exists x, y, z, t \in \mathbf{N},\ n = x^2 + y^2 + z^2 + t^2\}$,

we see that it is enough to show that every prime number which is congruent to 1 modulo 4 is in $\mathscr{C}_2$ and that every prime number is in $\mathscr{C}_4$.

We are going to construct the classical example of a noncommutative division ring, the ring of quaternions discovered by Hamilton, and elaborate on its arithmetical properties to establish a proof of the four-square theorem.

The most concrete of the constructions of the ring of quaternions undoubtedly consists of endowing the 4 dimensional real vector space with basis $\mathbf{1}, I, J, K$ and defining a bilinear multiplication on it, where $\mathbf{1}$ is the multiplicative inverse and which satisfies

$$\begin{aligned} &I^2 = J^2 = K^2 = -\mathbf{1}, \quad IJ = -JI = K, \quad JK = -KJ = I \\ &\text{and} \quad KI = -IK = J \end{aligned} \tag{3.1}$$

We should verify associativity "by hand": for example, $(IJ)K = K^2 = -\mathbf{1}$ and $I(JK) = I^2 = -\mathbf{1}$. To spare the 24 necessary verifications, we could also define $\mathbf{H}$ as the subalgebra of 2×2 complex matrices or 4×4 real matrices (associativity is immediate in this case, but one needs to check that these matrices satisfy formulas (3.1)). We could also define

$$\mathbf{H} = \left\{ \begin{pmatrix} \alpha & -\beta \\ \bar{\beta} & \bar{\alpha} \end{pmatrix} \middle|\ \alpha, \beta \in \mathbf{C} \right\}$$

with

$$\mathbf{1} = \begin{pmatrix} 1 & 0 \\ 0 & 1 \end{pmatrix}, \quad I = \begin{pmatrix} i & 0 \\ 0 & -i \end{pmatrix}, \quad J = \begin{pmatrix} 0 & 1 \\ -1 & 0 \end{pmatrix} \quad \text{and} \quad K = \begin{pmatrix} 0 & i \\ i & 0 \end{pmatrix}$$

or also

$$\mathbf{H} = \left\{ \begin{pmatrix} a & -b & c & -d \\ b & a & -d & -c \\ -c & -d & a & b \\ -d & c & -b & a \end{pmatrix} \middle| \; a, b, c, d \in \mathbf{R} \right\}$$

with

$$\mathbf{1} = \begin{pmatrix} 1 & 0 & 0 & 0 \\ 0 & 1 & 0 & 0 \\ 0 & 0 & 1 & 0 \\ 0 & 0 & 0 & 1 \end{pmatrix}, \; I = \begin{pmatrix} 0 & -1 & 0 & 0 \\ 1 & 0 & 0 & 0 \\ 0 & 0 & 0 & -1 \\ 0 & 0 & 1 & 0 \end{pmatrix},$$

$$J = \begin{pmatrix} 0 & 0 & -1 & 0 \\ 0 & 0 & 0 & 1 \\ 1 & 0 & 0 & 0 \\ 0 & -1 & 0 & 0 \end{pmatrix}, \; K = \begin{pmatrix} 0 & 0 & 0 & -1 \\ 0 & 0 & -1 & 0 \\ 0 & 1 & 0 & 0 \\ 1 & 0 & 0 & 0 \end{pmatrix}.$$

1.4. Remark. The construction of $\mathbf{H}$ endows it with the structure of an $\mathbf{R}$ algebra generated by two elements i and j, with the relations $i^2 = j^2 = -\mathbf{1}$ and $ij = -ji$. To see this, we let $k := ij$ and deduce the rest from the multiplication table since $k^2 = ijij = -iijj = -\mathbf{1}$ and $ik = iij = -j = (ii)j = -iji = -ki$, etc. The fact that $\mathbf{H}$ is noncommutative is already given in the multiplication table.

The *conjugate* of a quaternion $z = a\mathbf{1} + bI + cJ + dK$ is defined by $\bar{z} = a\mathbf{1} - bI - cJ - dK$, its *reduced trace* by $\mathrm{Tr}(z) = z + \bar{z}$ and its *reduced norm* by $\mathrm{N}(z) = z\bar{z}$ (from now on simply referred to as the trace and the norm).

1.5. Lemma. *If $z, w \in \mathbf{H}$, then $\overline{z+w} = \bar{z} + \bar{w}$ and $\overline{zw} = \bar{w} \cdot \bar{z}$, and if $z = a\mathbf{1} + bI + cJ + dK$, then $\mathrm{N}(z) = z\bar{z} = \bar{z}z = (a^2 + b^2 + c^2 + d^2)\mathbf{1}$ and $\mathrm{Tr}(z) = 2a\mathbf{1}$. Furthermore, $\mathrm{Tr}(z + z') = \mathrm{Tr}(z) + \mathrm{Tr}(z')$, $\mathrm{N}(zz') = \mathrm{N}(z)\,\mathrm{N}(z')$, and z is a root of the polynomial $X^2 - \mathrm{Tr}(z)X + \mathrm{N}(z) \in \mathbf{R}[X]$.*

Proof. These formulas can be checked by direct calculation (left to the reader). Take note that the conjugation is an *anti-isomorphism* of rings, i.e., it reverses the order of multiplication. □

It follows that $\mathbf{H}$ is a division ring, since if $z = a\mathbf{1} + bI + cJ + dK$ is a non-zero quaternion, then $\mathrm{N}(z) := a^2 + b^2 + c^2 + d^2 \in \mathbf{R}^*$ and $z\bar{z}/\mathrm{N}(z) = \mathbf{1}$, hence $z^{-1} = \bar{z}/\mathrm{N}(z)$.

We will now introduce the ring $\mathbf{Z}[i]$ (of Gaussian integers) and the two rings

$$A_0 = \mathbf{Z1} + \mathbf{Z}I + \mathbf{Z}J + \mathbf{Z}K \quad \text{and } A = A_0 + \mathbf{Z}\left(\frac{1 + I + J + K}{2}\right).$$

The set A is a subring of $\mathbf{H}$, because if we let $\delta := (1 + I + J + K)/2$, we have $\delta^2 = \delta - 1$ and $I\delta = \delta - 1 - I$, etc. The elements of the ring A are called *Hurwitz quaternions.*

It is clear that $\mathscr{C}_2 = \{\mathrm{N}(z) \mid z \in \mathbf{Z}[i]\}$ and $\mathscr{C}_4 = \{\mathrm{N}(z) \mid z \in A_0\}$. In fact, we also have $\mathscr{C}_4 = \{\mathrm{N}(z) \mid z \in A\}$, since if we assume the following elementary lemma, then we know that $\mathrm{N}(x\mathbf{1}+yI+zJ+tK) \in \mathbf{N}$ if $x, y, z, t \in \mathbf{Z}+1/2$.

1.6. Lemma. *Let* $\alpha = \dfrac{x\mathbf{1} + yI + zJ + tK}{2} \in A$, *where* x, y, z *and* t *are odd integers. Then there exists* $\epsilon = \dfrac{\pm\mathbf{1} \pm I \pm J \pm K}{2}$ *such that* $\epsilon\alpha$ *is in* A_0 *and* $\mathrm{N}(\alpha) = \mathrm{N}(\epsilon\alpha)$.

Proof. We write $x = 4x' + \epsilon_1$, $y = 4y' + \epsilon_2$, $z = 4z' + \epsilon_3$, $t = 4t' + \epsilon_4$, with $\epsilon_i = \pm 1$. If we set $\epsilon := \dfrac{\epsilon_1\mathbf{1} - \epsilon_2 I - \epsilon_3 J - \epsilon_4 K}{2}$, then $\mathrm{N}(\epsilon) = 1$, hence $\mathrm{N}(\alpha\epsilon) = \mathrm{N}(\alpha)$, and therefore

$$\begin{aligned}\alpha\epsilon &= 4\left(\frac{x'\mathbf{1} + y'I + z'J + t'K}{2}\right)\epsilon + \mathrm{N}(\epsilon)\\ &= (x'\mathbf{1} + y'I + z'J + t'K)\,(2\epsilon) + 1 \in A_0.\end{aligned}$$ □

The following lemma will also be useful.

1.7. Lemma. *In the rings* $\mathbf{Z}[i]$, A_0 *and* A, *an element is invertible if and only if its norm is 1.*

Proof. If α is invertible, then $1 = \mathrm{N}(\alpha\alpha^{-1}) = \mathrm{N}(\alpha)\,\mathrm{N}(\alpha^{-1})$, hence $\mathrm{N}(\alpha) = 1$. Conversely, if $\mathrm{N}(\alpha) = 1$, then $\alpha\bar{\alpha} = 1$. Since the rings that we are looking at are stable under conjugation, then $\bar{\alpha}$ is an element of the ring, and α is therefore invertible. □

Finally, since the norm is multiplicative, it is enough to show that every prime number p (resp. every prime number $\equiv 1 \bmod 4$) is the norm of a Hurwitz quaternion (resp. the norm of a Gaussian integer). Since $2 = 1^2 + 1^2$, it moreover suffices to show this for odd primes p. To do this, we will first prove that $\mathbf{Z}[i]$ is a *principal ring* and that A is left (or right) *principal.*

1.8. Proposition. *The ring* $\mathbf{Z}[i]$ *is Euclidean, hence principal. The ring* A *is left Euclidean, hence left principal (and also right Euclidean and right principal).*

Proof. We will use the symbol B for both of the rings A and $\mathbf{Z}[i]$. The statement means that for $\alpha \in B$ and $\beta \in B \setminus \{0\}$, there exist $q, r \in B$ such that $\alpha = q\beta + r$ with $\mathrm{N}(r) < \mathrm{N}(\beta)$ (when the ring is A, pay attention to the order of multiplication). Once this has been proven, we immediately have that $\mathbf{Z}[i]$ is principal. Actually, the "same" proof shows that A is (left) principal. Now let I be a non-zero left ideal of A (i.e., $A \cdot I \subset I$), so it contains an element $\beta \neq 0$ of minimal norm, and clearly $A\beta \subset I$. Conversely, let $\alpha \in I$, and write $\alpha = q\beta + r$ with $\mathrm{N}(r) < \mathrm{N}(\beta)$. We therefore have $r = \alpha - q\beta \in I$, hence r is zero and $I = A\beta$. Let us now prove that A and $\mathbf{Z}[i]$ are Euclidean. The proof is based on the following elementary lemma, whose proof is left to the reader.

1.9. Lemma. *Let $x \in \mathbf{R}$. Then there exists $m \in \mathbf{Z}$ such that $|x-m| \leqslant 1/2$, and there exists $n \in \mathbf{Z}$ such that $|x - n/2| \leqslant 1/4$.*

• Now let $\alpha \in \mathbf{Z}[i]$ and $\beta \in \mathbf{Z}[i] \setminus \{0\}$, hence $\alpha/\beta = x + iy \in \mathbf{Q}[i]$, and there exist $m, n \in \mathbf{Z}$ such that $|x - m| \leqslant 1/2$ and $|y - n| \leqslant 1/2$. Therefore

$$\mathrm{N}\left((x + iy) - (m + in)\right) = (x - m)^2 + (y - n)^2 \leqslant \frac{1}{4} + \frac{1}{4} = \frac{1}{2}.$$

The Gaussian integer $q := m + ni$ is the quotient obtained from the (obviously possible) division of α by β since

$$\mathrm{N}(\alpha - q\beta) \leqslant \frac{\mathrm{N}(\beta)}{2} < \mathrm{N}(\beta).$$

• If $\alpha \in A$ and $\beta \in A \setminus \{0\}$, then $\alpha\beta^{-1} = x + yI + zJ + tK \in \mathbf{H}$ and there exists $m \in \mathbf{Z}$ such that $|x - m/2| \leqslant 1/4$. We therefore choose $q = (m + nI + hJ + \ell K)/2$, where m, n, h and ℓ are integers with the same parity (and so that $q \in A$) and such that $|y - n/2|$, $|z - h/2|$ and $|t - \ell/2|$ are $\leqslant 1/2$. We therefore obtain

$$\begin{aligned}\mathrm{N}(\alpha\beta^{-1} - q) &= \left(x - \tfrac{m}{2}\right)^2 + \left(y - \tfrac{n}{2}\right)^2 + \left(z - \tfrac{h}{2}\right)^2 + \left(t - \tfrac{\ell}{2}\right)^2 \\ &\leqslant \frac{1}{16} + \frac{1}{4} + \frac{1}{4} + \frac{1}{4} < 1\end{aligned}$$

and hence the desired inequality,

$$\mathrm{N}(\alpha - q\beta) < \mathrm{N}(\beta). \qquad \square$$

We can now complete the proof of the two theorems.

Proof. (Sum of two squares.) The ring $\mathbf{Z}[i]$ is principal, hence factorial (a factorial ring is also called a *unique factorization domain* or *UFD*). It is also clear that $\mathbf{Z}[i]^* = \{\pm 1, \pm i\}$. Now let $p \equiv 1 \bmod 4$. We know that there exists $a \in \mathbf{Z}$ such that $a^2 \equiv -1 \bmod p$. Thus we have an equality of the form $(a + i)(a - i) = pm$. We can see that the Gaussian integer p

clearly divides neither $(a+i)$ nor $(a-i)$. It is therefore neither prime, nor irreducible since $\mathbf{Z}[i]$ is principal. We thus have the decomposition $p = \alpha\beta$ where α and β are non-invertible. Therefore, $\mathrm{N}(\alpha\beta) = \mathrm{N}(p) = p^2$ and $\mathrm{N}(\alpha) = \mathrm{N}(\beta) = p$, which proves the two-square theorem.

(Sum of four squares.) It is enough to show that if p is an odd prime number, then it is the norm of an element of A. The number of squares in $\mathbf{Z}/p\mathbf{Z}$ is $(p+1)/2$, and therefore the polynomial $-1-X^2$ equals a square for at least one X; in other words, there exist $a, b \in \mathbf{Z}$ such that $a^2+b^2+1 \in p\mathbf{Z}$. We see from this that $(\mathbf{1}+aI+bJ)(\mathbf{1}-aI-bJ) \in pA$. We therefore consider the (left) ideal $\mathscr{I}$ generated by p and $\mathbf{1} + aI + bJ$. On the one hand, we know $\mathscr{I} = A\beta$ because A is (left) principal, and on the other hand, we have the inclusions $pA = Ap \subset \mathscr{I} \subset A$. Thus $p = \alpha\beta$. Now we will check that β and α are not invertible and also that the above inclusions of rings are strict. If α were invertible, then p would divide $1 + aI + bJ$, and furthermore $(1 + aI + bJ) = p(x + yI + zJ + tK)/2$, so that $px = 2$, which is impossible (p is an odd prime). If β were invertible, we would have $\mathscr{I} = A$, hence $1 = q(1 + aI + bJ) + q'p$, and by multiplying (on the right) by $(1 - aI - bJ)$, we would get $(1 - aI - bJ) = q''p$, which is equally absurd. We can therefore conclude that $\mathrm{N}(p) = \mathrm{N}(\alpha)\,\mathrm{N}(\beta) = p^2$, where $\mathrm{N}(\alpha)$ and $\mathrm{N}(\beta)$ are different from 1, hence equal to p. □

Further down, we will give another proof of the two-square (resp. four-square) theorem, which uses the geometry of numbers.

2. Fermat's Equation ($n = 3$ and 4)

One of the most famous mathematical problems (called "Fermat's last theorem") was solved by Andrew Wiles [80], with the help of Taylor, in 1995:

2.1. Theorem. *Let $n \geqslant 3$, and let x, y and z be integers such that $x^n + y^n = z^n$. Then $xyz = 0$.*

Of course it is "enough" to prove the theorem for $n = 4$ and $n = p$, an odd prime. We will settle for proving it for $n = 3$ and 4, by using Fermat's principle of infinite descent. The proof proposed for $n = 4$ stays in $\mathbf{Z}$, but the one that we give for $n = 3$ takes place in $\mathbf{Z}[j]$ (with $j = \exp(2\pi i/3)$). The classical approach, due to Kummer, is based on the following factorization. Let $\zeta = \exp(2\pi i/p)$, so in the ring $\mathbf{Z}[\zeta]$ we have:

$$x^p + y^p = (x+y)(x+\zeta y)\cdots(x+\zeta^{p-1}y) = z^p.$$

We will do some calculations in the ring $\mathbf{Z}[\zeta]$, setting $\lambda = 1 - \zeta$.

2.2. Lemma. *The element* λ *is prime and* $\mathbf{Z}[\zeta]/\lambda\mathbf{Z}[\zeta] \cong \mathbf{F}_p$. *Furthermore, we have the decomposition*

$$p = \prod_{k=1}^{p-1}(1-\zeta^k) = \epsilon\lambda^{p-1}, \qquad \textit{where } \epsilon \in \mathbf{Z}[\zeta]^*.$$

The elements $\eta_k := \sin(k\pi/p)/\sin(\pi/p)$ *and* $\epsilon_k := (1-\zeta^k)/(1-\zeta)$ *are units in* $\mathbf{Z}[\zeta]$ *for* $1 \leqslant k \leqslant p-1$, *and* η_k/ϵ_k *is a root of unity.*

Proof. We will begin by factoring the pth cyclotomic polynomial (over $\mathbf{C}$), $\Phi_p(X) = X^{p-1} + X^{p-2} + \cdots + X + 1 = \prod_{k=1}^{p-1}(X-\zeta^k)$. From this, we get the formula $p = \Phi_p(1) = \prod_{k=1}^{p-1}(1-\zeta^k)$. Thus λ divides p and $p \in \lambda\mathbf{Z}[\zeta]$. Moreover, since $\zeta \equiv 1 \bmod \lambda$, every element of $\mathbf{Z}[\zeta]$ is congruent modulo λ to an integer between 0 and $p-1$ (inclusive), which proves that $\mathbf{Z}[\zeta]/\lambda\mathbf{Z}[\zeta] \cong \mathbf{F}_p$. Since $1-\zeta^k = (1-\zeta)(1+\cdots+\zeta^{k-1})$, we see that ϵ_k is in $\mathbf{Z}[\zeta]$. By the same reasoning and by using the inverse h of k modulo p and the equality $1-\zeta = (1-\zeta^k)(1+\cdots+\zeta^{k(h-1)})$, we see that ϵ_k^{-1} is also an integer and therefore that $\epsilon_k \in \mathbf{Z}[\zeta]^*$. Furthermore, if k is odd, then

$$\begin{aligned}\epsilon_k = \frac{1-\zeta^k}{1-\zeta} &= e^{\pi i(k-1)/p}\,\frac{e^{\pi ik/p} - e^{-\pi ik/p}}{e^{\pi i/p} - e^{-\pi i/p}}\\ &= \zeta^{\frac{k-1}{2}}\,\frac{\sin(\pi k/p)}{\sin(\pi/p)} = \zeta^{\frac{k-1}{2}}\eta_k,\end{aligned}$$

whereas if k is even, then $\epsilon_k = -\zeta^k\epsilon_{p-k}$. Finally, if $1-\zeta^k = \epsilon_k\lambda$, then $p = \epsilon\lambda^{p-1}$ where $\epsilon = \epsilon_1\cdots\epsilon_{p-1} \in \mathbf{Z}[\zeta]^*$. □

Remark. We could of course write other formulas which produce units such as:

$$2\cos\left(\frac{2\pi}{p}\right) = \zeta + \zeta^{-1} = \zeta^{-1}(1+\zeta^2) = \zeta^{-1}\frac{1-\zeta^4}{1-\zeta^2} = \zeta^{-1}\epsilon_4\epsilon_2^{-1}.$$

We will now return to Kummer's method for Fermat's equation in its factored form (where x, y, z are relatively prime in $\mathbf{Z}$):

$$(x+y)(x+\zeta y)\ldots(x+\zeta^{p-1}y) = z^p.$$

Let $\delta \in \mathbf{Z}[\zeta]$ be a number which divides two factors of the above equation, for example $x+\zeta^i y$ and $x+\zeta^j y$; then it divides $(\zeta^i-\zeta^j)y$ and $(\zeta^i-\zeta^j)x$, hence $(\zeta^i-\zeta^j)$, and therefore δ divides λ, so $\delta = 1$ or λ (up to a unit). If z is not divisible by p, then the factors are relatively prime, and *if we show that* $\mathbf{Z}[\zeta]$ *is factorial*, we can deduce that:

$$\text{for } i = 0, \ldots, p-1, \quad x+\zeta^i y = u_i\alpha_i^p, \quad \text{where } u_i \text{ is a unit and } \alpha_i \in \mathbf{Z}[\zeta].$$

If z is divisible by p, we also have, still assuming that $\mathbf{Z}[\zeta]$ is factorial, some similar identities with extra powers of λ. However, this approach is hindered by the fact that actually the ring $\mathbf{Z}[\zeta]$ is not factorial in general. In fact, if $n = p$ is prime, it is not factorial whenever $p \geqslant 23$. We should therefore try to find a substitute for the following lemma (where the proof is left as an exercise).

2.3. Lemma. *Let A be a factorial ring. If the elements $a_1, \ldots, a_m \in A$ are pairwise relatively prime and $a_1 \ldots a_m = a^p$, then, up to a unit, the a_i are pth powers.*

We will start by describing the solutions of Fermat's equation for $n = 2$.

2.4. Proposition. *Let x, y, z be (relatively prime) integers such that $x^2 + y^2 = z^2$, then (up to switching x and y) there exist (relatively prime) integers u and v such that*

$$x = u^2 - v^2, \qquad y = 2uv \qquad \textit{and} \qquad z = u^2 + v^2. \tag{3.2}$$

Proof. After having simplified by their gcd, we can assume that x, y, z are pairwise relatively prime. Notice that $(u^2 - v^2)^2 + (2uv)^2 = (u^2 + v^2)^2$. By considering congruences modulo 4, we know that z is odd and that x and y have different parity; we will therefore assume that x is odd and y is even. We write $y^2 = z^2 - x^2 = (z - x)(z + x)$. Now notice that if d divides $z - x$ and $z + x$, then it divides $2x$ and $2z$ and therefore also 2 (since x and z are relatively prime). Thus $\gcd(z - x, z + x) = 2$. The integers $(z - x)/2$ and $(z + x)/2$ being relatively prime and there product being a square, are themselves squares which gives us: $z - x = 2v^2$, $z + x = 2u^2$ and $y = 2uv$, and hence $x = u^2 - v^2$ and $z = u^2 + v^2$, as in the statement of the proposition. □

2.5. Theorem. *The equation $x^4 + y^4 = z^2$ does not have any integer solutions, except for $xyz = 0$. Consequently Fermat's equation for $n = 4$ does not have any nontrivial solutions.*

Proof. The main idea of the proof is Fermat's "infinite descent", which consists of proving that if the equation has a solution (x, y, z) where $xyz \neq 0$, then it has another solution (x_1, y_1, z_1) where $x_1 y_1 z_1 \neq 0$ and $|z_1| < |z|$. This will lead to a contradiction, because a decreasing sequence of positive integers is necessarily constant after a certain point.

So let (x, y, z) be a solution. We can assume that x, y and z are relatively prime. By the previous proposition, we know that $x^2 = u^2 - v^2$, $y^2 = 2uv$ and $z = u^2 + v^2$, where u and v are relatively prime. We see that u and v

have different parity, and hence u is odd, so $v = 2w$ (if not we would have $x^2 \equiv -1 \bmod 4$). By considering $y^2 = 4uw$, we see that u and w, which are relatively prime, need to be squares, $u = z_1^2$ and $w = a^2$. Furthermore, by again applying the previous proposition to $x^2+v^2 = u^2$, we have $x = b^2-c^2$, $v = 2bc$ and $u = b^2+c^2$ where b and c are relatively prime. However, recall that $v = 2w = 2a^2$, so we can see, as before, that b and c are squares, $b = x_1^2$ and $c = y_1^2$. It therefore follows that

$$z_1^2 = u = b^2 + c^2 = x_1^4 + y_1^4,$$

and we can check that $|z_1| < |z|$ by observing, for example, that $z = u^2 + v^2 = z_1^4 + 4a^4 > z_1$. □

2.6. Theorem. *The equation $x^3 + y^3 = z^3$ does not have any solutions, except for $xyz = 0$. More generally, there do not exist any algebraic integers $x, y, z \in \mathbf{Z}[j]$ such that $x^3 + y^3 = z^3$ and $xyz \neq 0$.*

Proof. It will be convenient to distinguish between the two cases, the easy one being when xyz does not have a factor of 3 and the more difficult one when, for example, z has a factor of 3. The idea of the proof in the second case is to show that if the equation has a solution, then it would have another "smaller one" (the principle of "infinite descent").

We can show, as in Proposition 3-1.8 for $\mathbf{Z}[i]$, that the ring $A := \mathbf{Z}[j]$ is principal, hence factorial and that the group of units is formed of $\pm 1, \pm j, \pm j^2$. In particular, we check directly[1] that if $u \in A^*$ and $u \equiv \pm 1 \bmod \lambda^2$, then $u = \pm 1$ (also recall that λ designates the prime element $1 - j$ and that $\mathrm{ord}_\lambda(u)$ designates the largest exponent such that $\lambda^{\mathrm{ord}_\lambda(u)}$ divides u).

2.7. Lemma. *If $x \in \mathbf{Z}[j]$ is not divisible by λ, then $x^3 \equiv \pm 1 \bmod \lambda^4$.*

Proof. We can assume $x \equiv 1 \bmod \lambda$ or moreover that $x = 1 + \lambda\alpha$. Then, $x^3 - 1 = (x-1)(x-j)(x-j^2) = \lambda^3\alpha(\alpha+1)(\alpha+1+j) \equiv 0 \bmod \lambda^4$, because the elements $0, 1, 1+j$ are distinct modulo λ and therefore constitute all of the elements of $\mathbf{Z}[j]/\lambda\mathbf{Z}[j]$. □

We will now return to the proof of the theorem.

First case: λ does not divide xyz. By the preceding lemma, we have $x^3 \equiv \pm 1 \bmod \lambda^4$ (and the same holds for y and z), and therefore a solution to Fermat's equation implies that $\pm 1 \pm 1 \pm 1 \equiv 0 \bmod \lambda^4$; such a congruence is obviously impossible (3 is only divisible by λ^2).

[1] This remark is a very special case of the famous "Kummer lemma", which says that a unit which is congruent modulo λ^p to a pth power is in fact the pth power of a unit in $\mathbf{Z}[\exp(2\pi i/p)]$, given that p is "regular" in the sense of Remark 3-4.25 (in particular, when the ring $\mathbf{Z}[\exp(2\pi i/p)]$ is factorial, which is the case for $p = 3$).

Second case: λ divides xyz. We can assume that λ divides z and not xy. We will show a slightly more general version, namely that the equation

$$x^3 + y^3 = uz^3, \tag{3.3}$$

where u is a unit (i.e., $u \in \mathbf{Z}[j]^*$) and $m := \operatorname{ord}_\lambda z > 0$, does not have a solution in $\mathbf{Z}[j]$. Observe that λ^2 must divide z since $\pm 1 \pm 1 \equiv uz^3 \bmod \lambda^4$, hence $z^3 \equiv 0 \bmod \lambda^4$, and therefore $\operatorname{ord}_\lambda(z) \geqslant 4/3$. We will therefore prove the descent statement:

if $x^3 + y^3 = uz^3$ where $x, y, z \in A$, $u \in A^*$ and $\operatorname{ord}_\lambda(z) = m \geqslant 2$, then there exist $x_1, y_1, z_1 \in A$ and $u' \in A^*$ where $\operatorname{ord}_\lambda(z_1) = m - 1$ and $x_1^3 + y_1^3 = u'z_1^3$.

We will of course begin by factoring:

$$(x + y)(x + jy)(x + j^2 y) = uz^3.$$

We can see that λ^2 must divide one of the factors on the left (because $\operatorname{ord}_\lambda(z^3) = 3m \geqslant 6$), say $x+y$, and therefore $\operatorname{ord}_\lambda(x+jy) = \operatorname{ord}_\lambda(x+j^2y) = 1$; for example $x + jy = x + y - \lambda y$ and λ does not divide y. Thus the gcd of two of the factors is exactly λ. Since A is factorial, we see that

$$\begin{cases} x + y &= u_1 X^3 \lambda^{3m-2} \\ x + jy &= u_2 Y^3 \lambda \\ x + j^2 y &= u_3 Z^3 \lambda \end{cases} \qquad \text{where } \gcd(X, Y, Z) = 1 \text{ and } u_1, u_2, u_3 \text{ are units.}$$

By multiplying the equations respectively by 1, j and j^2 and adding them, we obtain $0 = u_1 X^3 \lambda^{3m-2} + u_2 j Y^3 \lambda + u_3 j^2 Z^3 \lambda$. By simplifying by λ and letting $u_4 := ju_3/u_2$ and $u_5 := -j^2 u_1/u_2$, we obtain

$$Y^3 + u_4 Z^3 = u_5 \left(\lambda^{m-1} X\right)^3.$$

We finish by pointing out that $\pm 1 \pm u_4 \equiv 0 \bmod \lambda^2$, and therefore $u_4 = \pm 1$. We then let $x_1 = Y$, $y_1 = u_4 Z$, $z_1 = \lambda^{m-1} X$ and $u' = u_5$ so that we have $x_1^3 + y_1^3 = u' z_1^3$ and $\operatorname{ord}_\lambda(z_1) = m - 1$.

3. Pell's Equation $x^2 - dy^2 = 1$

In this section, we will always assume that $d > 0$, and we will discuss the solutions of the above equation by explaining how it is related to the units of the ring $\mathbf{Z}[\sqrt{d}]$ and "good" rational approximations of $\sqrt{d}$.

Let us point out that the equation always has as solutions $(x, y) = (\pm 1, 0)$; we will refer to these as trivial. We also point out that if d is a square, $d = a^2$, then $(x - ay)(x + ay) = 1$ implies $x + ay = x - ay = 1$ (or $= -1$), hence $2ay = 0$, and therefore there are no nontrivial solutions. The only interesting case is when d is not the square of an integer (hence $\sqrt{d} \notin \mathbf{Q}$). The main theorem can be stated as follows.

3.1. Theorem. *Let d be a positive integer which is not a square. Then there exists a nontrivial solution $(x_1, y_1) \in \mathbf{N}^* \times \mathbf{N}^*$ (called the* fundamental solution*) of the equation $x^2 - dy^2 = 1$ such that all positive integer solutions are given by (x_n, y_n) where $x_n + y_n\sqrt{d} := (x_1 + y_1\sqrt{d})^n$ and the general solutions are given by $(\pm x_n, \pm y_n)$.*

We can of course find solutions (x_n, y_n) by induction starting with (x_1, y_1) and observing that

$$(x_{n+1}, y_{n+1}) = (x_1 x_n + dy_1 y_n, y_1 x_n + x_1 y_n).$$

The connection to rational approximations of $\sqrt{d}$ is the following. Suppose that (x, y) is a nontrivial solution of the equation (where say $x, y > 0$), then

$$0 < \frac{x}{y} - \sqrt{d} = \frac{1}{y^2\left(\frac{x}{y} + \sqrt{d}\right)} < \frac{1}{2\sqrt{d}y^2}.$$

Conversely, if $x/y \in \mathbf{Q}$ is an approximation which satisfies the previous inequality, then

$$0 < x^2 - dy^2 = y^2\left(\frac{x}{y} - \sqrt{d}\right)\left(\frac{x}{y} + \sqrt{d}\right) < \frac{1}{2\sqrt{d}}\left(2\sqrt{d} + \frac{1}{2\sqrt{d}y^2}\right) < 2,$$

hence $x^2 - dy^2 = 1$ (because it is an integer). Thus a positive solution (x, y) of Pell's equation corresponds to a rational approximation x/y of $\sqrt{d}$ which satisfies $0 < \frac{x}{y} - \sqrt{d} < \frac{1}{2\sqrt{d}y^2}$.

To the ring $\mathbf{Z}[\sqrt{d}]$, we introduce the homomorphism $\sigma(a + b\sqrt{d}) = a - b\sqrt{d}$ (why is it a homomorphism?), as well as the norm

$$\mathrm{N}(\alpha) = \alpha\sigma(\alpha) = a^2 - db^2, \text{ if } \alpha = a + b\sqrt{d}.$$

The norm is multiplicative, and we have, as in $\mathbf{Z}[i]$, the following lemma, whose very similar proof is omitted.

3.2. Lemma. *In the ring $\mathbf{Z}[\sqrt{d}]$, an element is invertible if and only if its norm is ± 1.*

If we denote by $A^* = \mathbf{Z}[\sqrt{d}]^*$ and $U_1 = \{\alpha \mid \mathrm{N}(\alpha) = 1\}$, we see that the index $(A^* : U_1)$ is either 2 or 1, depending on whether there exists a unit with norm -1. Of course, the solutions (x, y) of Pell's equation correspond to the units $x + y\sqrt{d} \in U_1$, and Theorem 3-3.1 can be translated into the following statement.

3.3. Theorem. *There exists a unit $\epsilon_1 \in \mathbf{Z}[\sqrt{d}]^*$, called the fundamental unit, such that*

$$\mathbf{Z}[\sqrt{d}]^* = \{\pm\epsilon_1^n \mid n \in \mathbf{Z}\} \cong \{\pm 1\} \times \mathbf{Z}.$$

If $\mathrm{N}(\epsilon_1) = +1$, *then* $U_1 = \mathbf{Z}[\sqrt{d}]^*$ *and if* $\mathrm{N}(\epsilon_1) = -1$, *then we have*

$$U_1 = \{\pm\epsilon_1^{2n} \mid n \in \mathbf{Z}\} \cong \{\pm 1\} \times \mathbf{Z}.$$

In order to prove this theorem, we introduce the "logarithm" map, $L : \mathbf{Z}[\sqrt{d}]^* \to \mathbf{R}^2$ given by the formula $L(\alpha) = (\log|\alpha|, \log|\sigma(\alpha)|)$.

3.4. Proposition. *The map $L : \mathbf{Z}[\sqrt{d}]^* \to \mathbf{R}^2$ has the following properties.*

i) *The map L is a homomorphism, i.e., $L(\alpha\beta) = L(\alpha) + L(\beta)$.*
ii) *Its kernel is ± 1.*
iii) *Its image is a discrete subgroup.*
iv) *Its image contains the line $x + y = 0$.*

Proof. Property *i)* is immediate. Property *iv)* comes from the fact that $\log|\alpha| + \log|\sigma(\alpha)| = \log|\mathrm{N}(\alpha)| = 0$. To prove *ii)* and *iii)*, we will show that the preimage under L of a ball in $\mathbf{R}^2$ is finite, from which can we deduce, on the one hand, that the image is discrete and, on the other hand, that the kernel of L is finite, and therefore composed of roots of unity hence of ± 1 since $\mathbf{Z}[\sqrt{d}] \subset \mathbf{R}$. Now, an element $\alpha \in \mathbf{Z}[\sqrt{d}]^*$ is a root of $P := X^2 - t(\alpha)X + \mathrm{N}(\alpha) \in \mathbf{Z}[X]$, with $t(\alpha) = \alpha + \sigma(\alpha)$ (the "trace") and $\mathrm{N}(\alpha) = \pm 1$. If $L(\alpha)$ is in a ball of radius C, we have $|\alpha| = \exp(\log|\alpha|) \leqslant \exp(C)$ and the same for $|\sigma(\alpha)|$. It follows that $|t(\alpha)| \leqslant 2\exp(C)$. Therefore there are only a finite number of possible polynomials, and hence a finite number of α. □

We will now state a classical lemma.

3.5. Lemma. *Every discrete subgroup G of $\mathbf{R}$ is of the form $G = \mathbf{Z}\omega$.*

Proof. (Sketch) If $G = \{0\}$, we can choose $\omega = 0$, and if not, we choose $\omega := \inf\{x \in G \mid x > 0\}$. Since G is discrete, we have $\omega > 0$ and $\omega \in G$ (otherwise, there would be a sequence of elements of G which converge to ω, which contradicts the fact that G is discrete). Finally, if $x \in G$, we choose $m \in \mathbf{Z}$ such that $m\omega \leqslant x < (m+1)\omega$. Therefore, $0 \leqslant x - m\omega < \omega$ and $x - m\omega \in G$, hence $x = m\omega$. □

This lemma can be applied to $L(\mathbf{Z}[\sqrt{d}]^*)$, and provides a proof of the theorem under the condition that we prove the existence of a unit $\neq \pm 1$ or of a nontrivial solution to Pell's equation. These considerations show that it suffices to prove the following proposition.

3.6. Proposition. *Let d be a positive integer which is not a square. Then there exists a nontrivial solution (x_1, y_1) (i.e., with $y_1 \neq 0$) to the equation $x^2 - dy^2 = 1$.*

A good practical method for constructing this solution is the method of *continued fractions*, which is succinctly described later in this section. We are first going to prove the existence of a solution by showing, with an argument due to Dirichlet (and already used in the proof of Theorem 2-4.4) called the "pigeonhole principle", that there exist good rational approximations of $\sqrt{d}$, without actually explicitly constructing them, then give the continued fractions algorithm.

3.7. Lemma. *Let $\alpha \in \mathbf{R}$ and $N \geqslant 1$. Then there exists a rational number $p/q \in \mathbf{Q}$ such that*

$$\left|\alpha - \frac{p}{q}\right| \leqslant \frac{1}{qN} \qquad \text{and} \qquad 1 \leqslant q \leqslant N.$$

Proof. We cut the interval $[0,1]$ into N intervals of length $1/N$. Among the $N+1$ numbers $j\alpha - \lfloor j\alpha \rfloor$ (for $j = 0, \ldots, N$), there are therefore two in the same small interval and at a distance of at most $1/N$ from each other. In other words, there exist $0 \leqslant j < \ell \leqslant N$ such that $|(j\alpha - \lfloor j\alpha \rfloor) - (\ell\alpha - \lfloor \ell\alpha \rfloor)| \leqslant 1/N$. It follows that

$$\left|\alpha - \frac{\lfloor \ell\alpha \rfloor - \lfloor j\alpha \rfloor}{\ell - j}\right| \leqslant \frac{1}{(\ell - j)N} \cdot$$

The desired result follows by setting $p := \lfloor \ell\alpha \rfloor - \lfloor j\alpha \rfloor$ and $q := \ell - j$. □

Let us point out that the approximation provided by the lemma satisfies $|\alpha - p/q| \leqslant 1/q^2$.

3.8. Corollary. (Dirichlet) *Let $\alpha \in \mathbf{R} \setminus \mathbf{Q}$. Then there exist infinitely many rational numbers $p/q \in \mathbf{Q}$ such that*

$$\left|\alpha - \frac{p}{q}\right| \leqslant \frac{1}{q^2} \cdot$$

Proof. Let $N_1 \geqslant 1$ and p_1/q_1 be a rational number provided by the previous lemma such that $\left|\alpha - \frac{p_1}{q_1}\right| \leqslant \frac{1}{q_1 N_1} \cdot$ Since $\alpha \notin \mathbf{Q}$, the left-hand side of the inequality is non-zero. Therefore we can choose N_2 such that $1/N_2 < \left|\alpha - \frac{p_1}{q_1}\right|$. Now let p_2/q_2 be a rational number provided by the previous lemma such that $\left|\alpha - \frac{p_2}{q_2}\right| \leqslant \frac{1}{q_2 N_2}$. It follows that

$$\left|\alpha - \frac{p_2}{q_2}\right| \leqslant \frac{1}{q_2 N_2} \leqslant \frac{1}{N_2} < \left|\alpha - \frac{p_1}{q_1}\right|,$$

therefore $p_2/q_2 \neq p_1/q_1$. It is now clear that we can iterate this process indefinitely. □

3.9. Remarks. 1) If we remove the hypothesis that $\alpha \notin \mathbf{Q}$ in the statement of the corollary, the result would be false. To see this, if $\alpha = a/b$ and $a/b \neq p/q$ where $\left|\alpha - \frac{p}{q}\right| \leqslant \frac{1}{q^2}$, then

$$\frac{1}{bq} \leqslant \frac{|aq - bp|}{bq} = \left|\alpha - \frac{p}{q}\right| \leqslant \frac{1}{q^2},$$

and hence $q \leqslant b$. There would therefore only exist a finite number of p/q.
2) Let us look at the example $\alpha = \sqrt{d}$ where d is not a square. We can prove that the corollary is optimal in the following sense: there exists a constant $C > 0$ such that for every $p/q \in \mathbf{Q}$, we have

$$\left|\sqrt{d} - \frac{p}{q}\right| \geqslant \frac{C}{q^2}.$$

To do this, consider $P(X) = X^2 - d = (X - \sqrt{d})(X + \sqrt{d})$. It follows that $|P(p/q)| \geqslant 1/q^2$. Now, if for example $|\sqrt{d} - p/q| \leqslant 1$, we have $|p/q| \leqslant \sqrt{d}+1$, then $|p/q + \sqrt{d}| \leqslant 2\sqrt{d} + 1$, and thus

$$\left|\sqrt{d} - \frac{p}{q}\right| = \frac{|P(p/q)|}{\left|p/q + \sqrt{d}\right|} \geqslant \frac{1}{(2\sqrt{d} + 1)q^2}.$$

3) If α is an algebraic number of degree $d \geqslant 3$, the same proof shows that $\left|\alpha - \frac{p}{q}\right| \geqslant \frac{C}{q^d}$ *(Liouville's inequality).* In 1955, Roth proved-but the proof is much more difficult-that furthermore, for every $\epsilon > 0$ there exists a constant C, which depends on α and ϵ, such that for every $p/q \in \mathbf{Q}$ (see Chap. 6):

$$\left|\alpha - \frac{p}{q}\right| \geqslant \frac{C}{q^{2+\epsilon}}.$$

Proof. (of Proposition 3-3.6) We will apply Corollary 3-3.8 to $\sqrt{d} \notin \mathbf{Q}$. Thus there are infinitely many integers (x, y) such that $|\sqrt{d} - x/y| \leqslant 1/y^2$ and hence such that $|\sqrt{d} + x/y| \leqslant 2\sqrt{d} + 1$ and finally $|x^2 - dy^2| \leqslant 2\sqrt{d} + 1$.

In particular, there exists an integer c such that there are infinitely many solutions to the equation $x^2 - dy^2 = c$. Since there are only a finite number of classes modulo c, there even exist infinitely many pairwise congruent solutions modulo c. So we take (x_1, y_1) and (x_2, y_2) which are solutions to $x^2 - dy^2 = c$ and check that $x_1 \equiv x_2 \bmod c$ and $y_1 \equiv y_2 \bmod c$. We set

$$u + v\sqrt{d} := \frac{x_1 + y_1\sqrt{d}}{x_2 + y_2\sqrt{d}}.$$

We therefore have

$$u^2 - dv^2 = \mathrm{N}(u + v\sqrt{d}) = \frac{\mathrm{N}(x_1 + y_1\sqrt{d})}{\mathrm{N}(x_2 + y_2\sqrt{d})} = \frac{c}{c} = 1,$$

and it suffices to see that u and v are integers. Therefore, we compute

$$u + v\sqrt{d} = \frac{(x_1 + y_1\sqrt{d})(x_2 - y_2\sqrt{d})}{x_2^2 - dy_2^2} = \frac{x_1x_2 - dy_1y_2}{c} + \frac{y_1x_2 - x_1y_2}{c}\sqrt{d},$$

and notice that $x_1x_2 - dy_1y_2 \equiv x_1^2 - dy_1^2 \equiv 0 \bmod c$ and $y_1x_2 - x_1y_2 \equiv y_1x_1 - x_1y_1 \equiv 0 \bmod c$, which finishes the proof. □

Supplement. The slightly more general equation $x^2 - dy^2 = m$ does not always have a solution. For example, if $m = -1$ and p is a prime number congruent to 3 modulo 4 which divides d, then a solution would imply that $x^2 \equiv -1 \bmod p$, which is impossible. More generally, for every odd p which divides d but not m, it must be that $x^2 \equiv m \bmod p$, and hence $\left(\frac{m}{p}\right) = 1$. Conversely, if there existed a solution, there would exist infinitely many of them, since $\mathrm{N}(u\alpha) = m$ if $\mathrm{N}(\alpha) = m$ and $\mathrm{N}(u) = 1$. We have following proposition, which could be useful.

3.10. Proposition. *Let $m \in \mathbf{Z} \setminus \{0\}$, and let d be an integer which is not a square. Then there exist $\alpha_1, \ldots \alpha_r \in \mathbf{Z}[\sqrt{d}]$ such that:*

$$\left\{\alpha \in \mathbf{Z}[\sqrt{d}] \mid \mathrm{N}(\alpha) = m\right\} = \alpha_1 U_1 \cup \cdots \cup \alpha_r U_1.$$

Proof. It is clear that the set of solutions is the union of classes modulo U_1. We will show that there exists a finite union. If $\mathrm{N}(\alpha) = m$, it follows that α divides m and furthermore that $m\mathbf{Z}[\sqrt{d}] \subset \alpha\mathbf{Z}[\sqrt{d}]$. But the set of ideals which contain $m\mathbf{Z}[\sqrt{d}]$ is in bijection with the ideals of the quotient $\mathbf{Z}[\sqrt{d}]/m\mathbf{Z}[\sqrt{d}]$ and is consequently a finite set. However, $\alpha\mathbf{Z}[\sqrt{d}] = \alpha'\mathbf{Z}[\sqrt{d}]$ is equivalent to the fact that α and α' are equal up to a unit. The set of solutions is thus finite modulo the group of units, hence equal modulo the subgroup U_1. □

Continued fractions. We will now outline a procedure for calculating good rational approximations of a real number: the algorithm of *continued fractions.*

Notation. Let a_0 be a real number and $a_1, \ldots, a_n$ be a sequence of real numbers > 0. We set

$$[a_0, a_1, \ldots, a_n] := a_0 + \cfrac{1}{a_1 + \cfrac{1}{a_2 + \cdots + \cfrac{1}{a_n}}}$$

3.11. Definition. To a real number x, we can associate a sequence of integers a_n and an auxiliary sequence of real numbers x_n defined as follows: $a_0 := \lfloor x \rfloor$, $x_0 := x$, $x_{n+1} = 1/(x_n - a_n)$ and $a_{n+1} = \lfloor x_{n+1} \rfloor$. We conclude the sequence when x_n is an integer (which only happens when $x \in \mathbf{Q}$). We define the nth *convergent* as

$$\frac{p_n}{q_n} := [a_0, a_1, \ldots, a_n].$$

3.12. Lemma. *The following formulas hold.*

i) $x = [a_0, a_1, \ldots, a_{n-1}, x_n]$.

ii) $p_{n+1} = a_{n+1}p_n + p_{n-1}$ *(where* $p_0 = a_0$ *and* $p_1 = a_1a_0 + 1$*), while* $q_{n+1} = a_{n+1}q_n + q_{n-1}$ *(where* $q_0 = 1$ *and* $q_1 = a_1$*).*

iii) If $p_n/q_n = [a_0, \ldots, a_n]$*, then*

$$[a_0, \ldots, a_n, y] = \frac{p_n y + p_{n-1}}{q_n y + q_{n-1}}.$$

iv) $q_n p_{n-1} - p_n q_{n-1} = (-1)^n$.

v) $q_n p_{n-2} - p_n q_{n-2} = (-1)^{n-1} a_n$.

Proof. Let us point out right away that for all *real numbers* a_i, we have $[a_0, \ldots, a_{n-1}, a_n] = [a_0, \ldots, a_{n-1} + 1/a_n]$. The first formula can be proven by induction (the case $n = 0$ is satisfied by construction). Assume therefore that $x = [a_0, \ldots, a_{n-1}, x_n]$, hence $[a_0, \ldots, a_n, x_{n+1}] = [a_0, \ldots, a_{n-1}, a_n + 1/x_{n+1}] = [a_0, \ldots, a_{n-1}, x_n] = x$. Next,

$$[a_0, \ldots, a_{n-1}, a_n, a_{n+1}] = [a_0, \ldots, a_{n-1}, a_n + 1/a_{n+1}] = \frac{p'_n}{q'_n},$$

where we can assume (by induction) that the p'_n, q'_n are given by the formulas $p'_{m+1} = a'_{m+1}p'_m + p'_{m-1}$, where $a'_m = a_m$ for $m \leqslant n-1$ and $a'_n = a_n + 1/a_{n+1}$. Thus $p'_n = (a_n + 1/a_{n+1})p'_{n-1} + p'_{n-2} = (a_n + 1/a_{n+1})p_{n-1} + p_{n-2}$

and $q'_n = (a_n + 1/a_{n+1})q'_{n-1} + q'_{n-2} = (a_n + 1/a_{n+1})q_{n-1} + q_{n-2}$, hence

$$\begin{aligned}\frac{p'_n}{q'_n} &= \frac{(a_n + 1/a_{n+1})p_{n-1} + p_{n-2}}{(a_n + 1/a_{n+1})q_{n-1} + q_{n-2}} = \frac{a_{n+1}(a_n p_{n_1} + p_{n-2}) + p_{n-1}}{a_{n+1}(a_n q_{n_1} + q_{n-2}) + q_{n-1}} \\ &= \frac{a_{n+1}p_n + p_{n-1}}{a_{n+1}q_n + q_{n-1}}.\end{aligned}$$

Formula *iii*) can be proven similarly to the previous formulas. Formulas *iv*) and (v) can also be proven by induction, for example

$$\begin{aligned}p_{n+1}q_n - q_{n+1}p_n &= (a_{n+1}p_n + p_{n-1})q_n - (a_{n+1}q_n + q_{n-1})p_n \\ &= -(p_n q_{n-1} - q_n p_{n-1}),\end{aligned}$$

and the same for (v). □

3.13. Remark. We can also take as our initial values of the sequences (p_n) and (q_n) the values $p_{-2} = 0$, $p_{-1} = 1$ and $q_{-2} = 1$, $q_{-1} = 0$. Moreover, it often helpful to write the formulas in matrix form, for example:

$$\begin{pmatrix} p_n & p_{n-1} \\ q_n & q_{n-1} \end{pmatrix} = \begin{pmatrix} a_0 & 1 \\ 1 & 0 \end{pmatrix} \cdots \begin{pmatrix} a_n & 1 \\ 1 & 0 \end{pmatrix}.$$

3.14. Remark. These induction formulas allow us to calculate p_n and q_n, starting with the computation of the a_n; since $a_n \geqslant 1$, we also see that q_n grows at least as fast as a Fibonacci sequence and that the following lower bound holds: $q_n \geqslant \left(\frac{1+\sqrt{5}}{2}\right)^{n-1}$. Thus an approximation $|x - p/q| < 1/2q^2$, which we will show below must be a convergent of the continued fraction of x, can be computed in $O(\log q)$ steps; this remark is used in Exercise 3-6.12.

The following theorem can also be deduced from these formulas.

3.15. Theorem. *The sequence* (p_n/q_n) *converges to* x. *More precisely, the sequence of the* p_{2n}/q_{2n} *is increasing and converges to* x *and the sequence* p_{2n+1}/q_{2n+1} *is decreasing and converges to* x. *The following approximation holds:*

$$\frac{1}{q_n(q_n + q_{n+1})} < \left| x - \frac{p_n}{q_n} \right| < \frac{1}{q_n q_{n+1}}.$$

Furthermore, the convergents give the best approximations of x, *in the following sense. If* $q \leqslant q_n$ *and* $p/q \neq p_n/q_n$, *then*

$$q_n \left| x - \frac{p_n}{q_n} \right| < q \left| x - \frac{p}{q} \right|.$$

Furthermore, if $\left|x - \dfrac{p}{q}\right| < 1/2q^2$, *then there exists* n *such that* $p/q = p_n/q_n$.

Proof. We know that $a_n = \lfloor x_n \rfloor \leqslant x_n$. Now, the function $[a_0, \ldots, a_n]$ is clearly an increasing (resp. decreasing) function of a_m, for m even (resp. m odd). Therefore, if n is even, then $[a_0, \ldots, a_n] \leqslant [a_0, \ldots, x_n] = x$, and the converse if n is odd. By the lemma, we know

$$\frac{p_{n-1}}{q_{n-1}} - \frac{p_n}{q_n} = \frac{(-1)^n}{q_n q_{n-1}} \quad \text{and} \quad \frac{p_{n-2}}{q_{n-2}} - \frac{p_n}{q_n} = \frac{(-1)^{n-1} a_n}{q_n q_{n-2}}.$$

Hence we have the ordering

$$\frac{p_{2n}}{q_{2n}} < \frac{p_{2n+2}}{q_{2n+2}} < x < \frac{p_{2n+1}}{q_{2n+1}} < \frac{p_{2n-1}}{q_{2n-1}},$$

and therefore $|x - p_n/q_n| \leqslant |p_n/q_n - p_{n+1}/q_{n+1}| = 1/q_n q_{n+1}$, whereas $|x - p_n/q_n| \geqslant |p_n/q_n - p_{n+2}/q_{n+2}| = a_{n+2}/q_n q_{n+2} = a_{n+2}/q_n(a_{n+2}q_{n+1} + q_n) \geqslant 1/q_n(q_{n+1} + q_n)$. These approximations clearly show that the sequence (p_n/q_n) converges to x. Observe also that $1/q_{n+2} < |p_n - xq_n| < 1/q_{n+1}$ and that the sequence $(|p_n - xq_n|)$ is therefore strictly decreasing. Now let p/q be a fraction with $q \leqslant q_n$ and $p/q \neq p_n/q_n$. We can assume that $q_{n-1} < q$. If we solve the system of linear equations $up_n + vp_{n-1} = p$ and $uq_n + vq_{n-1} = q$, then we obtain $u = \pm(pq_n - qp_{n-1})$ and $v = \pm(pq_n - qp_n)$. In particular, u and v are non-zero integers. Since $q = uq_n + vq_{n-1} \leqslant q_n$, we see that u and v have opposite signs and that the two quantities $u(p_n - q_n x)$ and $v(p_{n-1} - q_{n-1}x)$ therefore have the same sign. We know that $p - qx = u(p_n - q_n x) + v(p_{n-1} - q_{n-1}x)$, hence

$$|p - qx| = |u(p_n - q_n x)| + |v(p_{n-1} - q_{n-1}x)| \geqslant |p_n - q_n x| + |p_{n-1} - q_{n-1}x|.$$

Finally, if $|x - p/q| < 1/2q^2$, we set $x - p/q = \epsilon\theta/q^2$ where $\epsilon = \pm 1$ and $0 < \theta < 1/2$. We will expand $p/q = [a_0, \ldots, a_m]$ as a finite continued fraction. By noticing that if $a_m > 1$, we see that $[a_0, \ldots, a_m] = [a_0, \ldots, a_m - 1, 1]$ and hence that we can choose[2] the parity of m. We choose the parity in such a way that $p_{m-1}q - pq_{m-1} = (-1)^m = \epsilon$. We now will define y by the equality $x = (yp_m + p_{m-1})/(yq_m + q_{m-1})$. Solving explicitly for y yields $y = (q - \theta q_{m-1})/\theta q$. By using $q_{m-1} < q$ and $\theta < 1/2$, we see that $y > 1$, and we can therefore write $y = [a_{m+1}, \ldots]$, where $a_{m+1} \geqslant 1$. By expanding the obtained continued fraction $x = [a_0, \ldots, a_m, a_{m+1}, \ldots]$, we have that $p/q = [a_0, \ldots, a_m]$ is a convergent. □

3.16. Remarks. 1) Whenever $x \in \mathbf{Q}$, its expansion as a continued fraction is finite (i.e., there exists n such that $a_n = 0$).

[2]It can also be shown that this is the only possible ambiguity in the expression of a continued fraction.

2) When $x \in \mathbf{R} \setminus \mathbf{Q}$, we therefore have that $x = \lim_{n\to\infty}[a_0, \ldots, a_n]$, which by convention is written $x = [a_0, \ldots, a_n, \ldots]$ and which is referred to as the continued fraction expansion of x.

3) A solution to Pell's equation $p^2 - dq^2 = 1$ provides, as we have seen, a good approximation p/q of $\sqrt{d}$. It should therefore appear as a convergent of the continued fraction expansion of $\sqrt{d}$. This is precisely how we find it, and in fact fairly rapidly, considering Remark 3-3.14.

3.17. Examples. The continued fraction expansion of $x = \sqrt{2}$ and of $y = \sqrt{7}$ are written respectively as

$$\sqrt{2} = [1, 2, 2, 2, \ldots] \qquad \text{and} \qquad \sqrt{7} = [2, 1, 1, 1, 4, 1, 1, 1, 4, \ldots].$$

It can be verified that these expansions are periodic. In the case of $\sqrt{2}$, the initial convergent p_0/q_0 gives $p_0^2 - 2q_0^2 = -1$ and $p_1/q_1 = 3/2$ gives $p_1^2 - 2q_1^2 = +1$. In the case of $\sqrt{7}$, the convergent $p_3/q_3 = 8/3$ gives $p_3^2 - 7q_3^2 = +1$. The fact that the continued fraction expansion is periodic is a very special case of Lagrange's theorem which says that the continued fraction expansion of the real number x is periodic if and only if x is quadratic, i.e., the root of a quadratic equation with integer coefficients (see, for example, Hardy and Wright's book [4]).

Let us give an example which illustrates the quality of the continued fraction algorithm: finding solutions of the equation $x^2 - 61y^2 = 1$ (try to find a solution to this one by guess and check!). The continued fraction expansion of $x = \sqrt{61}$ is written

$$\sqrt{61} = [7, 1, 4, 3, 1, 2, 2, 1, 3, 4, 1, 14, 1, \ldots],$$

and the expansion becomes periodic starting at $a_{12} = a_1 = 1$. The first convergents are

$$\frac{7}{1}, \frac{8}{1}, \frac{39}{5}, \frac{125}{16}, \frac{164}{21}, \frac{453}{58}, \frac{1070}{137}, \frac{1523}{195}, \frac{5639}{722},$$
$$\frac{24079}{3083}, \frac{29718}{3805}, \frac{440131}{56353}, \frac{469849}{60158}.$$

The tenth convergent, $p_{10}/q_{10} = 29718/3805$, provides the first solution to $x^2 - 61y^2 = -1$. The fundamental solution of $x^2 - 61y^2 = 1$ is from then on given by $x_1 + y_1\sqrt{61} = (p_{10} + q_{10}\sqrt{61})^2$, or

$$(x_1, y_1) = (1766319049, 226153980).$$

We will indicate, without proof (see for example the entertaining article [50] which describes, among other things and in detail, Archimedes' cattle problem, whose solution comes from the solution of a Pell equation), the following facts, which can be checked in the previous examples. We know that the expansion of $\sqrt{d}$ can be written as $\sqrt{d} = [a_0, \ldots, a_r, a_{r+1}, \ldots]$.

i) If r is the first subscript such that $a_{r+1} = 2a_0$, then the expansion of $\sqrt{d}$ becomes periodic starting at the latter coefficient (i.e., the sequence $a_1, \ldots, a_r, 2a_0$ repeats).
ii) If r is odd (which is the case for $d = 7$), then (p_r, q_r) provides the smallest solution to Pell's equation $x^2 - dy^2 = +1$ and there are no solutions to the equation $x^2 - dy^2 = -1$.
iii) If r is even (which is the case for $d = 2$ or 61), then (p_r, q_r) provides the smallest solution to the equation $x^2 - dy^2 = -1$ and the smallest solution to Pell's equation $x^2 - dy^2 = +1$ is given by (p_{2r+1}, q_{2r+1}).

4. Rings of Algebraic Integers

In this part, we will give you an idea of what some the general properties of ring extensions of **Z** *are. These properties will lead us to the notion of a Dedekind ring, which in turn generalizes some of the examples that we have already encountered with* $\mathbf{Z}[\sqrt{d}]$ *and* $\mathbf{Z}[\exp(2\pi i/n)]$. *The main tools are algebra and some geometry of numbers.*

We are familiar with the notion of an algebraic element (here "algebraic" is always taken to be in the sense "algebraic over the rationals"); this notion comes from field theory, and the corresponding notion for rings is as follows.

4.1. Definition. An *algebraic integer* is a complex number, α, which is a root of a monic polynomial with integer coefficients. More generally, an element α is called *integral* or an *algebraic integer* over a ring A if it is the root of a monic polynomial with coefficients in A.

Example. A rational number $\alpha = a/b$ is the root of $bX - a \in \mathbf{Z}[X]$ and, by making use of the fact that **Z** is factorial, we see that if α is an algebraic integer, then $\alpha \in \mathbf{Z}$.

4.2. Definition. An integral domain A is *integrally closed* if the only elements of $K := \mathrm{Frac}(A)$ which are algebraic over A are elements of A.

Examples. We can easily show that a principal or factorial ring is integrally closed. However, the ring $A = \mathbf{Z}[\sqrt{5}]$ is not integrally closed, since the number $\alpha := \dfrac{1+\sqrt{5}}{2}$, which is in $\mathbf{Q}(\sqrt{5})$, is a root $X^2 - X - 1$, which is integral over A (and even over **Z**) without being in A.

4.3. Lemma. *An element α is an algebraic integer over A if and only if $A[\alpha]$ is a finitely generated A-module, and also if and only if the ring $A[\alpha]$ is contained in a subring containing A and α, which is a finitely generated A-module.*

4.4. Corollary. *The sum, difference and product of two algebraic integers is an algebraic integer. If α is integral over B and every element of B is integral over A, then α is integral over A. In particular, if K is an extension of $\mathbf{Q}$, then the set:*

$$\mathcal{O}_K := \{\alpha \in K \mid \alpha \text{ is an algebraic integer}\}$$

is a ring.

Proof. If α is integral over A, then it satisfies an equation $\alpha^n = a_{n-1}\alpha^{n-1} + \cdots + a_0$ with $a_i \in A$. The A-module $A[\alpha] = A + A\alpha + \cdots + A\alpha^{n-1}$ is therefore finitely generated. Conversely, if $A[\alpha]$ is contained in a finitely generated A-module, $Au_1 + \cdots + Au_m$, we can write $\alpha u_i = \sum_{j=1}^{m} a_{i,j} u_j$, with $a_{i,j} \in A$. We will therefore let M be the $m \times m$ matrix of the coefficients $a_{i,j}$. The polynomial $P(X) := \det(XId - M)$ is monic with coefficients in A and $P(\alpha) = 0$ (think of the Cayley-Hamilton theorem, or redo its proof), and hence α is integral over A. For the corollary, observe that if α and β are algebraic integers, then $\mathbf{Z}[\alpha, \beta]$ is a finitely generated $\mathbf{Z}$-module (a generating set is given by a finite number of $\alpha^k \beta^l$), and hence its elements are all integral over $\mathbf{Z}$. □

4.5. Definition. A *number field* is a finite extension K of $\mathbf{Q}$ and $\mathcal{O}_K$ is the ring of integers of K.

4.6. Remark. We can always assume (by the primitive element theorem) that there exists an α such that $K = \mathbf{Q}(\alpha)$. It should also be noted that if α is algebraic over $\mathbf{Q}$, then there exists an integer $d \in \mathbf{Z}$ (a "denominator") such that $d\alpha$ is an algebraic integer. In particular, K is the field of fractions of $\mathcal{O}_K$.

4.7. Proposition. *The ring $\mathcal{O}_K$ is integrally closed.*

Proof. An element of K which is integral over $\mathcal{O}_K$ is integral over $\mathbf{Z}$, this is an immediate consequence of Lemma 3-4.3 above. It is therefore in $\mathcal{O}_K$.□

4.8. Definition. Let K be a number field and $\alpha \in K$; we define the *norm* $\mathrm{N}(\alpha) = \mathrm{N}_{\mathbf{Q}}^{K}(\alpha)$ (resp. the *trace* $\mathrm{Tr}(\alpha) = \mathrm{Tr}_{\mathbf{Q}}^{K}(\alpha)$) to be the determinant (resp. the trace) of multiplication by α, viewed as a $\mathbf{Q}$-linear map from K to K.

Afterwards, we will give a more concrete expression for the trace and the norm.

4.9. Lemma. *Let α be algebraic over* $\mathbf{Q}$ *and* $K = \mathbf{Q}(\alpha)$, *and let*

$$P(X) = X^d + a_{d-1}X^{d-1} + \cdots + a_0 = (X - \alpha_1)\cdots(X - \alpha_d)$$

be the minimal polynomial of α over $\mathbf{Q}$. *Then we have*

$$\mathrm{N}_{\mathbf{Q}}^{K}(\alpha) = \alpha_1 \cdots \alpha_d \qquad \textit{and} \qquad \mathrm{Tr}_{\mathbf{Q}}^{K}(\alpha) = \alpha_1 + \cdots + \alpha_d.$$

More generally, if $\alpha \in K$ *and* $m = [K : \mathbf{Q}(\alpha)]$, *then* $\mathrm{N}_{\mathbf{Q}}^{K}(\alpha) = (\alpha_1 \cdots \alpha_d)^m$ *and* $\mathrm{Tr}_{\mathbf{Q}}^{K}(\alpha) = m(\alpha_1 + \cdots + \alpha_d)$.

Proof. We are only going to prove the case $K = \mathbf{Q}(\alpha)$ and leave the general case as an exercise. It is sufficient to notice that the characteristic polynomial of multiplication by α, seen as a $\mathbf{Q}$-linear map from K to K, is nothing but the minimal polynomial of α. This is easily seen by taking the elements $1, \alpha, \ldots, \alpha^{d-1}$ as a basis for K over $\mathbf{Q}$. □

4.10. Remark. We can immediately deduce from the previous lemma that if $\alpha \in \mathscr{O}_K$, then $\mathrm{Tr}_{\mathbf{Q}}^{K}(\alpha)$ and $\mathrm{N}_{\mathbf{Q}}^{K}(\alpha)$ are in $\mathbf{Z}$. This follows from the fact that they are in $\mathbf{Q}$ and are algebraic integers.

4.11. Examples.

1. If $K = \mathbf{Q}(\sqrt{d})$, where d is square-free, then $\mathscr{O}_K = \mathbf{Z}[\sqrt{d}]$ if $d \equiv 2$ or $3 \bmod 4$, but $\mathscr{O}_K = \mathbf{Z}\left[\dfrac{1+\sqrt{d}}{2}\right]$ if $d \equiv 1 \bmod 4$. This follows from the fact that if $\alpha \in \mathscr{O}_K$, we can write $\alpha = x + y\sqrt{d}$, where a priori $x, y \in \mathbf{Q}$. We know that the trace and the norm are in $\mathbf{Z}$ and actually, since α is a root of $X^2 - \mathrm{Tr}(\alpha)X + \mathrm{N}(\alpha)$, this is equivalent to $\alpha \in \mathscr{O}_K$. Now, $\mathrm{Tr}(\alpha) = 2x$ and $\mathrm{N}(\alpha) = x^2 - dy^2$, hence $x = a/2$, $y = b/2$, where $a, b \in \mathbf{Z}$ and $a^2 - db^2 \in 4\mathbf{Z}$. If a is even, then b is also even and vice versa. If a and b are odd, we obtain $d \equiv 1 \bmod 4$, which proves the result.

2. If $K = \mathbf{Q}(\zeta)$, where $\zeta =: \exp(2\pi i/p)$, then $\mathscr{O}_K = \mathbf{Z}[\zeta]$. We have seen that $\lambda\mathbf{Z}[\zeta] \cap \mathbf{Z} = p\mathbf{Z}$ (recall that $\lambda := 1 - \zeta$). If $\alpha = a_0 + a_1\zeta + \cdots + a_{p-2}\zeta^{p-2}$ is an algebraic integer, where the a_i are a priori rational numbers, we can check that $\mathrm{Tr}(\lambda\alpha) = pa_0$. Now, $\mathrm{Tr}(\lambda\alpha)$ is in the ideal generated by λ, but also in $\mathbf{Z}$, so it is therefore an integer multiple of p. We can deduce from this that $a_0 \in \mathbf{Z}$. Then we start over again with $\alpha' := \zeta^{-1}(\alpha - a_0)$, and we then can conclude that $a_1 \in \mathbf{Z}$, and so on.

4.12. Proposition. *If $[K:\mathbf{Q}]=n$, then there exist $e_1,\dots,e_n \in \mathcal{O}_K$ such that $\mathcal{O}_K = \mathbf{Z}e_1 \oplus \cdots \oplus \mathbf{Z}e_n$ (as an abelian group or $\mathbf{Z}$-module).*

More generally, if I is a non-zero ideal of $\mathcal{O}_K$, then there exist $e'_1,\dots,e'_n \in \mathcal{O}_K$ such that $I = \mathbf{Z}e'_1 \oplus \cdots \oplus \mathbf{Z}e'_n$.

Let us point out that it is not always true that there exists an algebraic integer α such that $\mathcal{O}_K = \mathbf{Z}[\alpha]$.

Proof. The $\mathbf{Q}$-bilinear form $(x,y) := \mathrm{Tr}(xy)$ from $K \times K$ to $\mathbf{Q}$ is nondegenerate (because if $x \neq 0$, then $\mathrm{Tr}(xx^{-1}) = [K:\mathbf{Q}] \neq 0$, or see Exercise 3-6.15). If $f_1,\dots,f_n$ are vectors in a basis of K over $\mathbf{Q}$, we can assume, up to multiplication by a common denominator $d_0 \in \mathbf{Z}$, that they are in $\mathcal{O}_K$. Let $f_1^*,\dots,f_n^*$ be a dual basis (i.e., such that $\mathrm{Tr}(f_i f_j^*) = \delta_{ij}$) and let d be a common denominator of the f_j^*. For $x \in \mathcal{O}_K$, we can therefore write $x = x_1 f_1 + \cdots + x_n f_n$ where $x_i \in \mathbf{Q}$. We know that $\mathrm{Tr}\left(x(df_i^*)\right) = d\,\mathrm{Tr}(xf_i^*) = dx_i$ is in $\mathbf{Z}$, and therefore

$$\mathbf{Z}f_1 \oplus \cdots \oplus \mathbf{Z}f_n \subset \mathcal{O}_K \subset \frac{1}{d}\left(\mathbf{Z}f_1 \oplus \cdots \oplus \mathbf{Z}f_n\right),$$

which proves the first assertion.

If $I = \alpha\mathcal{O}_K$, then we can choose $e'_i = \alpha e_i$. If the ideal is not principal anymore, then we can nevertheless choose $\alpha \in I \setminus \{0\}$ such that, for a certain $d \geqslant 1$, we have

$$\mathbf{Z}\alpha e_1 \oplus \cdots \oplus \mathbf{Z}\alpha e_n = \alpha\mathcal{O}_K \subset I \subset \frac{1}{d}\alpha\mathcal{O}_K \subset \frac{1}{d}\left(\mathbf{Z}\alpha e_1 \oplus \cdots \oplus \mathbf{Z}\alpha e_n\right).$$

The second assertion follows from this. □

4.13. Definition. Let I be a non-zero ideal of $\mathcal{O}_K$. The *norm* of the ideal is defined as

$$\mathrm{N}(I) := \mathrm{card}\left(\mathcal{O}_K/I\right).$$

4.14. Proposition. *If $\alpha \in \mathcal{O}_K$, then*

$$\mathrm{N}(\alpha\mathcal{O}_K) = \left|\mathrm{N}_{\mathbf{Q}}^K(\alpha)\right|.$$

Furthermore, the norm is multiplicative on ideals: $\mathrm{N}(IJ) = \mathrm{N}(I)\,\mathrm{N}(J)$.

Proof. If M is a $\mathbf{Z}$-linear map from $\mathbf{Z}^n$ to $\mathbf{Z}^n$ with non-zero determinant, we have $\mathrm{card}(\mathbf{Z}^n/M\mathbf{Z}^n) = |\det(M)|$. If we denote by $M(\alpha)$ the multiplication by α from $\mathcal{O}_K$ to $\mathcal{O}_K$, we obtain

$$\mathrm{N}(\alpha\mathcal{O}_K) = \mathrm{card}\left(\mathcal{O}_K/\alpha\mathcal{O}_K\right) = |\det(M(\alpha))| = \left|\mathrm{N}_{\mathbf{Q}}^K(\alpha)\right|.$$

For the moment, we will settle for proving the second property in two special cases: the case where I and J are *comaximal*, i.e., $I + J = \mathcal{O}_K$, and the case where one of the two ideals is principal. The general case is more subtle and will be proven after the proof of Theorem 3-4.18 (we leave it to the reader to verify that no vicious circle is going on).

If $I+J=\mathcal{O}_K$, then there exist $i_0 \in I$ and $j_0 \in J$ such that $i_0+j_0=1$. We can deduce from this firstly that $I \cap J = IJ$ because $x \in I \cap J$ can be written $x = xi_0 + xj_0$ and secondly that the ring homomorphism $\mathcal{O}_K \to \mathcal{O}_K/I \times \mathcal{O}_K/J$, whose kernel is $I \cap J$, is surjective because $bi_0 + aj_0 \equiv a \bmod I$ and $bi_0 + aj_0 \equiv b \bmod J$. This homomorphism thus induces a ring isomorphism $\mathcal{O}_K/IJ \cong \mathcal{O}_K/I \times \mathcal{O}_K/J$ (by the generalized Chinese remainder theorem). Thus we have $\mathrm{N}(IJ) = \mathrm{N}(I)\,\mathrm{N}(J)$.

Now assume that $J = \alpha\mathcal{O}_K$. It follows from the exact sequence (of $\mathcal{O}_K$-modules or simply of abelian groups),

$$0 \to J/IJ \to \mathcal{O}_K/IJ \to \mathcal{O}_K/J \to 0,$$

that $\mathrm{N}(IJ) = \mathrm{N}(J)\,\mathrm{card}(J/IJ)$. The morphism $\phi : \mathcal{O}_K \to J/IJ$ given by $\phi(x) := \alpha x \bmod IJ$ is surjective, and its kernel is equal to $\{x \in \mathcal{O}_K \mid \alpha x \in \alpha I\} = I$. The desired equality $\mathrm{N}(I) = \mathrm{card}(\mathcal{O}_K/I) = \mathrm{card}(J/IJ)$ follows from this. □

Example of a non-principal ring. The ring $\mathbf{Z}[i\sqrt{3}]$ is neither principal nor factorial, because it is not integrally closed: $\mathbf{Z}[i\sqrt{3}]$ is strictly contained in $\mathbf{Z}[(1+i\sqrt{3})/2]$, which is principal, and has the same fraction field, namely $\mathbf{Q}(i\sqrt{3})$. More fundamentally, the rings $\mathbf{Z}[\sqrt{10}]$ and $\mathbf{Z}[i\sqrt{5}]$ are neither principal nor factorial. To see this, notice that

$$9 = 3^2 = (\sqrt{10}+1)(\sqrt{10}-1) \qquad \text{and} \qquad 6 = 2\cdot 3 = (1+i\sqrt{5})(1-i\sqrt{5})$$

give two essentially different decompositions into products of irreducible elements. In fact, we can show directly that the ideal generated by 3 and $\sqrt{10}+1$ in $\mathbf{Z}[\sqrt{10}]$ (resp. the ideal generated by 2 and $i\sqrt{5}+1$ in $\mathbf{Z}[i\sqrt{5}]$) is not principal, because the quotient by the ideal is $\mathbf{Z}/3\mathbf{Z}$ (resp. $\mathbf{Z}/2\mathbf{Z}$) and there is no element of norm 3 in $\mathbf{Z}[\sqrt{10}]$ (resp. of norm 2 in $\mathbf{Z}[i\sqrt{5}]$).

In order to measure how non-principal a ring is, we can introduce the following equivalence relation on ideals.

4.15. Definition. Two non-zero ideals I and J are *equivalent*, denoted $I \sim J$, if there exist two non-zero elements $\alpha, \beta \in \mathcal{O}_K$ such that $\alpha I = \beta J$.

The ring $\mathcal{O}_K$ is principal if and only if there is only one equivalence class. We will see that the set of classes is finite (Theorem 3-4.23, below) and forms a group, i.e., every ideal is *invertible*: for every non-zero ideal I, there exist a non-zero $\alpha \in \mathcal{O}_K$ and an ideal J such that $IJ = \alpha\mathcal{O}_K$.

4.16. Definition. A ring A is a *Dedekind ring* if it is Noetherian, integrally closed and if every non-zero prime ideal is maximal.

4.17. Examples. The fundamental example of a Dedekind ring is the ring of integers, $\mathcal{O}_K$, of a number field. To see this, it is integrally closed (cf. Proposition 3-4.7) and, since the quotient by a non-zero ideal is finite, the two other conditions can be easily checked. The set of ideals which contain a given non-zero ideal is finite, and a finite ring is integral if and only if it is a field.
If k is a field, the ring $k[T]$ is a Dedekind ring since it is principal. More generally, rings of the form $A = k[X,Y]/(f) = k[x,y]$ are Dedekind provided that they are integrally closed. (be careful: for example, if $f = Y^2 - X^3$, the element $\alpha = y/x$ is integral over A without being in A).

The fundamental property of Dedekind rings—that which in some sense replaces the notion of factoriality—is formulated in the following theorem.

4.18. Theorem. *Every non-zero ideal of $\mathcal{O}_K$ can be decomposed as a product of prime ideals; furthermore, this decomposition is unique (up to the order).*

Proof. We will start by stating some purely algebraic remarks. If $\beta \in K$ and I is an ideal in $\mathcal{O}_K$ which has the property that $\beta I \subset I$, then Lemma 3-4.3 shows that β is an algebraic integer. If I and J are two ideals such that $I = IJ$, then $J = \mathcal{O}_K$. To see this, if $\alpha_1, \ldots, \alpha_n$ is basis of I over $\mathbf{Z}$, then there exist $b_{ij} \in J$ such that $\alpha_i = \sum_j b_{ij}\alpha_j$, hence $\det(b_{ij} - \delta_{ij}) = 0$, and so $1 \in J$. From this, we can deduce the following assertion:

$$\text{if } \alpha I = JI, \text{ then } J = \alpha\mathcal{O}_K.$$

To see why this is true, for every $\beta \in J$ we have $\beta I \subset JI = \alpha I$, hence $(\beta\alpha^{-1})I \subset I$, and therefore $\beta\alpha^{-1}$ is an algebraic integer, and moreover $\beta \in \alpha\mathcal{O}_K$. Thus $\alpha^{-1}J$ is an ideal in $\mathcal{O}_K$ and $\alpha^{-1}JI = I$, therefore $\alpha^{-1}J = \mathcal{O}_K$ and $J = \alpha\mathcal{O}_K$. In the following section, we will use results from the geometry of numbers to show that there are a finite number of equivalence classes of ideals modulo principal ideals. If I is an ideal in $\mathcal{O}_K$, then there exist $m < n$ such that I^m and I^n are in the same class and, moreover, $\alpha I^m = \beta I^n$. We can deduce from this that $\alpha\mathcal{O}_K = \beta I^{n-m}$, and hence

$$\text{for every ideal } I \text{ of } \mathcal{O}_K, \text{ there exist } h \geqslant 1 \text{ and } \gamma \in \mathcal{O}_K \text{ such that } I^h = \gamma\mathcal{O}_K.$$

This allows us to prove the "cancellation" property of ideals:

$$\text{if } IJ = IJ', \text{ then } J = J'.$$

To see why this is true, by multiplying by I^{h-1}, we obtain $\gamma J = \gamma J'$, hence $J = J'$. We can also show that inclusion of ideals is equivalent to divisibility:

$$\text{if } I \subset J, \text{ then there exists an ideal } J' \text{ such that } I = JJ'.$$

This can be explained by noticing that if $J^h = \beta\mathscr{O}_K$, we obtain $J^{h-1}I \subset \beta\mathscr{O}_K$, thus $J' := \beta^{-1}J^{h-1}I$ is an ideal in $\mathscr{O}_K$, and $JJ' = \beta^{-1}J^hI = I$.

We will now prove the *existence* of the decomposition of an ideal. Let $I \neq \mathscr{O}_K$. If $\mathfrak{p}_1$ is a maximal ideal which contains I, $I \subset \mathfrak{p}_1$, then $I = \mathfrak{p}_1 I_1$; if $I_1 \neq \mathscr{O}_K$, we can still write $I = \mathfrak{p}_1\mathfrak{p}_2 I_2$. In this way, we iteratively construct a sequence of prime ideals such that $I = \mathfrak{p}_1 \cdots \mathfrak{p}_s I_s$, and since $\mathscr{O}_K$ is Noetherian, the process must eventually stop, i.e., there exists an n such that $I_n = \mathscr{O}_K$, and hence $I = \mathfrak{p}_1 \cdots \mathfrak{p}_n$.

We are now going to prove the *uniqueness* of the decomposition of an ideal. In order to do this, we will first point out that the previous results show that $\mathfrak{p}^{m+1}$ is included in $\mathfrak{p}^m$ and *distinct* from $\mathfrak{p}^m$. Hence we can define

$$\mathrm{ord}_{\mathfrak{p}}(I) := \max\{m \geqslant 0 \mid I \subset \mathfrak{p}^m\}.$$

We can easily check that $\mathrm{ord}_{\mathfrak{p}}(I)$ is zero for almost every $\mathfrak{p}$ and that

$$I = \prod_{\mathfrak{p}} \mathfrak{p}^{\mathrm{ord}_{\mathfrak{p}}(I)}.$$

This, together with the cancellation property, finishes the proof. □

Now we can move on to the general case of the formula $\mathrm{N}(IJ) = \mathrm{N}(I)\,\mathrm{N}(J)$.

Proof. (End of the proof of Proposition 3-4.14) It is enough, by the theorem on the decomposition of ideals stated above (Theorem 3-4.18), to show that the formula $\mathrm{N}(IJ) = \mathrm{N}(I)\,\mathrm{N}(J)$ holds when J is a non-zero prime ideal, in other words, for J maximal. Since $IJ \subset I$, we know that

$$\mathrm{card}\,(\mathscr{O}_K/IJ) = \mathrm{card}\,(\mathscr{O}_K/I)\,\mathrm{card}\,(I/IJ)\,.$$

Since J is maximal, $k := \mathscr{O}_K/J$ is a (finite) field. Since I/IJ is also an $\mathscr{O}_K$-module killed by J, we can consider it as a k-module or k-vector space. If we show that it has dimension 1, then we have proven that $\mathrm{card}(I/IJ) = \mathrm{card}\,(\mathscr{O}_K/J)$ which completes the proof. Now, a k-vector subspace $\{0\} \subset L \subset I/IJ$ is also an A-module, and therefore corresponds to an ideal I' such that $L = I'/IJ$ where $IJ \subset I' \subset I$. This gives us $I' = IJ'$ where $J \subset J' \subset \mathscr{O}_K$; since J is maximal, we must have that $J' = J$ or $\mathscr{O}_K$, and hence $I' = IJ$ or $I' = I$. Thus we can conclude that $L = \{0\}$ or I/IJ. □

4.19. Remark. The statement of the theorem (but not the proof) is also true for a general Dedekind ring—see for example P. Samuel's book *Théorie* *algébrique des nombres* [7] (Chap. 3). Another important property, which also serves as a definition for a Dedekind ring, is that fractional ideals are invertible. In fact, the key idea in the proof given in [7] is to show that if I is a maximal ideal in $\mathscr{O}_K$ and if we set $I^* := \{x \in K \mid xI \subset \mathscr{O}_K\}$, then $II^* = \mathscr{O}_K$. In particular, if we quotient the unitary monoid of ideals by the submonoid of prime ideals, we obtain a group, explicitly described below on page 104 (a fractional ideal I is a $\mathscr{O}_K$-submodule of K such that $dI \subset \mathscr{O}_K$ for some $d \in \mathscr{O}_K$ and which is invertible if there exists I' such that $II' = \mathscr{O}_K$).

We are now going to give a more thorough description of prime ideals. Let us first point out that if $\mathfrak{p}$ is a non-zero prime ideal in $\mathscr{O}_K$, then $\mathfrak{p} \cap \mathbf{Z}$ is a non-zero prime ideal in $\mathbf{Z}$, hence of the form $p\mathbf{Z}$ for some prime number p. Thus every prime ideal $\mathfrak{p}$ can be associated to a p, which is also the characteristic of the residue field $\mathscr{O}_K/\mathfrak{p}$. Conversely, if p is a prime number, then there is no reason that the ideal that it generates in $\mathscr{O}_K$ should still be prime and is therefore written, in light of the theorem above,

$$p\mathscr{O}_K = \mathfrak{p}_1^{e_1} \cdots \mathfrak{p}_s^{e_s} \quad \text{where } \mathfrak{p}_i \text{ are distinct prime ideals and } e_i \geqslant 1.$$

Let $f_i := [\mathscr{O}_K/\mathfrak{p}_i : \mathbf{F}_p]$, so that $\mathrm{N}\,\mathfrak{p}_i = p^{f_i}$. By taking the norms, we obtain

$$\mathrm{N}(p\mathscr{O}_K) = p^n = \mathrm{N}\,\mathfrak{p}_1^{e_1} \cdots \mathrm{N}\,\mathfrak{p}_s^{e_s} = p^{e_1 f_1 + \cdots + e_s f_s},$$

from which we have the relation

$$\sum_{i=1}^{s} e_i f_i = n. \tag{3.4}$$

By using the Chinese remainder theorem, we also have that

$$\mathscr{O}_K/p\mathscr{O}_K \cong (\mathscr{O}_K/\mathfrak{p}_1^{e_1}) \times \cdots \times (\mathscr{O}_K/\mathfrak{p}_s^{e_s}).$$

Describing the prime ideals in K thus boils down to describing the decomposition in $\mathscr{O}_K$ of primes in $\mathbf{Z}$.

4.20. Example. (Decomposition of primes in a quadratic field.) In the case where $K = \mathbf{Q}(\sqrt{d})$ and $[K : \mathbf{Q}] = 2$ (we can assume that d is square-free), we have three possibilities for its decomposition.

i) We can have $p\mathscr{O}_K = \mathfrak{p}_1\mathfrak{p}_2$ where $\mathrm{N}\,\mathfrak{p}_i = p$; we say then that p is *split* in K.
ii) We can have that $p\mathscr{O}_K = \mathfrak{p}_1$ where $\mathrm{N}\,\mathfrak{p}_1 = p^2$; we say then that p is *inert* in K.
iii) We can have $p\mathscr{O}_K = \mathfrak{p}_1^2$ where $\mathrm{N}\,\mathfrak{p}_1 = p$; we say then that p is *ramified* in K.

These cases correspond respectively to $s = 2$, $e_1 = e_2 = 1$ and $f_1 = f_2 = 1$; $s = 1$, $e_1 = 1$ and $f_1 = 2$; and $s = 1$, $e_1 = 2$ and $f_1 = 1$. We can also characterize them using the Legendre symbol.

4.21. Theorem. *Let $K = \mathbf{Q}(\sqrt{d})$ be a quadratic field, where d is square-free. If p is an odd prime number, then*

i) p is split in K if and only if $\left(\frac{d}{p}\right) = +1$;
ii) p is inert in K if and only if $\left(\frac{d}{p}\right) = -1$;
iii) p is ramified K if and only if $\left(\frac{d}{p}\right) = 0$, in other words if p divides d.

For the prime 2 the decomposition law is given by

i) 2 is split in K if and only if $d \equiv 1 \bmod 8$;
ii) 2 is inert K if and only if $d \equiv 5 \bmod 8$;
iii) 2 is ramified in K if and only if $d \equiv 2$ or $3 \bmod 4$.

Proof. If p is an odd prime, we have

$$\mathscr{O}_K/p\mathscr{O}_K \cong \mathbf{Z}[\sqrt{d}]/p\mathbf{Z}[\sqrt{d}].$$

This is a trivial remark if $d \equiv 2$ or $3 \bmod 4$, and for the case $d \equiv 1 \bmod 4$, it suffices to notice that if b is an odd integer,

$$a + b\left(\frac{1+\sqrt{d}}{2}\right) = a + \left(\frac{b-p}{2}\right)(1+\sqrt{d}) + p\left(\frac{1+\sqrt{d}}{2}\right),$$

hence $\mathscr{O}_K = \mathbf{Z}[\sqrt{d}] + p\mathscr{O}_K$. Next, we have the isomorphisms

$$\begin{aligned} A := \mathscr{O}_K/p\mathscr{O}_K &\cong \mathbf{Z}[\sqrt{d}]/p\mathbf{Z}[\sqrt{d}] \cong \mathbf{Z}[X]/(p, X^2 - d)\mathbf{Z}[X] \\ &\cong \mathbf{F}_p[X]/(X^2 - d)\mathbf{F}_p[X]. \end{aligned}$$

Therefore, we have the following three cases. Either $X^2 - d$ can be factored in $\mathbf{F}_p[X]$ into two distinct factors, which corresponds to $\left(\frac{d}{p}\right) = +1$, hence $A \cong \mathbf{F}_p \times \mathbf{F}_p$, and p is split; or $X^2 - d$ is irreducible in $\mathbf{F}_p[X]$, which corresponds to $\left(\frac{d}{p}\right) = -1$, hence $A \cong \mathbf{F}_{p^2}$, and p is inert; or finally $X^2 - d$ has a double root in $\mathbf{F}_p[X]$, which corresponds to $d = 0$ in $\mathbf{F}_p$ and $\left(\frac{d}{p}\right) = 0$, hence $A \cong \mathbf{F}_p[X]/X^2\mathbf{F}_p[X]$, and p is ramified.

If $p = 2$ and $d \equiv 2$ or $3 \bmod 4$, then we have

$$\begin{aligned} \mathscr{O}_K/2\mathscr{O}_K &\cong \mathbf{Z}[\sqrt{d}]/2\mathbf{Z}[\sqrt{d}] \cong \mathbf{Z}[X]/(2, X^2 - d)\mathbf{Z}[X] \\ &\cong \mathbf{F}_2[X]/(X^2 - d)\mathbf{F}_2[X] \cong \mathbf{F}_2[X]/(X - d)^2\mathbf{F}_2[X], \end{aligned}$$

and hence 2 is ramified. Now, if $d \equiv 1 \bmod 4$, since the minimal polynomial of $\dfrac{1+\sqrt{d}}{2}$ is $X^2 - X - \dfrac{d-1}{4}$, then we have

$$\begin{aligned}\mathscr{O}_K/2\mathscr{O}_K &\cong \mathbf{Z}\left[\frac{1+\sqrt{d}}{2}\right]/2\mathbf{Z}\left[\frac{1+\sqrt{d}}{2}\right]\\ &\cong \mathbf{Z}[X]/(2, X^2 - X - \frac{d-1}{4})\mathbf{Z}[X]\\ &\cong \mathbf{F}_2[X]/(X^2 - X - \frac{d-1}{4})\mathbf{F}_2[X].\end{aligned}$$

Thus if $\dfrac{d-1}{4} \equiv 0 \bmod 2$, in other words $d \equiv 1 \bmod 8$, then $X^2 - X - \dfrac{d-1}{4} = X(X-1)$ in $\mathbf{F}_2[X]$, hence $\mathscr{O}_K/2\mathscr{O}_K \cong \mathbf{F}_2 \times \mathbf{F}_2$, and 2 is split. But if $\dfrac{d-1}{4} \equiv 1 \bmod 2$, in other words $d \equiv 5 \bmod 8$, since $X^2 - X - \dfrac{d-1}{4} = X^2 + X + 1$ is irreducible in $\mathbf{F}_2[X]$, then $\mathscr{O}_K/2\mathscr{O}_K \cong \mathbf{F}_4$, and 2 is inert. □

4.22. Remark. We will now consider an odd prime number p and $K = \mathbf{Q}(\sqrt{d})$. We can see that there exists a prime *ideal* $\mathfrak{p}$ in $\mathscr{O}_K$ such that $\mathrm{N}\,\mathfrak{p} = p$ if and only if p is ramified or split, in other words if and only if the Legendre symbol $\left(\frac{d}{p}\right)$ equals 0 or 1. Thus we recover the congruence conditions for the solvability of the equation $x^2 - dy^2 = p$ established in Example 1-3.6. We then know that this equation or the equation $\mathrm{N}_{\mathbf{Q}}^K(x + y\frac{1+\sqrt{d}}{2}) = x^2 + xy - \frac{d-1}{4}y^2 = p$, where $d \equiv 1 \bmod 4$, has a solution if and only if the congruence conditions are satisfied and the associated ideal $\mathfrak{p}$ is principal. From this, we can deduce another proof of the two-square theorem.

We have shown that every ideal is invertible modulo the equivalence relation given in 3-4.15 and can therefore talk about the *ideal class group*, denoted $C\ell_K$ or $\mathrm{Pic}(\mathscr{O}_K)$. Thus the ring $\mathscr{O}_K$ is principal if and only if the group $C\ell_K$ is reduced to one element. We have seen that the rings $\mathscr{O}_K$ are in general not principal, but we can nevertheless say that they are "almost principal" by the *finiteness theorem*, which we will now state.

4.23. Theorem. *The class group of ideals $C\ell_K$ of a number field K is finite.*

4.24. Corollary. *Let $h_K := \mathrm{card}(C\ell_K)$, which is called the class number of K. For every non-zero ideal I in the ring $\mathscr{O}_K$, the ideal I^{h_K} is principal. Conversely, if $\gcd(h_K, m) = 1$ and I^m is principal, then I is principal.*

4.25. Remark. The corollary above is essentially due to Kummer, who used the following variation of it: if $K = \mathbf{Q}(\exp(2i\pi/p))$ has the property that p does not divide the number of classes h_K (we say then that p is *regular*), then an ideal whose pth power is principal is itself principal. This property, in addition to another one related to units in the always regular case, allowed Kummer to prove "Fermat's last theorem" for all regular prime exponents. The smallest non-regular prime exponent is 37. Kummer's proof follows the outline of the proof of Theorem 3-2.6; the first case is handled with the aid of congruences modulo λ^p and the second case with the aid of a descent where Kummer's lemma on the units of a cyclotomic field plays a crucial role (for the details, see the book by Borevich-Shafarevich [2], or also [58] or [77]).

In the following section, we will go over the main points of the proof of Theorem 3-4.23, as well as the structure of the group of units of $\mathcal{O}_K$. These last two properties (finiteness of the class group and finite generation of the group of units) are not purely algebraic (they are moreover false for Dedekind rings in general), and the proof of these properties will rely on the geometry of numbers.

5. Geometry of Numbers

We will start with the following statement from topology, which generalizes Lemma 3-3.5.

5.1. Proposition. *A discrete subgroup G in $\mathbf{R}^n$ has a basis over $\mathbf{Z}$ formed of r linearly independent vectors over $\mathbf{R}$ (where $r \leqslant n$); in particular, $G \cong \mathbf{Z}^r$.*

Proof. Let $e_1, \ldots, e_r$ be a maximal system of vectors of G which are linearly independent over $\mathbf{R}$; it suffices to prove that $\mathbf{Z}e_1 + \cdots + \mathbf{Z}e_r$ is a subgroup of finite index in G. The intersection of G and the compact set $K_0 := \{x_1e_1 + \cdots + x_re_r \mid x_i \in [0,1]\}$ is a finite set. Now let $x \in G$. It can be naturally written as $x = x_1e_1 + \cdots + x_re_r$, where x_i are a priori in $\mathbf{R}$. The vectors $y^{(m)} := (mx_1 - \lfloor mx_1 \rfloor)e_1 + \cdots + (mx_r - \lfloor mx_r \rfloor)x_re_r$ are all in $G \cap K_0$. If we let m vary from 0 to $M := \operatorname{card}(G \cap K_0)$, two of them will of course be equal, say $y^{(m_1)} = y^{(m_2)}$. This gives us $x_i = (\lfloor m_1x_i \rfloor - \lfloor m_2x_i \rfloor)/(m_1 - m_2)$, and hence, by letting $d := M!$, we have

$$\mathbf{Z}e_1 + \cdots + \mathbf{Z}e_r \subset G \subset \frac{1}{d}\left(\mathbf{Z}e_1 + \cdots + \mathbf{Z}e_r\right),$$

which finishes the proof. □

5.2. Definition. Whenever $r = n$, we say that G is a *lattice* in $\mathbf{R}^n$; this boils down to requiring that G be discrete and that $\mathbf{R}^n/G$ be compact. We therefore define the *volume* or *determinant* of a lattice G to be the absolute value of the determinant of a basis of G (with respect to the canonical basis of $\mathbf{R}^n$).

5.3. Theorem. (Minkowski) *Let $K \subset \mathbf{R}^n$ be a compact, convex, and symmetric (i.e., $x \in K$ implies that $-x \in K$) set. Assume that* $\mathrm{vol}(K) \geqslant 2^n$. *Then there exists a non-zero x in $K \cap \mathbf{Z}^n$.*

More generally, if Λ is a lattice and $\mathrm{vol}(K) \geqslant 2^n \det(\Lambda)$, *then there exists a non-zero x in $K \cap L$.*

Remark. The statement is optimal because, for example, the open cube defined by $\max_i |x_i| < 1$ is convex and symmetric and has volume equal to 2^n, while the compact cube defined by $\max_i |x_i| \leqslant 1 - \epsilon$ is convex and symmetric and has volume equal to $2^n(1-\epsilon)^n$.

Proof. The second statement follows from the first by making a linear variable change which takes the lattice Λ to $\mathbf{Z}^n$.

We set $C := [0, 1[^n$. Let $T \subset \mathbf{R}^n$, and suppose that $(T + \lambda) \cap (T + \mu) = \emptyset$ for $\lambda \neq \mu \in \mathbf{Z}^n$. This gives us $T = \cup_{\lambda \in \mathbf{Z}^n} (T \cap (C + \lambda))$, hence

$$\begin{aligned} \mathrm{vol}(T) &= \sum_{\lambda \in \mathbf{Z}^n} \mathrm{vol}\,(T \cap (C + \lambda)) = \sum_{\lambda \in \mathbf{Z}^n} \mathrm{vol}\,((T - \lambda) \cap C) \\ &= \mathrm{vol}\,((\cup_{\lambda \in \mathbf{Z}^n} (T - \lambda)) \cap C) \leqslant \mathrm{vol}(C) = 1. \end{aligned}$$

Conversely, if $\mathrm{vol}(T) > 1$, then there exists $x \in T \cap (T + \lambda)$ with $0 \neq \lambda \in \mathbf{Z}^n$, and also there exists such a λ in $T - T$. We now return to the proof by letting $T := \frac{1}{2}K = \left\{\frac{x}{2} \mid x \in K\right\}$. Then we have $K = T - T$ and $\mathrm{vol}(T) = 2^{-n}\mathrm{vol}(K)$. If $\mathrm{vol}(K) > 2^n$, then $K \cap (\mathbf{Z}^n \setminus \{0\})$ is nonempty.

If $\mathrm{vol}(K) = 2^n$ and K is compact, we get the same result. For every $m > 0$, the set $K_m = (1 + 1/m)K$ contains a non-zero element x_m in the lattice $\mathbf{Z}^n$. The sequence (x_m), with values in the intersection of the compact set K_1 and the lattice $\mathbf{Z}^n$, contains an eventually constant subsequence, whose limit is $x \in \mathbf{Z}^n \setminus \{0\}$. Furthermore, the point x is in $\cap_{m>0} K_m$, which coincides with the compact set K. □

Applications. We can use Minkowski's theorem above to give other proofs of the two-square and four-square theorems.

A prime number $p \equiv 1 \bmod 4$ is the sum of two squares.

Proof. Let $a \in \mathbf{Z}$ such that $a^2 + 1 \equiv 0 \bmod p$, and we define the lattice

$$\Lambda := \{(x, y) \in \mathbf{Z}^2 \mid y \equiv ax \bmod p\}.$$

Then $\det(\Lambda) = p$ and $\mathrm{vol}(B(0,r)) = \pi r^2$, hence whenever $\pi r^2 \geqslant 4p$, there exists a non-zero vector in $B(0,r) \cap \Lambda$. We can choose $r := \sqrt{4p/\pi}$. Then there exists a non-zero $(x, y) \in \Lambda$ such that

$$0 < x^2 + y^2 \leqslant r^2 = 4p/\pi < 2p.$$

So we have that $x^2 + y^2 \equiv (1 + a^2)x^2 \equiv 0 \bmod p$, and hence $x^2 + y^2 = p$.□

Every positive integer is the sum of four squares.

Proof. Let $n = p_1 \cdots p_r$ be square-free; it suffices to show that n is the sum of four squares. As in the first proof of Lagrange's theorem, choose a_i and b_i such that

$$a_i^2 + b_i^2 + 1 \equiv 0 \bmod p_i.$$

Consider the lattice given by

$$\Lambda := \{x \in \mathbf{Z}^4 \mid x_3 \equiv a_i x_1 + b_i x_2 \bmod p_i \text{ and } x_4 \equiv b_i x_1 - a_i x_2 \bmod p_i,\ 1 \leqslant i \leqslant r\}.$$

The volume of Λ is $\leqslant (p_1 \cdots p_r)^2 = n^2$. We can choose ρ such that $\mathrm{vol}(B(0,\rho)) = \dfrac{1}{2}\pi^2\rho^4 = 2^4 \det(\Lambda)$, and then by Minkowski's theorem, we have $0 \neq x \in \Lambda$ such that $0 < x_1^2 + x_2^2 + x_3^2 + x_4^4 \leqslant \rho^2 < 2n$. However, $x_1^2 + x_2^2 + x_3^2 + x_4^4 \equiv x_1^2 + x_3^2 + (a_i x_1 + b_i x_2)^2 + (b_i x_1 - a_i x_2)^2 \equiv 0 \bmod p_i$. Thus n divides $x_1^2 + x_2^2 + x_3^2 + x_4^4$ and hence $x_1^2 + x_2^2 + x_3^2 + x_4^4 = n$. □

5.4. Remark. We could also prove the four-square theorem by proving Jacobi's formula (see Exercise 3-6.11).

If we denote by $r_4(n) := \mathrm{card}\{(x, y, z, t) \in \mathbf{Z}^4 \mid x^2 + y^2 + z^2 + t^2 = n\}$, then

$$r_4(n) = 8 \sum_{\substack{d \mid n \\ 4 \nmid d}} d = \begin{cases} 8\sum_{d \mid n} d & \text{if } n \text{ is odd,} \\ 24\sum_{d \mid n, 2 \nmid d} d & \text{if } n \text{ is even,} \end{cases} \tag{3.5}$$

where $n > 0$. To see why this is true, the right hand side of the equality is clearly positive. This formula can be written in terms of generating functions, where we denote by $r_k(n)$ the number of ways to write n as the sum of k squares, i.e., $r_k(n) := \mathrm{card}\{(x_1, \ldots, x_k) \in \mathbf{Z}^k \mid x_1^2 + \cdots + x_k^2 = n\}$.

We will define the following series (formal or convergent if $|q| < 1$):

$$\Theta(q) = \sum_{n \in \mathbf{Z}} q^{n^2},$$

so that

$$\Theta(q)^k = \left(\sum_{n \in \mathbf{Z}} q^{n^2} \right)^k = \sum_{n \in \mathbf{N}} r_k(n) q^n,$$

and likewise

$$Z(q) = \sum_{n=1}^{\infty} \frac{nq^n}{1-q^n} = \sum_{n \in \mathbf{N}^*} \sigma(n) q^n, \qquad \text{where } \sigma(n) = \sum_{d|n} d.$$

Then Jacobi's formula can be written as

$$\Theta(q)^4 = 1 + 8\left(Z(q) - 4Z(q^4) \right).$$

The following theorem of Hermite can also be proven using Minkowski's theorem (although Hermite's method provides a better constant γ_n).

5.5. Theorem. *There exists a constant γ_n such that if $Q : \mathbf{R}^n \to \mathbf{R}$ is a positive-definite quadratic form, then*

$$m(Q) := \min_{x \in \mathbf{Z}^n \setminus \{0\}} Q(x) \leqslant \gamma_n (\det Q)^{1/n}.$$

Proof. Consider the ellipsoid $B_Q(r) := \{x \in \mathbf{R}^n \mid Q(x) \leqslant r^2\}$; its volume is $v_n r^n / \sqrt{\det(Q)}$, where v_n is the volume of the unit ball for the usual Euclidean norm. We can choose r in such a way that this volume is equal to 2^n. Therefore, there exists $x \in \mathbf{Z}^n \setminus \{0\}$ in $B_Q(r)$ which satisfies

$$Q(x) \leqslant r^2 = \frac{4}{v_n^{2/n}} (\det Q)^{1/n}.$$ □

Let us now proceed to some applications to general number field theory.

If $K = \mathbf{Q}(\alpha)$ and P is the minimal polynomial of α, then $n = [K : \mathbf{Q}] = \deg(P)$, and P has r_1 real roots $\alpha_1, \ldots, \alpha_{r_1}$ and r_2 pairs of complex roots $\alpha_{r_1+1}, \bar{\alpha}_{r_1+1}, \ldots, \alpha_{r_1+r_2}, \bar{\alpha}_{r_1+r_2}$ (so $n = r_1 + 2r_2$). The embeddings from K into $\mathbf{R}$ are therefore given by $\sigma_i(\alpha) = \alpha_i$ (for $1 \leqslant i \leqslant r_1$) and the complex (nonreal) embeddings by $\sigma_{r_1+i}(\alpha) = \alpha_{r_1+i}, \bar{\sigma}_{r_1+i}(\alpha) = \bar{\alpha}_{r_1+i}$ (for $1 \leqslant i \leqslant r_2$).

5.6. Theorem. (Dirichlet's unit theorem) *Let K be a number field with r_1 real embeddings and r_2 pairs of complex conjugate embeddings. The group* *of units $\mathcal{O}_K^*$ is a finitely generated abelian group, which is isomorphic to the direct product of the finite group of roots of unity of K and the free group $\mathbf{Z}^r$, where $r := r_1 + r_2 - 1$.*

Proof. We are going to prove this theorem for an integer r, where $r \leqslant r_1 + r_2 - 1$ (see the remarks below and Exercise 3-6.23 for the complete proof). We will proceed as we did when we looked at the units of $\mathbf{Q}(\sqrt{d})$. To do this, we introduce the homomorphism $L : \mathcal{O}_K^* \to \mathbf{R}^{r_1+r_2}$ defined by

$$L(\alpha) := (\log|\sigma_1(\alpha)|, \ldots, \log|\sigma_{r_1}(\alpha)|, 2\log|\sigma_{r_1+1}(\alpha)|, \ldots, 2\log|\sigma_{r_1+r_2}(\alpha)|)\,,$$

and we will show that there are only a finite number of elements $\alpha \in \mathcal{O}_K^*$ such that $L(\alpha)$ is contained in a given ball in $\mathbf{R}^{r_1+r_2}$, so that the kernel turns out to be finite, consisting of roots of unity contained in K, and the image $L(\mathcal{O}_K^*)$ turns out to be discrete. We can observe that the image is contained in the hyperplane $x_1 + \cdots + x_{r_1+r_2} = 0$ because $\log|\mathrm{N}_{\mathbf{Q}}^K(\alpha)| = 0$. Finally, since a discrete subgroup of $\mathbf{R}^m$ is isomorphic to $\mathbf{Z}^r$, where $r \leqslant m$, we have the statement of the theorem when $r \leqslant r_1 + r_2 - 1$. □

5.7. Examples. An imaginary quadratic field satisfies $r_1 = 0$ and $r_2 = 1$, hence $r = 0$, i.e., $\mathcal{O}_K^*$ is finite. A real quadratic field satisfies $r_1 = 2$ and $r_2 = 0$, hence $r \leqslant 1$; actually $r = 1$ and $\mathcal{O}_K^* \cong \{\pm 1\} \times \mathbf{Z}$ as we have seen in Theorem 3-3.3. In the case $K = \mathbf{Q}(\sqrt[3]{2})$, we have $r_1 = r_2 = 1$ hence $r \leqslant 1$, and so by setting $\alpha := \sqrt[3]{2}$ to lighten the notation, we see that $(1 + \alpha + \alpha^2)(\alpha - 1) = 1$, hence $1 + \alpha + \alpha^2$ is a unit and $r = 1$. In the case of $K = \mathbf{Q}(\zeta)$ where $\zeta := \exp(2\pi i/p)$, we see that $r_1 = 0$, $r_2 = (p-1)/2$, hence $r \leqslant (p-3)/2$. It can be shown directly that $r = (p-3)/2$ by checking that the units $\eta_k := \sin(k\pi/p)/\sin(\pi/p)$, for $k = 2, \ldots, (p-1)/2$, are independent and therefore generate a subgroup of rank $(p-3)/2$ hence of finite index in $\mathcal{O}_K^*$.

To prove the finiteness of the class group, we use the embedding of K into $E := \mathbf{R}^{r_1} \times \mathbf{C}^{r_2} \cong \mathbf{R}^{[K:\mathbf{Q}]}$ given by

$$\Phi(\alpha) := (\sigma_1(\alpha), \ldots, \sigma_{r_1}(\alpha), \sigma_{r_1+1}(\alpha), \ldots, \sigma_{r_1+r_2}(\alpha))\,.$$

We can show, as we did previously, that the image of $\mathcal{O}_K$ is discrete. Since $\mathcal{O}_K \cong \mathbf{Z}^n$, the image is therefore a lattice with volume $V = V_K$.

Now, we will prove that if $r_1 + r_2 - 1 \geqslant 1$, then the group of units is infinite. We consider the convex set in $\mathbf{R}^n = \mathbf{R}^{r_1} \times \mathbf{C}^{r_2}$ given by

$$B(t_1, \ldots, t_{r_1+r_2}) := \{x \in \mathbf{R}^{r_1} \times \mathbf{C}^{r_2} \mid |x_i| \leqslant t_i\}\,,$$

whose volume is $2^{r_1}\pi^{r_2} t_1 \cdots t_{r_1}(t_{r_1+1} \cdots t_{r_1+r_2})^2$. We can choose t_i with volume equal to $2^n V_K$. Thus $t_1 \cdots t_{r_1}(t_{r_1+1} \cdots t_{r_1+r_2})^2$ equals a fixed constant, say A_K. Minkowski's theorem therefore guarantees the existence of a non-zero algebraic integer α such that $\Phi(\alpha) \in B(t_1, \ldots, t_{r_1+r_2})$ and hence such that $|\mathrm{N}(\alpha)| \leqslant A_K$. By choosing, for example, t_1 smaller and smaller (and of course letting one of the t_i get bigger and bigger), we can even obtain infinitely many elements of norm smaller than A_K. Since there are only a finite number of ideals of norm $\leqslant A_K$, we obtain infinitely many elements which generate the same ideal and whose quotients are therefore units. This argument is refined in Exercise 3-6.23 to provide a complete proof of the theorem with $r_1 + r_2 - 1$ independent units.

Minkowski's theorem also allows us to prove the following lemma.

5.8. Lemma. *There exists $c_1 > 0$ (which depends on K) such that if I is a non-zero ideal in $\mathcal{O}_K$, then there exists a non-zero element $\alpha \in I$ such that $\left|\mathrm{N}_{\mathbf{Q}}^K(\alpha)\right| \leqslant c_1 \mathrm{N}(I)$.*

Proof. Let K_t be the compact, convex, symmetric set in E defined by $|x_i| \leqslant t$. Its volume is proportional to t^n (where $n = [K : \mathbf{Q}]$), more precisely $\mathrm{vol}(K_t) = 2^{r_1}\pi^{r_2}t^n$. The lattice $\Phi(I)$ has volume $V_K \mathrm{N}(I)$. Therefore, if $2^{r_1}\pi^{r_2}t^n = 2^n V_K \mathrm{N}(I)$, we would have $\Phi(I) \cap K_t \neq \{0\}$. It follows that there exists a non-zero $\alpha \in I$ such that $\Phi(\alpha) \in K_t$, and hence $\left|\mathrm{N}_{\mathbf{Q}}^K(\alpha)\right| \leqslant t^n \leqslant c_1 \mathrm{N}(I)$ where $c_1 := (4/\pi)^{r_2} V_K$. □

The constant "c_1" (or sometimes optimal value of this constant) is often called "Minkowski's constant".

5.9. Corollary. *The set of ideal classes of $\mathcal{O}_K$ is finite.*

Proof. Let c_1 be the constant given in the preceding lemma and set $m := \lfloor c_1 \rfloor !$. For a non-zero ideal I in $\mathcal{O}_K$, choose α to be a non-zero element in I with $|\mathrm{N}(\alpha)| \leqslant c_1 \mathrm{N}(I)$. It follows that $(I : \alpha\mathcal{O}_K) \leqslant c_1$, and consequently $mI \subset \alpha\mathcal{O}_K$. Then we set $J := \frac{m}{\alpha} I$, which is an ideal in $\mathcal{O}_K$ in the same equivalence class as I since $\alpha J = mI$. Furthermore, since $\alpha \in I$, we have $m\alpha \in \alpha J$, and hence $m \in J$. We already know that there exist a finite number of ideals in $\mathcal{O}_K$ which contain m (they are in bijection with the ideals in $\mathcal{O}_K / m\mathcal{O}_K$). □

By using the fact that every ideal (or every class of ideals) is invertible, as well as the multiplicativity of the norm of ideals, we can control the finite set of classes. We will use the constant c_1 from the previous lemma (3-5.8) to do this.

5.10. Corollary. *Every ideal class in $\mathcal{O}_K$ contains an ideal with norm smaller than c_1.*

Proof. Let $\mathscr{C}$ be an ideal class and I an integral ideal in the inverse class. By the lemma, there exists $\alpha \in I$ such that $\left|\mathrm{N}_{\mathbf{Q}}^{K}(\alpha)\right| \leqslant c_1 \mathrm{N}(I)$. We therefore have $\alpha\mathscr{O}_K \subset I$, and hence there exists an ideal J such that $\alpha\mathscr{O}_K = IJ$. The ideal J is an integral ideal which belongs to the class $\mathscr{C}$, and we have $\mathrm{N}(J) = \left|\mathrm{N}_{\mathbf{Q}}^{K}(\alpha)\right| \mathrm{N}(I)^{-1} \leqslant c_1$. □

5.11. Remarks. We could give a more explicit expression for the constant c_1 which appears in Corollary 3-5.10. We need to first define the *discriminant*, denoted Δ_K, of a number field K. To do this, we introduce a basis $\alpha_1, \dots, \alpha_n$ over $\mathbf{Z}$ of $\mathscr{O}_K$ as well as the set $\sigma_1, \dots, \sigma_n$ of embeddings of K into $\mathbf{R}$ or $\mathbf{C}$. Then,

$$\Delta_K := \left(\det(\sigma_i(\alpha_j)\right)^2 \in \mathbf{Z}. \tag{3.6}$$

An $\mathbf{R}$-linear variable change $(z, \bar{z}) \mapsto (\mathrm{Re}(z), \mathrm{Im}(z))$ over the complex coordinates shows that

$$V_K = 2^{-r_2}\sqrt{|\Delta_K|}. \tag{3.7}$$

We can often compute the absolute value of this discriminant in the following manner (see Exercise 3-6.13): if α is an algebraic integer of K, whose minimal polynomial is $f(X) \in \mathbf{Z}[X]$, and if the index $u := (\mathscr{O}_K : \mathbf{Z}[\alpha])$ is finite, then

$$u^2|\Delta_K| = \left|\mathrm{N}_{\mathbf{Q}}^{K}(f'(\alpha))\right|. \tag{3.8}$$

This gives us that every ideal class of $\mathscr{O}_K$ contains an ideal of norm $\leqslant (2/\pi)^{r_2}\sqrt{|\Delta_K|}$. By looking at the decomposition of small prime numbers, we can, at least if the discriminant is not too large, deduce what the structure of the class group $C\ell_K$ is. We can of course improve the bounds—see Samuel [7], for example—and obtain the following value for "Minkowski's constant":

$$c_1 = \left(\frac{4}{\pi}\right)^{r_2} \frac{n!}{n^n}\sqrt{\Delta_K}. \tag{3.9}$$

This improvement allows us to establish *Hermite's inequality*:

$$[K : \mathbf{Q}] \leqslant c_2 \log|\Delta_K|, \quad \text{for } K \neq \mathbf{Q}, \tag{3.10}$$

The bound that we have obtained lets us determine what the class group is in the following examples and in Exercises 3-6.16, 3-6.17 and 3-6.18.

5.12. Examples.

1. Take $K = \mathbf{Q}(i\sqrt{19})$, then $|\Delta_K| = 19$, and every ideal class contains an ideal with norm $\leqslant 2\sqrt{19}/\pi < 3$; however, we can check that 2 is prime

in $\mathcal{O}_K = \mathbf{Z}[\frac{1+i\sqrt{19}}{2}]$ (cf. Theorem 3-4.21). The only ideal with norm < 3 is the unit ideal and $\mathbf{Z}[\frac{1+i\sqrt{19}}{2}]$ is thus principal.

2. Take $K = \mathbf{Q}(i\sqrt{23})$, then $|\Delta_K| = 23$, and every ideal class contains an ideal with norm $\leqslant 2\sqrt{23}/\pi < 4$. We can check that 2 and 3 are split in $\mathcal{O}_K = \mathbf{Z}[\frac{1+i\sqrt{23}}{2}]$, but no element has 2 or 3 as a norm. However, $\mathrm{N}(\frac{1+i\sqrt{23}}{2}) = 6$. Thus $2\mathcal{O}_K = \mathfrak{p}_1\mathfrak{p}_2$, $3\mathcal{O}_K = \mathfrak{p}'_1\mathfrak{p}'_2$ and $(\frac{1+i\sqrt{23}}{2})\mathcal{O}_K = \mathfrak{p}_1\mathfrak{p}'_1$. The classes of $\mathfrak{p}_1$ and $\mathfrak{p}_2$ are distinct since the only elements of norm 4 are ± 2. It follows that $C\ell_K = \{1, [\mathfrak{p}_1], [\mathfrak{p}_2]\} \cong \mathbf{Z}/3\mathbf{Z}$. In particular, $\mathbf{Z}[\frac{1+i\sqrt{23}}{2}]$ is not principal.

3. Take $K = \mathbf{Q}(\sqrt{13})$, then $|\Delta_K| = 13$ and every ideal class contains an ideal with norm $\leqslant \sqrt{13} < 4$; however, we can check that 2 is prime in $\mathcal{O}_K = \mathbf{Z}[\frac{1+\sqrt{13}}{2}]$ (cf. Theorem 3-4.21) and next we check that $\mathrm{N}(\frac{1+\sqrt{13}}{2}) = -3$ hence $3\mathcal{O}_K = \mathfrak{p}_1\mathfrak{p}_2$ with $\mathfrak{p}_1, \mathfrak{p}_2$ principal ideals generated by $\frac{1 \pm \sqrt{13}}{2}$. The ring $\mathbf{Z}[\frac{1+\sqrt{13}}{2}]$ is thus principal.

4. Take $K = \mathbf{Q}(\sqrt{10})$, then $|\Delta_K| = 40$ and every ideal class contains an ideal with norm $\leqslant \sqrt{40} < 7$; the primes 2 and 5 are ramified and 3 is split in $\mathcal{O}_K = \mathbf{Z}[\sqrt{10}]$ (cf. Theorem 3-4.21) thus $2\mathcal{O}_K = \mathfrak{p}^2$, $5\mathcal{O}_K = \mathfrak{q}^2$ and $3\mathcal{O}_K = \mathfrak{p}_1, \mathfrak{p}_2$ and the class group is generated by these four prime ideals. We can check that no element has norm ± 2 or ± 3 (the equations $x^2 \pm 2 = 10y^2$ or $x^2 \pm 3 = 10y^2$ have no solution modulo 10) hence $\mathfrak{p}$, $\mathfrak{p}_1, \mathfrak{p}_2$ are not principal. Notice that $\sqrt{10}\mathcal{O}_K = \mathfrak{p}\mathfrak{q}$ hence we may omit $\mathfrak{q}$; next $\mathrm{N}(2+\sqrt{10}) = -6$ thus $(2+\sqrt{10})\mathcal{O}_K = \mathfrak{p}\mathfrak{p}_i$ thus the class group is generated by $\mathfrak{p}$ which is of order 2. It follows that $C\ell_K = \{1, [\mathfrak{p}]\} \cong \mathbf{Z}/2\mathbf{Z}$ and the ring $\mathbf{Z}[\sqrt{10}]$ is not principal.

We will end this chapter by pointing out an often useful generalization about the ring of algebraic integers, $\mathcal{O}_K$. For this, we make use of a finite set S of prime ideals in $\mathcal{O}_K$.

5.13. Definition. Let K be a number field and S a finite set of prime ideals in $\mathcal{O}_K$. An element $\alpha \in K$ is called an *S-algebraic integer* if for every prime ideal $\mathfrak{p} \notin S$, $\mathrm{ord}_{\mathfrak{p}}(\alpha) \geqslant 0$. We denote by $\mathcal{O}_{K,S}$ the ring of S-integers and $\mathcal{O}^*_{K,S}$ the set of *S-units*.

5.14. Remark. We can prove, with a variation on the proof of Theorem 3-5.6, that the group $\mathcal{O}_{K,S}^*$ has rank $\leqslant r_1 + r_2 - 1 + |S|$ (and actually this is always an equality). Moreover, by using Theorem 3-4.23, we see that we can always choose S such that $\mathcal{O}_{K,S}$ is principal. In fact, it suffices that the ideals whose classes are generators of $C\ell_K$ can be written as a product of ideals in S.

6. Exercises

6.1. Exercise. *In this and the following two exercises, we denote by A and A_0 the rings of quaternions defined on page 78. Prove that*

$$\mathbf{Z}[i]^* = \{\pm 1, \pm i\}, \qquad A_0^* = \{\pm 1, \pm I, \pm J, \pm K\} \qquad \textit{and}$$

$$A^* = A_0^* \cup \left\{ \frac{\pm 1 \pm I \pm J \pm K}{2} \right\}$$

(The group A_0^* is the quaternion group of order 8. The group A^*, whose order is 24, is isomorphic to the group $\mathrm{SL}(2, \mathbf{F}_3)$*—see Exercise 3-6.3 below). Prove that A_0 is not (left) principal. Prove also that an element with norm equal to a prime number is irreducible.*

6.2. Exercise. *If B is a commutative ring, we define the ring $\mathbf{H}_B$ as the additive group*

$$\mathbf{H}_B := \{x\mathbf{1} + yI + zJ + tK \mid x, y, z, t \in B\}$$

endowed with the B-bilinear multiplication which has the same multiplication table as $\mathbf{H}$. *Notice that $A_0 = \mathbf{H}_\mathbf{Z}$ and $A \subset \mathbf{H}_{\mathbf{Z}\left[\frac{1}{2}\right]}$. (We could also consider $\mathbf{H}_B$ to be the "tensor product" $\mathbf{H}_B = A_0 \otimes_\mathbf{Z} B$.)*

1) Let F be a field of characteristic $\neq 2$ which contains two elements a and b such that $a^2 + b^2 + 1 = 0$. Prove that the map from $\mathbf{H}_F$ to the algebra of 2×2 matrices with coefficients in F given by

$$\mathbf{1} \mapsto \begin{pmatrix} 1 & 0 \\ 0 & 1 \end{pmatrix}, \; I \mapsto \begin{pmatrix} a & b \\ b & -a \end{pmatrix}, \; J \mapsto \begin{pmatrix} 0 & 1 \\ -1 & 0 \end{pmatrix} \text{ and } K \mapsto \begin{pmatrix} -b & a \\ a & -b \end{pmatrix},$$

is an isomorphism of F-algebras.

2) If p is an odd prime, deduce from the previous question that $\mathbf{H}_{\mathbf{F}_p}$ is isomorphic to the algebra of 2×2 matrices with coefficients in $\mathbf{F}_p$.

6.3. Exercise. *We will use the same notation as in the previous exercise. Our goal is to show that the group A^*, formed of the invertible elements of the Hurwitz quaternion algebra, is isomorphic to* $\mathrm{SL}(2, \mathbf{F}_3)$.

1) The homomorphism of reduction modulo 3 from $\mathbf{Z}\left[\frac{1}{2}\right]$ *to* $\mathbf{F}_3$ *induces a ring homomorphism from* $\mathbf{H}_{\mathbf{Z}\left[\frac{1}{2}\right]}$ *to* $\mathbf{H}_{\mathbf{F}_3}$. *Deduce from this a group homomorphism* $\phi : A^* \to \mathbf{H}^*_{\mathbf{F}_3} \cong \mathrm{GL}(2, \mathbf{F}_3)$.

2) Let $m \in A^*$ *such that* $m^2 = \mathbf{1}$ *(resp.* $m^3 = \mathbf{1}$*). Prove that if* $m \equiv \mathbf{1} \bmod 3$, *then* $m = \mathbf{1}$. *Conclude that* $\mathrm{Ker}(\phi) = \{\mathbf{1}\}$.

Hint.– We can assume $m \neq \mathbf{1}$. *Write* m *as* $m = \mathbf{1} + 3^h x$ *where* $x \in A$, $h \geqslant 1$ *and* $x \not\equiv 0 \bmod 3$, *which leads to a contradiction.*

3) Conclude from this that A^* *is isomorphic to a subgroup of index 2 of* $\mathrm{GL}(2, \mathbf{F}_3)$, *so it must be equal to* $\mathrm{SL}(2, \mathbf{F}_3)$.

6.4. Exercise. *Let* $\alpha, \beta \in k^*$, *and let* $\sqrt{\alpha}$ *be a root (in an extension of* k*) of* $X^2 - \alpha = 0$. *We set:*

$$\mathbf{H} = \mathbf{H}_{\alpha,\beta} := \left\{ \begin{pmatrix} a + b\sqrt{\alpha} & \beta(c + d\sqrt{\alpha}) \\ c - d\sqrt{\alpha} & a - b\sqrt{\alpha} \end{pmatrix} \middle| \, a, b, c, d \in k \right\}.$$

i) Prove that $\mathbf{H}$ *is a subalgebra of* $\mathrm{Mat}(2 \times 2, k(\sqrt{\alpha}))$ *and that a basis is given by the identity* $\mathbf{1} = I_2$ *and the three matrices*

$$I := \sqrt{\alpha}\begin{pmatrix} 1 & 0 \\ 0 & -1 \end{pmatrix}, \; J := \begin{pmatrix} 0 & \beta \\ 1 & 0 \end{pmatrix}, \; K := \sqrt{\alpha}\begin{pmatrix} 0 & \beta \\ -1 & 0 \end{pmatrix}.$$

Then check that the following multiplication identities hold: $I^2 = \alpha$, $J^2 = \beta$, $K^2 = -\alpha\beta$ *and*

$$IJ = -JI = K, \; JK = -KJ = -\beta I, \; KI = -IK = -\alpha J.$$

ii) Prove that an element $q = \begin{pmatrix} a + b\sqrt{\alpha} & \beta(c + d\sqrt{\alpha}) \\ c - d\sqrt{\alpha} & a - b\sqrt{\alpha} \end{pmatrix}$ *of* $\mathbf{H} \setminus \{0\}$ *is invertible if and only if*

$$\det(q) = N(a + b\sqrt{\alpha}) - \beta N(c + d\sqrt{\alpha}) = a^2 - \alpha b^2 - \beta c^2 + \alpha\beta d^2 \neq 0.$$

(If $[k(\sqrt{\alpha}) : k] = 2$, *the quantity* $N(a + b\sqrt{\alpha}) := a^2 - \alpha b^2$ *is the norm of* $a + b\sqrt{\alpha}$ *for the extension* $k(\sqrt{\alpha})/k$.*) Deduce from this that* $\mathbf{H}$ *is a division ring if and only if* $[k(\sqrt{\alpha}) : k] = 2$ *and* β *is not a norm of* $k(\sqrt{\alpha})/k$.

6.5. Exercise. *Modify the proof of Minkowski's theorem (Theorem 3-5.3) in order to obtain* $\mathrm{card}\,(K \cap \Lambda) \geqslant 2^n \mathrm{vol}(K)/\det(\Lambda)$.

6.6. Exercise. *Prove that there exist constants* C_n *(and explain what they are) such that if* L *is a lattice in* $\mathbf{R}^n$ *endowed with the Euclidean norm, then* *there exists a basis, $e_1, \ldots, e_n$, of L which satisfies*

$$\det(L) \leqslant \|e_1\| \cdots \|e_n\| \leqslant C_n \det(L).$$

Hint.– The first inequality is true for every basis. For the second, look for a basis which is as close as possible to an orthogonal basis.

6.7. Exercise. *Let v_n be the volume of the unit ball in $\mathbf{R}^n$. For $s > 0$, we define the function $\Gamma(s) := \int_0^\infty e^{-t} t^{s-1} dt$.*

1) Prove that $\Gamma(1) = 1$ and $\Gamma(1/2) = \sqrt{\pi}$. Prove that the values of the Γ function at positive integers (resp. half-integers) can be computed with the formula $\Gamma(s+1) = s\Gamma(s)$.

2) By computing the integral $\int_{\mathbf{R}^n} \exp[-(x_1^2 + \cdots + x_n^2)] dx_1 \cdots dx_n$ in two different ways, prove that

$$v_n = \frac{\pi^{n/2}}{\Gamma\left(\frac{n}{2} + 1\right)}.$$

and in particular that $v_{2m} = \pi^m / m!$ and $v_{2m+1} = \pi^m \alpha$ where $\alpha \in \mathbf{Q}^*$ (also specify α).

6.8. Exercise. *Let q be a positive-definite quadratic form with integer coefficients and such that for all $x \in \mathbf{Q}^n$ there exists $y \in \mathbf{Z}^n$ such that $q(x-y) < 1$. Let $m \in \mathbf{N}$; prove that there exists $y \in \mathbf{Z}^n$ such that $q(x) = m$ if and only if there exists $z \in \mathbf{Q}^n$ such that $q(z) = m$.*

Hint.– If $x \in \mathbf{Z}^n$ such that $q(x) = \ell^2 m$ and $\ell \geqslant 2$, choose $y \in \mathbf{Z}^n$ such that $q(\frac{x}{\ell} - y) < 1$, and then $x' := ax + by$ where $a = q(y) - m$ and $b = 2(m\ell - B(x, y))$. Then check that $q(x') = \ell'^2 m$, where $\ell' = \ell q(\frac{x}{\ell} - y) < \ell$.

By using this property for $q(x_1, x_2, x_3) = x_1^2 + x_2^2 + x_3^2$ and the Hasse-Minkowski theorem (Theorem 6-3.18 and Corollary 6-3.19), reprove the three-square theorem (Theorem 3-1.2).

Deduce from this the following theorem due to Gauss: Every number $m \in \mathbf{N}$ can be written as the sum of three triangular numbers (i.e., of the form $x(x-1)/2$).

Hint.– *Write $8m + 3$ as the sum of three squares $x_1^2 + x_2^2 + x_3^2$ and observe that the x_i must be odd.*

6.9. Exercise. *We write the vectors of $\mathbf{R}^n$ as column vectors. The group of $n \times n$ square matrices with coefficients in the ring A whose determinant* $\det(A) \in A^*$ *is denoted by* $\mathrm{GL}_n(A)$ *(in other words, the set of invertible matrices in the ring* $\mathrm{Mat}(n \times n, A)$*). If A is a subring of $\mathbf{R}$, we denote by*

$\mathscr{S}_n(A)$ the set of symmetric, positive-definite matrices with coefficients in A. If moreover $Q[x] = {}^t xQx$ is a quadratic form with integer coefficients, we say that it represents *an integer m if there exists $x \in \mathbf{Z}^n$ such that $Q[x] = m$. Finally, recall that two quadratic forms Q and Q' with integer coefficients are called equivalent if there exists $U \in \mathrm{GL}_n(\mathbf{Z})$ such that $Q' = Q[U] := {}^t UQU$.*

a) Prove that two equivalent forms represent the same set of integers.

b) Let $x \in \mathbf{Z}^n$ such that $\gcd(x_1, \ldots, x_n) = 1$. Prove that there exists a matrix $U \in \mathrm{GL}_n(\mathbf{Z})$ whose first column is x.

c) Let Q be a matrix in $\mathscr{S}_n(\mathbf{R})$, and let

$$m(Q) := \min_{x \in \mathbf{Z}^n \setminus \{0\}} Q[x].$$

Let $x \in \mathbf{Z}^n \setminus \{0\}$ be the minimal vector in the definition of $m(Q)$. Prove that a matrix $U \in \mathrm{GL}_n(\mathbf{Z})$ can be constructed so that $Q' = Q[U]$ satisfies

$$Q'[e_1] = m(Q') = m(Q).$$

d) Let Q be a matrix in $\mathscr{S}_n(\mathbf{R})$ such that $Q'[e_1] = m(Q')$. Prove that a matrix $U \in \mathrm{GL}_n(\mathbf{Z})$ of the form $U = \begin{pmatrix} 1 & {}^t b \\ 0 & \\ \vdots & V \\ 0 & \end{pmatrix}$ (where $b \in \mathbf{Z}^{n-1}$ and V is an $(n-1) \times (n-1)$ square matrix) can be constructed so that the matrix $Q'' = Q'[U]$ satisfies

$$Q''[e_1] = m(Q'') = m(Q') \quad \text{and} \quad Q''[e_2] = \min_{\substack{x \in \mathbf{Z}^n \\ \gcd(x_2, \ldots, x_n) = 1}} Q''[x].$$

e) A matrix $Q \in \mathscr{S}_n(\mathbf{R})$ is called reduced *if it satisfies the following property:*

$$\forall k \in [1, n], \qquad Q[e_k] = \min_{\substack{x \in \mathbf{Z}^n \\ \gcd(x_k, \ldots, x_n) = 1}} Q[x].$$

By iterating the procedure from the previous questions, prove that every matrix in $\mathscr{S}_n(\mathbf{R})$ is equivalent to a reduced matrix.

f) Let $Q \in \mathscr{S}_n(\mathbf{R})$ be a reduced matrix with coefficients $q_{i,j}$. Prove that $0 < q_{1,1} \leqslant q_{2,2} \leqslant \ldots \leqslant q_{n,n}$ and that $2|q_{i,j}| \leqslant q_{i,i}$.

g) Let $Q \in \mathscr{S}_n(\mathbf{R})$. Prove that there exist $D = \mathrm{diag}(d_1, \ldots, d_n)$ and T, an upper triangular matrix whose coefficients on the diagonal are all equal *to 1, such that $Q = D[T]$. Deduce from this Hadamard's inequality:*

$$\det(Q) \leqslant q_{1,1}q_{2,2}\cdots q_{n,n}.$$

h) Let $Q \in \mathscr{S}_n(\mathbf{R})$ be a reduced matrix. Prove the existence of a constant C_n such that

$$\det(Q) \leqslant q_{1,1}q_{2,2}\cdots q_{n,n} \leqslant C_n \det(Q).$$

(We can afterwards take $C_2 = 4/3$ and $C_3 = 2$ to be the first admissible initial values.)

Hint.– If $q_{n,n} \ll q_{1,1}$, then the proof is fairly easy; if not, there exists $k \leqslant n-1$ such that $q_{n,n} \ll q_{k+1,k+1}$ but $M_k q_{k,k} \leqslant q_{k+1,k+1}$. We then have the decomposition $Q = \begin{pmatrix} Q_1 & 0 \\ 0 & Q_2 \end{pmatrix} \begin{bmatrix} I & U \\ 0 & I \end{bmatrix}$, where Q_1 is a $k \times k$ matrix extracted from Q which will give us an inequality of the form $q_{k+1,k+1} \leqslant \dfrac{k^2}{4} q_{k,k} + m(Q_2)$. Finish the proof by applying Hermite's theorem (3-5.5) to Q_2 and an induction hypothesis to Q_1.

i) We denote by $\mathscr{H}_n(D)$ the set of equivalence classes of matrices of $\mathscr{S}_n(\mathbf{Z})$ whose determinant equals D. Prove that $h_n(D) := \operatorname{card} \mathscr{H}_n(D)$ is finite.

j) Prove that $h_2(1) = h_3(1) = 1$.

Hint.– *It can be shown that any 2×2 or 3×3 reduced matrix which has determinant 1 is the identity matrix.*

k) Prove that a form $Q = ((q_{i,j}))_{1 \leqslant i,j \leqslant n}$ is positive-definite if and only if

$$\forall k \in [1, n], \quad \det((q_{i,j}))_{1 \leqslant i,j \leqslant k}) > 0.$$

l) Application. Let n be a positive integer. Suppose that you know a positive integer d such that $-d$ is a square modulo $r := dn - 1$. Deduce from this that n is the sum of three squares. First show that if $-d = m^2 - \ell r$ (where ℓ has to be $\geqslant 1$), then the matrix

$$Q := \begin{pmatrix} \ell & m & 1 \\ m & r & 0 \\ 1 & 0 & n \end{pmatrix}$$

is positive-definite and has determinant 1. Then show that n is simply $Q(0,0,1)$.

6.10. Exercise. *Use the following hints and the previous exercise (3-6.9) to prove the three-square theorem (Theorem 3-1.2); we will also need* Dirichlet's theorem: *"If a and b are relatively prime, then there exists a prime number $p \equiv a \bmod b$."*

a) If $n = 2(2m+1) \equiv 2 \bmod 4$, prove that we can find a prime number p of the form $p = (4u+1)n - 1$. Let $d = 4u + 1$, and conclude that n is the sum of three squares.

b) If $n \equiv 1 \bmod 8$ *(resp.* $\equiv 3 \bmod 8$*, resp.* $\equiv 5 \bmod 8$*), we set* $c = 3$ *(resp.* $c = 1$*, resp.* $c = 3$*). Prove that we can find a prime number* p *of the form* $p = 4un + \frac{cn-1}{2}$*. Let* $d = 8u + c$ *(so that* $2p = nd - 1$*), and conclude that* n *is the sum of three squares.*

Finally, prove that the three-square theorem follows from these statements.

6.11. Exercise. (Jacobi's four-square formula, see [79] and the book by Ireland-Rosen [5]) *We would like to compute* $r_k(m) := \operatorname{card}\{(x_1, \ldots, x_k) \in \mathbf{Z}^k \mid x_1^2 + \cdots + x_k^2 = m\}$ *for* $k = 2$ *and* 4*. To help us do this, we introduce the quantity*

$$N_k(m) := \operatorname{card}\{(x_1, \ldots, x_k) \in \mathbf{N}^k \mid x_i \text{ odd and } x_1^2 + \cdots + x_k^2 = m\}.$$

a) Let χ *be the character modulo 4 which equals* $+1$ *(resp.* -1*, resp.* 0*) if* $x \equiv 1 \bmod 4$ *(resp. if* $x \equiv -1 \bmod 4$*, resp. if* x *is even). Prove that*

$$N_2(m) = \sum_{d \mid m} \chi(d),$$

and deduce that

$$r_2(m) = 4 \sum_{d \mid m} \chi(d).$$

b) Prove that the following equalities hold for $m \equiv 4 \bmod 8$*:*

$$N_4(m) = \sum_R N_2(m_1) N_2(m_2) = \sum_S (-1)^{\frac{a-c}{2}},$$

where R *is the set of pairs of natural numbers* (m_1, m_2) *such that* $m_1 + m_2 = m$ *and* $m_1 \equiv m_2 \equiv 2 \bmod 4$*, and* S *is the set of quadruples of odd natural numbers* (a, b, c, d) *such that* $2ab + 2cd = m$*.*

c) By setting $a = x + y$*,* $b = z - t$*,* $c = x - y$ *and* $d = z + t$*, prove that*

$$N_4(m) = \sum_{S'} (-1)^y,$$

where S' *is now the set of quadruples* $(x, y, z, t) \in \mathbf{Z}^4$ *such that* $|y| < x$*,* $|t| < z$*,* $m = 4(xz - yt)$ *and* x *and* y *(resp.* z *and* t*) have different parity. We denote by* $\mathscr{N}_0$ *(resp.* $\mathscr{N}_1$*, resp.* $\mathscr{N}_2$*) the sum restricted to* $y = 0$ *(resp.* $y > 0$*, resp.* $y < 0$*), so that* $N_4(m) = \mathscr{N}_0 + \mathscr{N}_1 + \mathscr{N}_2$*.*

d) Prove that $\mathscr{N}_0 = \sum_{d \mid m} d$ *(where the sum is restricted to being over* odd *divisors) and that* $\mathscr{N}_1 = \mathscr{N}_2$*. Then show that* $\mathscr{N}_1 = 0$ *by using the variable change (and checking that it indeed defines a bijection from* S' *to* S'*)* $x' = 2uz - t$*,* $y' = z$*,* $z' = y$ *and* $t' = 2uy - x$ *where* u *is chosen as the unique integer such that* $2u - 1 < x/y < 2u + 1$*.*

e) Prove Jacobi's formula (3.5) by successively showing that the following identities hold. For every m, we have $r_4(4m) = r_4(2m)$; if m is odd, then $r_4(4m) = 16N_4(4m) + r_4(m)$; if m is even, then $r_4(2m) = 3r_4(m)$. You can use the following "obvious" identity:

$$(x_1+x_2)^2+(x_1-x_2)^2+(x_3+x_4)^2+(x_3-x_4)^2 = 2x_1^2+2x_2^2+2x_3^2+2x_4^2.$$

6.12. Exercise. *Let $N = pq$ be an RSA number gotten from two very large prime numbers. Let d be the public exponent and e the secret one, so that $de \equiv 1 \bmod \phi(N)$.*

a) Observe that $\phi(N) \sim N$ (size-wise), and show that there exists an integer $k \sim ed/\phi(N)$ such that

$$\frac{d}{N} - \frac{k}{e} = \frac{1}{eN} - \frac{k}{e}\left(\frac{\phi(N)}{N} - 1\right).$$

b) Use Theorem 3-3.15 to prove that if the absolute value of the right hand side is smaller than $1/2e^2$, then the continued fraction expansion algorithm of d/N gives a fast computation of k/e and hence of e.

c) Prove that if $e \leqslant \frac{1}{3}N^{1/4}$, the previous condition holds.

6.13. Exercise. *Let K be a number field of degree $n = [K : \mathbf{Q}]$ and $\sigma_1, \ldots, \sigma_n$ its real or complex embeddings. For every lattice $L = \mathbf{Z}\alpha_1 \oplus \cdots \oplus \mathbf{Z}\alpha_n$, we set*

$$\Delta_L := \left(\det(\sigma_j(\alpha_i))\right)^2.$$

1) Prove that if $L \subset \mathcal{O}_K$, then $\Delta_L \in \mathbf{Z}$ and that if L' is a sublattice of index u in L, then $\Delta_{L'} = u^2\Delta_L$.

2) Let α be an algebraic integer such that $K = \mathbf{Q}(\alpha)$, and let $f(X)$ be its minimal polynomial. We set $L = \mathbf{Z}[\alpha]$. Prove that

$$\Delta_L = (-1)^{\frac{n(n-1)}{2}} \mathrm{N}_{\mathbf{Q}}^{K}(f'(\alpha)).$$

6.14. Exercise. *Let K be a number field such that $\mathcal{O}_K = \mathbf{Z}[\alpha]$ for some α. We denote by $f(X)$ the minimal polynomial of α. Prove that p is ramified in $K/\mathbf{Q}$ if and only if $f(X)$ has a double root in $\bar{\mathbf{F}}_p$. Deduce from this that p is ramified in $K/\mathbf{Q}$ if and only if p divides Δ_K.*

Remark. It can be shown (but this is more difficult) that this last conclusion is still true even if we do not assume the existence of α such that $\mathcal{O}_K = \mathbf{Z}[\alpha]$.

6.15. Exercise. ("Computational" proof of the nondegeneracy of the bilinear form $(x, y) \mapsto \mathrm{Tr}(xy) = \mathrm{Tr}_k^K(xy)$.) *Let K be a number field of* *degree* $n := [K : \mathbf{Q}]$. *We set, for* $x_1, \ldots, x_n \in K$,

$$D(x_1, \ldots, x_n) = \det\left(\operatorname{Tr}(x_i x_j)\right).$$

a) Let $\sigma_1, \ldots, \sigma_n$ *be the real and complex embeddings of* K. *Check that*

$$D(x_1, \ldots, x_n) = \det\left(\sigma_i(x_j)\right)^2.$$

b) Suppose that the elements y_i *are gotten from the* x_i *by a* $\mathbf{Q}$*-linear transformation, A. Verify that therefore*

$$D(y_1, \ldots, y_n) = (\det A)^2 D(x_1, \ldots, x_n).$$

c) Let α *be a primitive element of* K *(i.e., such that* $K = \mathbf{Q}(\alpha)$*) and* $F(X)$ *its minimal polynomial. Prove that*

$$D(1, \alpha, \ldots, \alpha^{n-1}) = (-1)^{\frac{n(n-1)}{2}} \mathrm{N}_{\mathbf{Q}}^{K}(F'(\alpha)) \neq 0.$$

Deduce from this that $D(x_1, \ldots, x_n) \neq 0$ *if and only if* $x_1, \ldots, x_n$ *form a basis (over* $\mathbf{Q}$*) of* K *and that the bilinear form given by the trace is non-degenerate. (This result is in fact valid for every finite separable extension* L/K.*)*

6.16. Exercise. *In this exercise, you are asked to show that the ring of integers of* $K = \mathbf{Q}(i\sqrt{d})$, *for* $d = 1, 2, 3, 7, 11, 19, 43, 67, 163$, *is principal.*

i) For the first five values of d, *the ring is Euclidean for the norm.*

ii) For $d = 19, 43, 67, 163$, *the integer* $n = \mathrm{N}(\frac{1 + i\sqrt{d}}{2})$ *is prime and the prime numbers* $\leqslant n$ *are inert in* K.

iii) Check that every ideal of norm $< 2\sqrt{d}/\pi$ *is principal and finish the proof by using Corollary 3-5.10 and the remarks that follow it.*

(Note: it is more difficult to prove—but true—that these are the only rings of quadratic imaginary integers that are principal.)

6.17. Exercise. *Let* $\zeta := \exp(2\pi i/5)$ *and* $K = \mathbf{Q}(\zeta)$, *hence* $\mathcal{O}_K = \mathbf{Z}[\zeta]$. *Prove that* $\Delta_K = 125$ *by using Exercise 3-6.13. Check that 2 and 3 stay prime in* $\mathcal{O}_K$ *and that* $5\mathcal{O}_K = ((1 - \zeta)\mathcal{O}_K)^4$. *Deduce from this that every ideal of norm* < 6 *is principal, and conclude that* $\mathcal{O}_K$ *is principal by using Corollary 3-5.10 and the remarks that follow it.*

6.18. Exercise. *Let* $\omega := \sqrt[3]{2}$ *and* $K = \mathbf{Q}(\omega)$. *Prove that the discriminant (in the sense of Exercise 3-6.13) of* $\mathbf{Z}[\omega]$ *equals* $\pm 3^3 2^2$, *and deduce that* $\mathcal{O}_K = \mathbf{Z}[\omega]$ *or* $(\mathcal{O}_K : \mathbf{Z}[\omega]) = 3$. *By considering the norm, trace, etc. of* $a + b\omega + c\omega^2$, *conclude that* $\mathcal{O}_K = \mathbf{Z}[\omega]$.

Check that $2 = \omega^3$, $3 = (1+\omega)^3(\omega - 1)$ *and* $5 = (1+\omega^2)(1+2\omega-\omega^2)$. *Prove that the elements* ω, $1+\omega$, $1+\omega^2$ *and* $1+2\omega-\omega^2$ *are prime and that* $\omega - 1$ *is invertible.*

Conclude that $\mathbf{Z}[\sqrt[3]{2}]$ *is principal by using Corollary 3-5.10 and the remarks that follow it.*

6.19. Exercise. *In this exercise, you are asked to study the integer valued solutions of the equation*

$$F(x,y,z) = x^3 + 2y^3 + 4z^3 - 6xyz = m, \qquad (*)$$

where m *is a non-zero integer.*

We set $\omega := \sqrt[3]{2}$ *and* $K = \mathbf{Q}(\omega)$. *We will assume (see Exercise 3-6.18) that the ring of integers of* K *is equal to* $\mathcal{O}_K = \mathbf{Z}[\omega]$ *and that it is principal.*

1) Prove that $1, \omega, \omega^2$ *form a basis for* K *over* $\mathbf{Q}$, *and compute* $\mathrm{N}_{\mathbf{Q}}^{K}(x + y\omega + z\omega^2)$.

2) Let p *be a prime number. Prove that either* p *stays prime in* $\mathcal{O}_K$ *or there exists an ideal* $\mathfrak{p}$ *in* $\mathcal{O}_K$ *with norm* p.

Hint.– *Find the decomposition of* $p\mathcal{O}_K$ *into products of prime ideals and enumerate the possibilities.*

3) Find $\alpha \in \mathcal{O}_K$ *(resp.* $\beta \in \mathcal{O}_K$*) such that* $\mathrm{N}_{\mathbf{Q}}^{K}(\alpha) = 2$ *(resp.* $\mathrm{N}_{\mathbf{Q}}^{K}(\beta) = 3$*).*

4) Let $p \neq 2, 3$. *Suppose that there exists an ideal* $\mathfrak{p}$ *in* $\mathcal{O}_K$ *of norm* p; *prove that there exists* $a \in \mathbf{F}_p^*$ *such that* $a^3 = 2$.

5) Let $p \neq 2, 3$. *Suppose that there exists* $a \in \mathbf{Z}$ *such that* $a^3 \equiv 2 \bmod p$; *prove that there exists an ideal* $\mathfrak{p}$ *in* $\mathcal{O}_K$ *of norm* p.

Hint.– *You can look at the factorization* $a^3 - 2 = (a - \omega)(a^2 + a\omega + \omega^2)$ *and deduce that* p *is not prime in* $\mathcal{O}_K$.

6) Check that if p *stays prime in* $\mathcal{O}_K$, *then* $p \equiv 1 \bmod 3$. *Prove by counterexample that the converse is false. (Check the case* $p = 31$.*)*

7) We can write $m = \pm \prod_p p^{m_p}$. *Prove that equation* $(*)$ *has an integer solution* (x,y,z) *if and only if for every prime number* $p \neq 2, 3$ *such that the congruence* $a^3 \equiv 2 \bmod p$ *does not have a solution, the integer* m_p *is divisible by 3.*

8) Prove that if equation $(*)$ *has an integer solution* (x,y,z), *then it has infinitely many.*

6.20. Exercise. *1) Let* α *be an algebraic integer with minimal polynomial* $P = X^d + p_{d-1}X^{d-1} + \cdots + p_0 \in \mathbf{Z}[X]$ *which generates the number field* $K = \mathbf{Q}(\alpha)$. *Let* p *be an odd prime number. Suppose now that* $\mathcal{O}_K = \mathbf{Z}[\alpha]$ *(or more generally that* p *does not divide* $(\mathcal{O}_K : \mathbf{Z}[\alpha])$*). Prove that if the* *reduction modulo p of the polynomial P can be factored in* $\mathbf{F}_p[X]$ *into*

$$\bar{P} = P_1^{e_1} \cdots P_r^{e_r}, \qquad \text{where the } P_i \text{ are irreducible and distinct in } \mathbf{F}_p[X],$$

then $\mathscr{O}_K/p\mathscr{O}_K \cong \mathbf{F}_p[X]/(P_1^{e_1}) \times \cdots \times \mathbf{F}_p[X]/(P_r^{e_r})$ *and* $p\mathscr{O}_K = \mathfrak{p}_1^{e_1} \cdots \mathfrak{p}_r^{e_r}$, *where the* $\mathfrak{p}_i$ *are prime ideals of norm* $\mathrm{N}(\mathfrak{p}_i) = p^{\deg(P_i)}$. *(Recall that p is said to be ramified in the extension* $K/\mathbf{Q}$ *if one of the* e_i *is* $\geqslant 2$.*)*

2) Let $\Phi_m \in \mathbf{Z}[X]$ *be the mth cyclotomic polynomial. Recall how to factor* $\bar{\Phi}_m$ *in* $\mathbf{F}_p[X]$. *(Treat the case where p divides m separately.)*

3) Let ζ *be a primitive mth root of unity and* $K = \mathbf{Q}(\zeta)$. *Since* $\mathbf{Q}(\zeta) = \mathbf{Q}(-\zeta)$, *we can assume that either m is odd or 4 divides m; we will also assume that* $\mathscr{O}_K = \mathbf{Z}[\zeta]$. *Prove that p is ramified in the extension* $K/\mathbf{Q}$ *if and only if p divides m. Assuming that p is relatively prime to m, we let r be the order of p modulo m. Prove that*

$$p\mathscr{O}_K = \mathfrak{p}_1 \cdots \mathfrak{p}_{\phi(m)/r}, \qquad \text{where } N\mathfrak{p}_i = p^r.$$

We will now assume that $m = 5$, $\zeta := \exp(2\pi i/5)$ *and* $K = \mathbf{Q}(\zeta)$.

4) Give a necessary and sufficient condition for an integer $n \in \mathbf{N}^*$ *to be the norm of an ideal in* $\mathscr{O}_K$.

Hint.– *Look for a condition in terms of the factorization of n.*

5) We will consider the Gauss sum

$$\tau := \sum_{x \in \mathbf{F}_5} \exp\left(\frac{2\pi i x^2}{5}\right).$$

Find, up to sign, the value of τ, *and deduce from this that K contains the real quadratic field* $\mathbf{Q}(\sqrt{5})$.

6) Prove that $\epsilon := \dfrac{1+\sqrt{5}}{2}$ *generates a subgroup of finite index of the group of units* $\mathscr{O}_K^*$.

7) Supposing that $\mathscr{O}_K$ *is principal (see Exercise 3-6.17), describe the set of solutions* $(x, y, z, t) \in \mathbf{Z}^4$ *of the equation*

$$N_{\mathbf{Q}}^K\left(x + y\zeta + z\zeta^2 + t\zeta^3\right) = m$$

for $m = 2$, $m = 5$, $m = 2^4 \cdot 11 = 176$. *In particular, specify whether the set of solutions is empty, finite or infinite.*

6.21. Exercise. *In this exercise, you are asked to determine which natural numbers can be written in the form* $x^2 + 3y^2$ *or (equivalently) in the form* $x^2 - xy + y^2$.

1) We let $j := \dfrac{-1 + i\sqrt{3}}{2}$. *Consider the rings* $A_0 = \mathbf{Z}[i\sqrt{3}]$ *and* $A = \mathbf{Z}[j]$ *Prove that A is principal and factorial, but that* A_0 *is not even factorial. Specify the group of units of* A_0^* *and* A^*.

2) Let the norm be defined by $N : \mathbf{Q}(i\sqrt{3}) \to \mathbf{Q}$*. Check that* $N(a+bi\sqrt{3}) = a^2+3b^2$, $N(x+yj) = x^2-xy+y^2$*, and prove that an integer* n *is the norm of an element of* A_0 *if and only if it is the norm of an element in* A*.*

Hint.– *You could show that if* α *is in* A *but not in* A_0*, then* $j\alpha$ *or* $j^2\alpha$ *is in* A_0*.*

3) Let p *be a prime number not equal to 2 or 3. Prove that if* p *is the norm of an element of* A_0 *(or* A*), then* -3 *is a square modulo* p *and deduce from this that* $p \equiv 1 \bmod 3$*.*

4) Let p *be a prime other than 2. Prove that if* $p \equiv 2 \bmod 3$ *and* $n = mp$ *is the norm of an element of* A_0 *(or of* A*), then* $m = n'p$ *and* n' *is also a norm.*

5) Prove that 2 is an irreducible element in A*. Deduce from this that if* $n = 2m$ *is the norm of an element of* A_0 *(or of* A*), then* $m = 2n'$ *and* n' *is also a norm.*

6) Now assume that $p \equiv 1 \bmod 3$*. Prove that* -3 *is a square modulo* p *and deduce from this that* p *is not irreducible in* A *and, consequently, that it is a norm.*

7) By using the previous questions, prove the following result.

An integer $n \geqslant 1$ can be written as x^2+3y^2 or, equivalently, as x^2-xy+y^2 where $x, y \in \mathbf{Z}$ if and only if for every prime $p \equiv 2 \bmod 3$, $\mathrm{ord}_p(n)$ is even.

8) State and prove a similar result with the norm associated to a quadratic field whose ring of integers is principal.

6.22. Exercise. *Prove that the only integer solutions of the equation*

$$y^2 = x^3 - 2$$

are $(x, y) = (3, \pm 5)$*.*

Hint.– *First show that* x, y *must be odd. By working in* $A := \mathbf{Z}[i\sqrt{2}]$*, prove that* $y + i\sqrt{2}$ *must be a cube in* A *and conclude by identifying the real and imaginary parts.*

6.23. Exercise. *In this exercise, you are asked to finish the proof of the unit theorem, by constructing* $r_1 + r_2 - 1$ *independent units. (You can assume that* $r_1 + r_2 - 1 \geqslant 1$*.)*

a) Let $A = (a_{i,j})$ *be an* $r \times r$ *matrix such that* $\forall i$, $|a_{i,i}| > \sum_{j \neq i} a_{i,j}$*. Prove that* A *is invertible.*

Let $r = r_1 + r_2 - 1$ *and* $\Phi : \mathscr{O}_K \hookrightarrow \mathbf{R}^{r_1} \times \mathbf{C}^{r_2} =: E$ *be the usual embedding. We denote by* $|x|_i := |\sigma_i(x)|$*, for* $i = 1, \ldots, r_1 + r_2$*, the absolute values associated to the different embeddings (resp. pairs of conjugate embeddings).*

b) Let $C > 1$*. By repeating the proof of the existence of infinitely many units, prove that a unit* ϵ_j *can be constructed for every absolute value such that*

$$|\epsilon_j|_j > C, \quad \text{and} \quad |\epsilon_j|_i \leqslant 1/4, \quad \text{for } i \neq j.$$

Hint.– *Construct, using Minkowski's theorem, algebraic integers such that* $|\alpha|_i \leqslant 1/4$ *for* $i \neq j$ *and* $|\mathrm{N}_{\mathbf{Q}}^K(\alpha)| \leqslant C$*, and use this to find a subset which generates the same ideal.*

c) Prove that $\epsilon_1, \ldots, \epsilon_r$ *are independent.*

6.24. Exercise. *Let* $\alpha = \sqrt[3]{2}$ *and* $K = \mathbf{Q}(\alpha)$*. In this exercise, you are asked to prove that the equation*

$$\mathrm{N}_{\mathbf{Q}}^K\left(x + 4y + z\alpha + w\alpha^2\right) - 6(x+y)(x^2 + xy + 7y^2) = 0, \tag{3.11}$$

which has nontrivial solutions modulo N *for every* $N \geqslant 2$ *(cf. Exercise 1-6.26 by taking* $w = 0$*), does not have any nontrivial integer solutions.*

1) Let $(x, y, z, w) \in \mathbf{Z}^4$ *be a primitive solution. We let* $d = \gcd(x, y)$*. Prove that* $\gcd(6, d) = 1$ *and that* 3 *does not divide* $f(x, y) := x^2 + xy + 7y^2$*. Check that, in particular,* $x \not\equiv y \bmod 3$*.*

2) i) Assume that p *does not divide* d*. Prove that if* $p \equiv 2 \bmod 3$*, then* p *does not divide* $f(x, y)$ *and that if* $p \equiv 1 \bmod 3$ *and* 2 *is not a cube modulo* p*, then* p *does not divide* $f(x, y)$*.*

Hint.– *If not,* p *would divide* $x + 4y$ *and this would force* p *to be equal to* 19.

ii) Finally show that if $p \equiv 1 \bmod 3$ *and* 2 *is a cube modulo* p*, then there exist integers* a *and* b *such that* $p = a^2 + 27b^2 = (a + 3bi\sqrt{3})(a - 3bi\sqrt{3})$*.*

Hint.– *We know* $(a + bj)^3 = A + 3Bj$*. A norm of* $K(i\sqrt{3})/\mathbf{Q}(i\sqrt{3})$ *can therefore be written as* $A + 3Bj$*.*

3) Let $\rho := \dfrac{1 + i\sqrt{3}}{2}$*, so we have the factorization*

$$f(x, y) = (x + y(3\rho - 1))(x + y(2 - 3\rho)).$$

Deduce from the previous arguments that there exist integers a, b *and* m *such that*

$$x + y(3\rho - 1) = \rho^m = a + 3bi\sqrt{3}.$$

4) By reducing modulo 3, prove that $\rho^m = \pm 1$*, then that* y *is even and* x *is odd. Conclude by referring back to the equation.*

Note: this example is due to Birch and Swinnerton-Dyer and was taken from [13].

Chapter 4

Analytic Number Theory

"– Eh! qu'aimes-tu donc, extraordinaire étranger?
– J'aime les nuages... les nuages qui passent... là-bas... là-bas...
les merveilleux nuages!"
Charles Baudelaire

The theme of this chapter is the distribution of prime numbers. We will begin by giving some statements and relatively elementary proofs, before introducing the key tool: the classical theory of functions of a complex variable, of which we will give a brief overview. The two following sections contain proofs of Dirichlet's "theorem on arithmetic progressions" and the "prime number theorem". Dirichlet series and in particular the Riemann zeta function play a fundamental role. We will illustrate this by additionally proving the functional equation of the zeta function and by formulating the famous Riemann hypothesis.

1. Elementary Statements and Estimates

The (written) history of prime numbers generally begins with the following theorem.

1.1. Theorem. (Euclid) *The set of prime numbers is infinite.*[1]

Given a finite list of prime numbers, $p_1, \ldots, p_r$, Euclid's argument consists of constructing a new prime number by considering possible prime factors of $N := p_1 \cdots p_r + 1$.

[1]Euclid's statement of course does not mention infinity; it says that given a finite collection of prime numbers, one can deduce another one from it.

M. Hindry, *Arithmetics*, Universitext,
DOI 10.1007/978-1-4471-2131-2_4,

There are many ways to expand on Euclid's proposition.

1.2. Proposition. *The series with terms* $\log(p)p^{-1}$ *and* p^{-1} *are divergent. To be more precise,*

$$\sum_{p\leqslant x}\frac{\log(p)}{p}=\log x+O(1)\quad and\quad \sum_{p\leqslant x}\frac{1}{p}=\log\log x+C+O\left(\frac{1}{\log x}\right). \tag{4.1}$$

These statements can be refined as follows.

1.3. Theorem. (Prime number theorem) *As* x *tends to infinity, we have the following asymptotic behavior:*

$$\pi(x):=\mathrm{card}\{p\ prime\ ,\ p\leqslant x\}\sim\frac{x}{\log x}. \tag{4.2}$$

We could state various equivalent forms of this theorem, for example,

$$\theta(x)\sim x;\qquad \psi(x)\sim x\qquad \text{or also}\qquad p_n\sim n\log n,$$

where we let (following Tchebychev)

$$\theta(x)=:\sum_{p\leqslant x}\log p,\qquad \psi(x)=\sum_{p^m\leqslant x}\log p \tag{4.3}$$

and where p_n denotes the nth prime number. We will also prove the following theorem.

1.4. Theorem. (Dirichlet's theorem on arithmetic progressions) *Let* $a,b\geqslant 1$ *be two relatively prime integers. Then there exist infinitely many primes* p *of the form* $a+bn$.

We could make this statement more precise by showing that prime numbers are distributed more or less uniformly over the congruence classes modulo b.

1.5. Theorem. *With the same hypotheses as before, we have*

$$\sum_{\substack{p\leqslant x\\ p\equiv a \bmod b}}\frac{1}{p}=\frac{\log\log x}{\phi(b)}+C_{a,b}+O\left(\frac{1}{\log x}\right).$$

We will now give a more refined statement but will not however prove it.

1.6. Theorem. *As x tends to infinity, we have the following asymptotic behavior:*

$$\pi(x; a, b) := \operatorname{card}\{p \text{ prime }, \ p \leqslant x, \ p \equiv a \bmod b\} \sim \frac{x}{\phi(b) \log x}. \qquad (4.4)$$

In this section, we will expand on some so-called "elementary" methods (which in this context means that they do not involve complex variables) which can be used to prove the previous assertions, except for Dirichlet's theorem on arithmetic progressions and the prime number theorem. They will however allow us to prove a partial version: there exist two constants, $c_1, c_2 > 0$, such that $c_1 x/\log x \leqslant \pi(x) \leqslant c_2 x/\log x$.

1.7. Lemma. *The following estimate holds:* $n \log 2 \leqslant \log \binom{2n}{n} \leqslant n \log 4$.

Proof. From the binomial theorem, we know that $\binom{2n}{n} \leqslant \sum_{k=0}^{2n} \binom{2n}{k} = (1+1)^{2n} = 4^n$. Next, we have the following lower bound: $\displaystyle\binom{2n}{n} = \frac{(2n)!}{(n!)^2} = \frac{2n(2n-1)\cdots(n+1)}{n(n-1)\cdots 1} \geqslant 2^n$. □

1.8. Lemma. *The following formula holds:* $\operatorname{ord}_p(n!) = \sum_{m \geqslant 1} \left\lfloor \frac{n}{p^m} \right\rfloor$; *furthermore, the sum can be restricted to* $m \leqslant \log n/\log p$.

Proof. Write $n! = 1 \cdot 2 \cdot 3 \cdots n = \prod_{k=1}^{n} k$. The number of integers $\leqslant n$ which are divisible by p is $\lfloor n/p \rfloor$, and the number of integers $\leqslant n$ divisible by p^2 is $\lfloor n/p^2 \rfloor$, etc. Thus $\operatorname{ord}_p(n!)$ is the sum of the $\lfloor n/p^m \rfloor$. Finally, $p^m \leqslant n$ is equivalent to $m \leqslant \log n/\log p$, hence the first statement is proved. □

We can therefore write

$$\log \binom{2n}{n} = \sum_{p \leqslant 2n} \operatorname{ord}_p \binom{2n}{n} \log p = \sum_{p \leqslant 2n} \left(\sum_{m \geqslant 1} \left\lfloor \frac{2n}{p^m} \right\rfloor - 2 \left\lfloor \frac{n}{p^m} \right\rfloor \right) \log p. \qquad (4.5)$$

To find a lower bound, we only keep the terms that satisfy $n < p \leqslant 2n$. In fact, such a p clearly divides $\binom{2n}{n} = (2n)!/(n!)^2$, and thus we obtain

$$n \log 4 \geqslant \log \binom{2n}{n} \geqslant \sum_{n < p \leqslant 2n} \log p = \theta(2n) - \theta(n).$$

From this, we obtain an upper bound of the form $\theta(x) \leqslant Cx$. This is true because

$$\theta(2^m) = \sum_{k=0}^{m-1} \theta(2^{k+1}) - \theta(2^k) \leqslant \sum_{k=0}^{m-1} 2^k \log 4 = (2^m - 1)\log 4.$$

Therefore, if $2^m \leqslant x < 2^{m+1}$, then

$$\theta(x) \leqslant \theta(2^{m+1}) \leqslant 2^{m+1} \log 4 \leqslant (2\log 4)x. \tag{4.6}$$

To obtain an upper bound, we could notice that $\lfloor 2u \rfloor - 2\lfloor u \rfloor$ always equals 0 or 1 and equals 0 whenever $u < 1/2$. Thus

$$n \log 2 \leqslant \log \binom{2n}{n} = \sum_{p \leqslant 2n} \left(\sum_{m \geqslant 1} \left\lfloor \frac{2n}{p^m} \right\rfloor - 2 \left\lfloor \frac{n}{p^m} \right\rfloor \right) \log p$$

$$\leqslant \sum_{p \leqslant 2n} \left(\frac{\log(2n)}{\log p} \right) \log p = \log(2n)\pi(2n).$$

From this, we have a lower bound of the form $\pi(x) \geqslant Cx/\log x$. This is true because if $2n \leqslant x < 2(n+1)$, then

$$\pi(x) \geqslant \pi(2n) \geqslant \frac{n \log 2}{\log(2n)} \geqslant \left(\frac{x}{2} - 1\right) \frac{\log 2}{\log x}. \tag{4.7}$$

Furthermore, we can easily see that

$$\theta(x) = \sum_{p \leqslant x} \log p \leqslant \log x \sum_{p \leqslant x} 1 = \pi(x) \log x. \tag{4.8}$$

Next, notice that for $2 \leqslant y < x$,

$$\pi(x) - \pi(y) = \sum_{y < p \leqslant x} 1 \leqslant \frac{1}{\log y} \sum_{y < p \leqslant x} \log p = \frac{1}{\log y} \left(\theta(x) - \theta(y)\right).$$

It follows that

$$\pi(x) \leqslant \frac{\theta(x)}{\log y} + \pi(y) \leqslant \frac{\theta(x)}{\log y} + y.$$

By choosing $y = x/(\log x)^2$ and by recalling the previous inequality (4.8), we have

$$\frac{\theta(x)}{\log x} \leqslant \pi(x) \leqslant \frac{\theta(x)}{\log x + 2\log\log x} + \frac{x}{(\log x)^2}. \tag{4.9}$$

To summarize, it is easy to see from inequalities (4.6), (4.7), (4.8) and (4.9) that $(\theta(x) \sim x)$ is equivalent to $(\pi(x) \sim x/\log x)$ and that

$$C_1 x \leqslant \theta(x) \leqslant C_2 x \qquad \text{and} \qquad C_3 x/\log x \leqslant \pi(x) \leqslant C_4/\log x. \tag{4.10}$$

Furthermore, the following comparison of the function $\theta(x)$ to the function $\psi(x)$ is not difficult to see:

$$\begin{aligned}\theta(x) \leqslant \psi(x) := \sum_{p^m \leqslant x} \log p &= \theta(x) + \theta(\sqrt{x}) + \theta(\sqrt[3]{x}) + \dots \\ &\leqslant \theta(x) + \frac{\log(x)}{\log 2}\theta(\sqrt{x}) \leqslant \theta(x) + C \log x \sqrt{x}.\end{aligned}$$

Finally, if we denote by p_n the nth prime number, we have $\pi(p_n) = n$ by definition. The prime number theorem therefore implies that $n \sim p_n/\log(p_n)$ and that $p_n \sim n \log n$. We can check that the latter statement is in fact equivalent to the prime number theorem.

1.9. Lemma. (Abel's formula) *Let* $A(x) := \sum_{n \leqslant x} a_n$ *and* f *be a function of class* $\mathscr{C}^1$. *Then,*

$$\sum_{y<n\leqslant x} a_n f(n) = A(x)f(x) - A(y)f(y) - \int_y^x A(t)f'(t)dt. \tag{4.11}$$

Proof. We first point out that $\int_n^{n+1} A(t)f'(t)dt = A(n)\int_n^{n+1} f'(t)dt = A(n)\left(f(n+1) - f(n)\right)$. Therefore, setting $N = \lfloor x \rfloor$ and $M = \lfloor y \rfloor$ yields

$$\begin{aligned}\int_M^N A(t)f'(t)dt &= \sum_{n=M}^{N-1} \int_n^{n+1} A(t)f'(t)dt = \sum_{n=M}^{N-1} A(n+1)\left(f(n) - f(n)\right) \\ &= \sum_{n=M+1}^{N} f(n)(A(n-1) - A(n)) + f(N)A(N) - A(M)f(M) \\ &= -\sum_{n=M+1}^{N} f(n)a_n + f(N)A(N) - A(M)f(M).\end{aligned}$$

This proves the formula when x and y are integers. For the general formula, observe that

$$\int_{\lfloor x \rfloor}^x A(t)f'(t)dt = A(\lfloor x \rfloor)\left(f(x) - f(\lfloor x \rfloor)\right) = A(x)f(x) - A(\lfloor x \rfloor)f(\lfloor x \rfloor). \square$$

Applications. 1) The formula gives a fairly precise comparison between the "sum" and the "integral" (see Exercise 4-6.10 for some refinements). To be more precise, if we take $a_n = 1$ and integrate by parts, we have:

$$\sum_{n=M+1}^{N} f(n) = \int_M^N f(t)dt + \int_M^N (t - \lfloor t \rfloor)f'(t)dt. \tag{4.12}$$

If we choose $f(t) = 1/t$, we obtain

$$\sum_{n=1}^{N} \frac{1}{n} = 1 + \int_1^N \frac{dt}{t} - \int_1^N (t - \lfloor t \rfloor) \frac{dt}{t^2}$$
$$= \log N + \left(1 - \int_1^\infty (t - \lfloor t \rfloor) \frac{dt}{t^2}\right) + \int_N^\infty (t - \lfloor t \rfloor) \frac{dt}{t^2}$$
$$= \log N + \gamma + O\left(\frac{1}{N}\right),$$

where $\gamma := 1 - \int_1^\infty (t - \lfloor t \rfloor) \frac{dt}{t^2}$ is *Euler's constant.*

2) Take $a_n = 1$, so $A(t) = \lfloor t \rfloor$, $y = 1$ and $f(t) = \log t$. We therefore have

$$\log(\lfloor x \rfloor !) = \lfloor x \rfloor \log(x) - \int_1^x \frac{\lfloor t \rfloor dt}{t}$$
$$= x \log x - \int_1^x dt + (\lfloor x \rfloor - x) \log x - \int_1^x \frac{\lfloor t \rfloor - t}{t} dt$$
$$= x \log x - x + O(\log x).$$

We should point out that Stirling's formula gives a slightly more precise statement, namely $n! \sim n^n e^{-n} \sqrt{2\pi n}$, and hence $\log(n!) = n \log n - n + \frac{1}{2} \log n + \frac{1}{2} \log(2\pi) + \epsilon(n)$ where $\lim_{n\to\infty} \epsilon(n) = 0$.

Furthermore, we see that

$$\log(\lfloor x \rfloor !) = \sum_{p \leqslant x} \mathrm{ord}_p(\lfloor x \rfloor !) \log p$$
$$= \sum_{p \leqslant x} \sum_{m \geqslant 1} \left\lfloor \frac{x}{p^m} \right\rfloor \log p$$
$$= x \sum_{p \leqslant x} \frac{\log p}{p} + \sum_{p \leqslant x} \log p \left(\left\lfloor \frac{x}{p} \right\rfloor - \frac{x}{p} \right) + \sum_{p \leqslant x} \sum_{m \geqslant 2} \left\lfloor \frac{x}{p^m} \right\rfloor \log p$$
$$= x \sum_{p \leqslant x} \frac{\log p}{p} + O(x),$$

where the last estimate comes from the upper bound $\theta(x) = \sum_{p \leqslant x} \log p = O(x)$, from (4.6) and the estimate

$$\sum_{p \leqslant x} \sum_{m \geqslant 2} \left\lfloor \frac{x}{p^m} \right\rfloor \log p \leqslant x \sum_{p \leqslant x} \sum_{m \geqslant 2} \frac{\log p}{p^m} = x \sum_{p \leqslant x} \frac{\log p}{p(p-1)} = O(x).$$

From this, we can deduce the first formula in Proposition 4-1.2,

$$\sum_{p \leqslant x} \frac{\log p}{p} = \log x + O(1). \qquad (4.13)$$

To get the second, we apply Abel's formula (Lemma 4-1.9) letting $f(t) = 1/\log t$ and $a_n = \log p/p$ if $n = p$ is prime and $a_n = 0$ otherwise. By setting $A(x) = \sum_{p \leqslant x} \frac{\log p}{p}$, we have that

$$\begin{aligned}\sum_{p\leqslant x}\frac{1}{p} &= \sum_{n\leqslant x} a_n f(n)\\ &= \frac{A(x)}{\log x} + \int_2^x \frac{A(t)dt}{t(\log t)^2}\\ &= 1 + O(1/\log x) + \int_2^x \frac{dt}{t\log t} + \int_2^x \frac{(A(t)-\log t)}{t(\log t)^2}dt\\ &= \log\log x - \log\log 2 + 1 + \int_2^\infty \frac{(A(t)-\log t)}{t(\log t)^2}dt + O(1/\log x).\end{aligned}$$

2. Holomorphic Functions (Summary/Reminders)

This section, without proofs, is a summary of some of the fundamental properties of functions of a complex variable that we will be using. It could be helpful to use [74] as a reference.

Concerning series, we will use the product rule for calculating the product of two *absolutely* convergent series:

$$\left(\sum_{n=0}^{\infty} a_n\right)\left(\sum_{n=0}^{\infty} b_n\right) = \sum_{n=0}^{\infty}\left(\sum_{k=0}^{n} a_k b_{n-k}\right),$$

as well as rearrangement of the order of summation in a series with *positive* terms $a_{m,n}$:

$$\sum_{n=0}^{\infty}\left(\sum_{m=0}^{\infty} a_{m,n}\right) = \sum_{m=0}^{\infty}\left(\sum_{n=0}^{\infty} a_{m,n}\right).$$

A power series $S(z) = \sum_{n=0}^{\infty} a_n z^n$ is said to have a radius of convergence $R \geqslant 0$ (possibly $R = 0$ or $R = +\infty$) if the series converges for all $|z| < R$ and diverges for all $|z| > R$; furthermore, the convergence is absolute in the interior of the disc of convergence and the function is of class $\mathscr{C}^\infty$ with $S^{(k)}(z) = \sum_{n=k}^{\infty} n(n-1)\cdots(n-k+1)a_n z^{n-k}$. In fact, the function S can be expanded as a power series around every point $z_0 \in D(0,R)$, in other words, for every $z \in D(z_0, r) \subset D(0,R)$, we have $S(z) = \sum_{n=0}^{\infty} b_n(z-z_0)^n$ (with $b_n = S^{(n)}(z_0)/n!$). Such a function only has a finite number of zeros in every closed disc (or compact set) which is contained in $D(0,R)$. We define the multiplicity of a zero, z_0, as the integer k such that $S(z) =$

$(z-z_0)^k \sum_{n=0}^{\infty} b_n(z-z_0)^n$ where $b_0 \neq 0$. A function which can be expressed a power series in a neighborhood of every point is called *analytic.*

2.1. Definition. A *holomorphic* function $f : U \to \mathbf{C}$ on an open set U is a function which is (complex) differentiable at every point in U, i.e.,

$$\lim_{z \to z_0} \frac{f(z) - f(z_0)}{z - z_0} = f'(z_0) \in \mathbf{C} \quad \text{exists.}$$

If F is a closed set in the complex plane, f is said to be holomorphic over F if it is holomorphic over an open set U which contains F.

Of course the notions of "differentiable" and "analytic" are very different in a real variable; in a complex variable, however, they are equivalent.

2.2. Proposition. *Let $f : U \to \mathbf{C}$ be a holomorphic function and assume that $D(z_0, r) \subset U$. Then for every $z \in D(z_0, r)$, we have $f(z) = \sum_{n=0}^{\infty} a_n(z - z_0)^n$, where $a_n = f^{(n)}(z_0)/n!$.*

2.3. Proposition. *Let $f : U \to \mathbf{C}$ be a holomorphic function and assume that U is connected and f is not identically zero. Then the set of zeros of f is discrete in U.*

2.4. Corollary. *Let $f, g : U \to \mathbf{C}$ be two holomorphic functions and assume that U is connected. If the set $\{z \in U \mid f(z) = g(z)\}$ is not discrete in U, then $f = g$. In particular, a holomorphic function on a disc $D(z_0, r) \subset U$ admits at most one analytic continuation to all of U.*

Next, we will define *meromorphic* functions as functions which are holomorphic on an open set U except for at the *poles.* At a pole z_0, a meromorphic function has the following behavior: there exists an integer m, called *the order of the pole*, such that the function $(z - z_0)^m f(z)$ has a holomorphic continuation in a neighborhood of z_0 and is not equal to zero at z_0. This is the same as saying that $f(z)$ can be written, in a neighborhood of z_0, as

$$f(z) = \frac{a_m}{(z - z_0)^m} + \frac{a_{m-1}}{(z - z_0)^{m-1}} + \cdots + \frac{a_1}{z - z_0} + \begin{array}{l}\text{a holomorphic function}\\ \text{at} z_0.\end{array}$$

The coefficient a_1 is called the *residue* of f at z_0 and is denoted by $\operatorname{Res}(f; z_0)$. Its importance comes from its usefulness in calculating integrals.

We define the integral along a path as follows: for $\gamma : [a, b] \to \mathbf{C}$ of class $\mathscr{C}^1$, we set

$$\int_\gamma f(z)dz := \int_a^b f(\gamma(t))\gamma'(t)dt.$$

The variable change formula shows that the value of the integral does not depend on the parametrization of the path but does, however, depend on the direction in which you integrate along it. For convenience sake, we will call a *simple contour* a path $\gamma : [a, b] \to \mathbf{C}$ such that $\gamma(a) = \gamma(b)$, but γ is injective on $[a, b[$ and travels in the counterclockwise direction. A theorem due to Camille Jordan shows that any such a contour partitions the plane into two connected parts, the interior and the exterior.

2.5. Theorem. (Residue theorem) *Let f be a meromorphic function on U. Let γ be a simple contour which does contain any poles of f and S the set of poles of f in the interior of γ. Then,*

$$\int_\gamma f(z)dz = 2\pi i \sum_{a \in S} \operatorname{Res}(f; a).$$

If U is *simply connected*, i.e., "without holes", then if f is holomorphic on U and γ_1, γ_2 are two paths in U, both of which join a and b, then $\int_{\gamma_1} f(z)dz = \int_{\gamma_2} f(z)dz$. Thus we can define an antiderivative of a holomorphic function $f(z)$ on such an open set by the formula $F(b) = \int_\gamma f(z)dz$, where γ is a path in U which joins a and b.

2.6. Proposition. *Let $f_n(z)$ be a sequence of holomorphic functions from U to $\mathbf{C}$. If the sequence converges uniformly on all compact sets in U to a function f, then f is holomorphic, and the kth derivatives $f_n^{(k)}$ converge uniformly on every compact set in U to the function $f^{(k)}$.*

We will expand on this point with the example of series of functions. Let $(u_n(z))$ be a sequence of holomorphic functions such that the series $S(z) := \sum_{n=0}^{\infty} u_n(z)$ converges; suppose moreover that it converges uniformly on every compact set, in other words, $\left|\left|\sum_{n=M}^{N} u_n(z)\right|\right|_{K,\infty} \to 0$ when M and N tend to infinity, and $K \subset U$ is any compact subset. Then the function $S(z)$ is holomorphic, and

$$S^{(k)}(z) = \sum_{n=0}^{\infty} u_n^{(k)}(z).$$

2.7. Example. (The complex logarithm). The function $\exp(z) = e^z = \sum_{n=0}^{\infty} z^n/n!$ is holomorphic on $U = \mathbf{C}$, and the series converges uniformly on every disk centered at 0 with radius R. We define

$$F(z) = \sum_{n=1}^{\infty} (-1)^{n+1} \frac{(z-1)^n}{n} = -\sum_{n=1}^{\infty} \frac{(1-z)^n}{n}.$$

The series converges normally at every point in the open disk $D(1,1) = \{z \in \mathbf{C} \mid |1-z| < 1\}$, and the convergence is uniform (and also normal) on every closed disk with center 1 and radius $r < 1$. Therefore, $F(z)$ is holomorphic on $D(1,1)$. If z is a real number in the interval $]0,2[$, we can see that $F(z) = \log z$ (ordinary logarithm), and in particular,

$$\exp(F(z)) = z.$$

The previous formula indicates that the two functions, the identity and $\exp \circ F$, which are analytic on the disk $D(1,1)$, coincide on the segment $]0,2[$ and hence on the whole disk. Thus F defines a complex logarithm on the disk $|z-1| < 1$.

2.8. Definition. Let $f(z)$ be a holomorphic function on U. We say that $F(z)$ is a *branch of the logarithm* of f on U (and we write, with a slight abuse of notation, $F(z) = \log f(z)$) if $F(z)$ is holomorphic and if $\exp(F(z)) = f(z)$.

2.9. Remark. If $F(z)$ is a branch of the logarithm of f, then f is never 0 on U, we have $|\exp(F(z))| = \exp(\operatorname{Re} F(z)) = |f(z)|$, and hence

$$\operatorname{Re}\log f(z) = \log|f(z)|.$$

Likewise, $f'(z)/f(z) = F'(z)\exp(F(z))\,/\exp(F(z)) = F'(z)$, and also

$$\frac{d}{dz}\log f(z) = \frac{f'(z)}{f(z)}.$$

Finally, if F_1 and F_2 are two logarithms, then $F_2(z) = F_1(z) + 2k\pi i$ on any connected set U.

This remark suggests that we should construct the logarithm of $f(z)$ as an antiderivative $f'(z)/f(z)$, with the condition that f is not zero. We have seen that this is possible if U is simply connected.

2.10. Proposition. *Let U be a simply connected open subset of the complex plane and $f(s)$ a holomorphic function without any zeros in U. Then there exists a holomorphic branch $F(s) = \log f(s)$ on U. Two such branches differ by an integer multiple of $2\pi i$.*

We will finish this summary by explaining the notion of an infinite product. The first idea consists of saying that a product is convergent if $\lim_N \prod_{n=0}^{N} p_n$ exists. This could be confusing because it is not true that such a product is zero if and only if one of the factors is zero. For example, it can be easily checked that

$$\lim_{N\to\infty}\prod_{n=1}^{N}\left(1-\frac{1}{n+1}\right)=0.$$

We can overcome this inconvenience by defining infinite products a little differently. Observe first that a necessary condition for the convergence of a non-zero product $\prod_n p_n$ is to have $\lim_n p_n = 1$; it therefore does not hurt to assume that $p_n = 1 + u_n$ where u_n tends to zero. In particular, $\log(1+u_n) = \sum_{k=1}^{\infty}(-1)^k u_n^k/k$ is well-defined when $|u_n| < 1$ and hence when $n \geqslant n_0$, which justifies the following definition.

2.11. Definition. A product $\prod_{n=0}^{\infty}(1+u_n)$ is *convergent* (resp. *absolutely convergent*) if there exists n_0 such that $|u_n| < 1$ for all $n \geqslant n_0$ and the series $\sum_{n=n_0}^{\infty}\log(1+u_n)$ is convergent (resp. absolutely convergent). A product of functions $\prod_{n=0}^{\infty}(1+u_n(z))$ is *uniformly convergent* on K if there exists n_0 such that $|u_n(z)| < 1$ for all $n \geqslant n_0$ and $z \in K$ and the series $\sum_{n=n_0}^{\infty}\log(1+u_n(z))$ is uniformly convergent (on K).

2.12. Lemma. *A product $P := \prod_{n=0}^{\infty}(1+u_n)$ is absolutely convergent if and only if the series $\sum_{n=0}^{\infty}|u_n|$ is convergent. If P is convergent, it is zero if and only if one of the factors $1+u_n$ is zero.*

2.13. Proposition. *Let $(u_n(z))$ be a sequence of holomorphic functions on an open set U such that the series $\sum_n \log(1+u_n(z))$ converges uniformly on every compact subset of U.*

i) Then the function defined by the infinite product

$$P(z) := \prod_{n=0}^{\infty}(1+u_n(z))$$

is holomorphic on U.

ii) For every $z_0 \in U$, only a finite number of $p_n(z) := 1+u_n(z)$ are zero at z_0, and hence

$$\operatorname{ord}_{z_0} P(z) = \sum_{n=0}^{\infty}\operatorname{ord}_{z_0} p_n(z).$$

3. Dirichlet Series and the Function $\zeta(s)$

We call a *Dirichlet series* a series of the form $F(s) = \sum_{n=1}^{\infty}\frac{a_n}{n^s}$. We will now state its first important property.

3.1. Proposition. *Let* $F(s) = \sum_{n=1}^{\infty} \frac{a_n}{n^s}$ *be a Dirichlet series that we will assume to be convergent at* s_0. *Then it converges uniformly on the sets* $E_{C,s_0} = \{s \in \mathbf{C} \mid \operatorname{Re}(s - s_0) \geqslant 0, |s - s_0| \leqslant C \operatorname{Re}(s - s_0)\}$.

Proof. For $M \geqslant 1$, set $A_M(x) := \sum_{M<n\leqslant x} a_n n^{-s_0}$. By the hypothesis, we then have $|A_M(x)| \leqslant \epsilon(M)$ where $\epsilon(M)$ tends to zero as M tends to infinity. Abel's formula gives

$$\sum_{M<n\leqslant N} a_n n^{-s} = \sum_{M<n\leqslant N} a_n n^{-s_0} n^{-(s-s_0)}$$
$$= A_M(N) N^{-(s-s_0)} + (s - s_0) \int_M^N A_M(t) t^{-(s-s_0+1)} dt.$$

We can find an upper bound for the integral as follows:

$$\left| \int_M^N A_M(t) t^{-(s-s_0+1)} dt \right| \leqslant \epsilon(M) \int_M^N t^{-(\sigma-\sigma_0+1)} dt$$
$$= \epsilon(M) \frac{M^{-(\sigma-\sigma_0)} - N^{-(\sigma-\sigma_0)}}{(\sigma - \sigma_0)}.$$

By restricting to an angular sector E_{C,s_0} bounded by $\sigma - \sigma_0 = \operatorname{Re}(s) - \operatorname{Re}(s_0) \geqslant 0$ and $|s - s_0| = C(\sigma - \sigma_0)$, we obtain

$$\left| \sum_{M<n\leqslant N} a_n n^{-s} \right| \leqslant \epsilon(M)(1 + C),$$

which suffices to show the uniform convergence on this sector (cf. the figure below). □

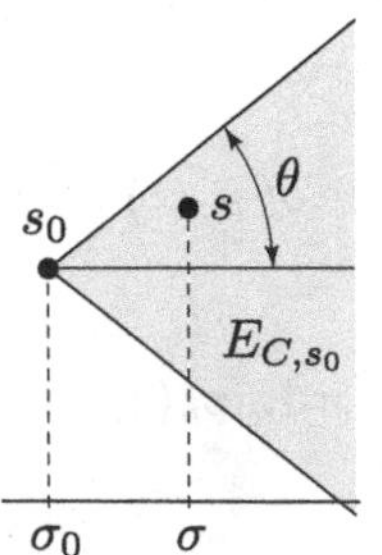

the domain $|s - s_0| \leqslant \dfrac{\sigma - \sigma_0}{\cos\theta}$

The following corollary is a result of the general theorems recalled in the previous section (in particular, Proposition 4-2.6).

3.2. Corollary. *Every Dirichlet series* $F(s) = \sum_{n=1}^{\infty} \frac{a_n}{n^s}$ *has an abscissa of convergence, say* σ_0, *such that the series converges for* $\operatorname{Re}(s) > \sigma_0$ *and* *diverges for* $\mathrm{Re}(s) < \sigma_0$. *Furthermore, the function F defined by the series is holomorphic in the half-plane of convergence* $\mathrm{Re}(s) > \sigma_0$, *and its derivatives are given by* $F^{(k)}(s) = \sum_{n=1}^{\infty} a_n(-\log n)^k n^{-s}$.

Proof. It suffices to let $\sigma_0 = \inf\{\sigma \in \mathbf{R} \mid \text{ the series converges at } \sigma\}$, then to observe that every compact set in the (open) half-plane of convergence is contained in a sector, as above, where the convergence is uniform. □

3.3. Remarks. 1) If a Dirichlet series converges at $s_0 = \sigma_0 + it_0$ to the number S, then the proof of Proposition 4-3.1 (above) shows that $S = \lim_{\epsilon \to 0^+} F(\sigma_0 + \epsilon + it_0)$.
2) Set $A(t) := \sum_{n \leqslant t} a_n$. The previous proof allows us to establish the formula

$$\sum_{n=1}^{\infty} a_n n^{-s} = s \int_1^{\infty} \left(\sum_{n \leqslant t} a_n \right) t^{-s-1} dt = s \int_1^{\infty} A(t) t^{-s-1} dt. \tag{4.14}$$

In particular, if $A(t) = \sum_{n \leqslant t} a_n$ is bounded, then the series converges whenever $\mathrm{Re}(s) > 0$.

The most famous Dirichlet series is the Riemann *zeta function*, defined by the series $\sum_{n=1}^{\infty} n^{-s}$. It is well-known, at least for real values and the general case follows from the real case, that the abscissa of convergence is $+1$.

3.4. Theorem. (Euler Product) *If* $\mathrm{Re}(s) > 1$, *then the following formula holds*

$$\zeta(s) = \sum_{n=1}^{\infty} \frac{1}{n^s} = \prod_p \left(1 - \frac{1}{p^s}\right)^{-1}. \tag{4.15}$$

Proof. Notice that $\sum_p |p^{-s}| = \sum_p p^{-\sigma} \leqslant \sum_n n^{-\sigma}$, hence the product is absolutely convergent. If $\mathrm{Re}(s) > 0$, the convergence of geometric series allows us to write $(1 - p^{-s})^{-1} = \sum_{m=0}^{\infty} p^{-ms}$. By taking the product over the prime numbers $p_1, \ldots, p_r$ which are $\leqslant T$, we obtain

$$\begin{aligned}
\prod_{p \leqslant T} (1 - p^{-s})^{-1} &= \prod_{p \leqslant T} \left(\sum_{m=0}^{\infty} p^{-ms} \right) \\
&= \sum_{\substack{m_1, \ldots, m_r \geqslant 1 \\ p_1, \ldots, p_r \leqslant T}} (p_1^{m_1} \cdots p_r^{m_r})^{-s} \\
&= \sum_{n \in \mathscr{N}(T)} n^{-s},
\end{aligned}$$

where we denote by $\mathscr{N}(T)$ the set of integers all of whose prime factors are $\leqslant T$. Thus whenever $\mathrm{Re}(s) > 1$, we have

$$\left|\sum_{n=1}^{\infty}\frac{1}{n^s} - \prod_{p\leqslant T}\left(1-\frac{1}{p^s}\right)^{-1}\right| = \left|\sum_{n\notin\mathscr{N}(T)}\frac{1}{n^s}\right| \leqslant \sum_{n>T}\left|\frac{1}{n^s}\right| = \sum_{n>T}\frac{1}{n^\sigma}.$$

The last sum is the tail-end of a real convergent series (when $\sigma := \mathrm{Re}(s) > 1$) and thus tends to zero, which proves both the convergence of the product and Euler's formula. □

3.5. Corollary. *The function $\zeta(s)$ does not have any zeros in the open half-plane* $\mathrm{Re}(s) > 1$. *A holomorphic branch of* $\log\zeta(s)$ *for* $\mathrm{Re}(s) > 1$ *can be constructed by setting*

$$\log\zeta(s) = \sum_p\sum_{m\geqslant 1}\frac{p^{-ms}}{m}. \tag{4.16}$$

Furthermore, if we define the von Mangoldt function *by*

$$\Lambda(n) = \begin{cases}\log p & \textit{if } n = p^m,\\ 0 & \textit{if not,}\end{cases}$$

then

$$-\frac{\zeta'(s)}{\zeta(s)} = \sum_p\sum_{m\geqslant 1}\frac{\log p}{p^{ms}} = \sum_{n=1}^{\infty}\frac{\Lambda(n)}{n^s}. \tag{4.17}$$

Proof. We know $1 - p^{-s} \neq 0$ and that the product is convergent. The first assertion is therefore obvious. The second formula can be deduced from Euler's formula by taking the series expansion of the logarithm (valid for $|x| < 1$) and summing:

$$\log\left((1-x)^{-1}\right) = \sum_{m=1}^{\infty}\frac{x^m}{m}.$$

The second formula is therefore gotten by differentiating the first. □

Interlude (I). These formulas can be generalized by replacing $\mathbf{Z}$ and $\mathbf{Q}$ by $\mathscr{O}_K$ and K, and the uniqueness of the decomposition into prime factors by the uniqueness of the decomposition into prime ideals (Theorem 3-4.18). We denote by $\mathscr{I}_K$ the set of non-zero ideals and $\mathscr{P}_K$ the set of non-zero (maximal) prime ideals in $\mathscr{O}_K$. Now we can introduce the Dedekind zeta function and prove that

$$\zeta_K(s) := \sum_{I\in\mathscr{I}_K}\mathrm{N}(I)^{-s} = \prod_{\mathfrak{p}\in\mathscr{P}_K}\left(1-\mathrm{N}(\mathfrak{p})^{-s}\right)^{-1} \text{ for } \mathrm{Re}(s) > 1.$$

3.6. Proposition. *The function* $\zeta(s)$ *can be analytically continued to a meromorphic function to the half-plane* $\mathrm{Re}(s) > 0$, *with a unique pole at* $s = 1$ *with residue equal to* $+1$.

Proof. The statement says that $\zeta(s) - 1/(s-1)$, originally defined for $\mathrm{Re}(s) > 1$, has a holomorphic continuation to the half-plane $\mathrm{Re}(s) > 0$. To prove this, we can write $\zeta(s)$, using the expression $\lfloor t \rfloor = \sum_{n \leqslant t} 1$, as

$$\begin{aligned}\zeta(s) = \sum_{n=1}^{\infty} \frac{1}{n^s} &= s \int_1^{\infty} \lfloor t \rfloor t^{-s-1} dt \\ &= s \int_1^{\infty} t^{-s} dt + s \int_1^{\infty} (\lfloor t \rfloor - t) t^{-s-1} dt \\ &= \frac{1}{s-1} + 1 + s \int_1^{\infty} (\lfloor t \rfloor - t) t^{-s-1} dt.\end{aligned}$$

We know that $|\lfloor t \rfloor - t| \leqslant 1$. The last integral is hence convergent and defines a holomorphic function for $\mathrm{Re}(s) > 0$. □

3.7. Remark. Actually, the function $\zeta(s) - 1/(s-1)$ can be extended to the whole complex plane, and moreover, $\zeta(s)$ satisfies a functional equation (see Theorem 4-5.6 further down).

4. Characters and Dirichlet's Theorem

4.1. Definition. If G is a finite abelian group, a homomorphism from G to $\mathbf{C}^*$ is called a *character*. The set of characters of G forms a group denoted by $\hat{G}$.

4.2. Proposition. *The group* $\hat{G}$ *is isomorphic to the group* G *(but not canonically).*

Proof. If $G = \mathbf{Z}/n\mathbf{Z}$, a character satisfies $\chi(1) \in \mu_n$ (where μ_n denotes as usual the group of nth roots of unity), and the map $\chi \mapsto \chi(1)$ provides an isomorphism between $\hat{G}$ and μ_n. The latter is isomorphic to $\mathbf{Z}/n\mathbf{Z}$ and hence to G. We will now show that $\widehat{G_1 \times G_2} \cong \hat{G}_1 \times \hat{G}_2$; to see why this is true, a character χ of $G_1 \times G_2$ can be written as $\chi(g_1, g_2) = \chi(g_1, e_2)\chi(e_1, g_2)$, and, by setting $\chi_1 = \chi(\cdot, e_2)$ and $\chi_2 = \chi(e_1, \cdot)$, we obtain an isomorphism $\chi \mapsto (\chi_1, \chi_2)$ from $\widehat{G_1 \times G_2}$ to $\hat{G}_1 \times \hat{G}_2$. The general case is now easy: we have $G \cong \mathbf{Z}/n_1\mathbf{Z} \times \cdots \times \mathbf{Z}/n_r\mathbf{Z}$, hence

$$\begin{aligned}\hat{G} &\cong (\mathbf{Z}/n_1\mathbf{Z} \times \cdots \times \mathbf{Z}/n_r\mathbf{Z})\widehat{\ } \\ &\cong \widehat{\mathbf{Z}/n_1\mathbf{Z}} \times \cdots \times \widehat{\mathbf{Z}/n_r\mathbf{Z}} \cong \mathbf{Z}/n_1\mathbf{Z} \times \cdots \times \mathbf{Z}/n_r\mathbf{Z} \cong G.\end{aligned}$$ □

4.3. Lemma. *If x is an element of order r in G, then for every ξ, an rth root of unity, there exist $|G|/r$ characters χ such that $\chi(x) = \xi$. In particular, in the ring of polynomials $\mathbf{C}[T]$, the following formula holds:*

$$\prod_{\chi\in\hat{G}} (1 - \chi(x)T) = (1 - T^r)^{|G|/r}.$$

Proof. We can immediately see that $\chi(x) \in \mu_r$ since $\chi(x)^r = \chi(x^r) = \chi(e_G) = 1$. Consider the map $\chi \mapsto \chi(x)$ from $\hat{G}$ to μ_r, which is a homomorphism whose kernel we will now identify. Let H be the subgroup generated by x (so that $H \cong \mathbf{Z}/r\mathbf{Z}$). The kernel of the previous homomorphism consists of characters which satisfy $\chi(x) = 1$ and also of characters which are trivial on H. The latter are in bijection with the characters of G/H, and their cardinality is therefore $\mathrm{card}(G/H) = |G|/r$. We see that the image has cardinality r, thus the homomorphism is surjective, which completes the proof of the first part of the lemma. For the last formula, it suffices to notice that

$$\prod_{\chi\in\hat{G}} (1 - \chi(x)T) = \left(\prod_{\xi\in\mu_r} (1 - \xi T)\right)^{|G|/r} = (1 - T^r)^{|G|/r}. \qquad \square$$

We will essentially use this lemma in the form of the following corollary.

4.4. Corollary. *Let p be a prime number which does not divide n and r the order of p modulo n. Consider the set of Dirichlet characters modulo n (see below). Then the following formula holds:*

$$\prod_{\chi \bmod n} \left(1 - \frac{\chi(p)}{p^s}\right)^{-1} = \left(1 - \frac{1}{p^{rs}}\right)^{-\phi(n)/r}. \tag{4.18}$$

4.5. Proposition. *If G is a finite commutative group, we have the following relations:*

$$\forall g \in G \setminus \{e\},\ \sum_{\chi\in\hat{G}} \chi(g) = 0 \qquad \textit{and} \qquad \forall \chi \in G \setminus \{1\},\ \sum_{g\in G} \chi(g) = 0.$$

Proof. If $g = e$, we clearly have $\sum_{\chi\in\hat{G}} \chi(g) = |G|$. If $g \neq e$, there exists, by the previous lemma, a character χ_1 such that $\chi_1(g) \neq 1$, and hence

$$\sum_{\chi\in\hat{G}} \chi(g) = \sum_{\chi\in\hat{G}} (\chi\chi_1)(g) = \chi_1(g) \sum_{\chi\in\hat{G}} \chi(g),$$

which gives us the first equality. We will handle the other sum similarly by observing that if $\chi = 1$, then $\sum_{g\in G} \chi(g) = |G|$, and if $\chi \neq 1$, then there exists g_1 such that $\chi(g_1) \neq 1$, hence

$$\sum_{g\in G} \chi(g) = \sum_{g\in G} \chi(gg_1) = \chi(g_1) \sum_{g\in G} \chi(g),$$

which gives us the second formula. □

4.6. Lemma. *Let $a \in G$. Then,*

$$\frac{1}{|G|} \sum_{\chi\in\hat{G}} \overline{\chi}(a)\chi(x) = \begin{cases} 1 & \textit{if } x = a, \\ 0 & \textit{if not.} \end{cases}$$

Proof. This statement follows from the previous formulas since $\chi(a)$ is a root of unity, so $\overline{\chi}(a) = \chi(a)^{-1} = \chi(a^{-1})$, and therefore $\sum_{\chi\in\hat{G}} \overline{\chi}(a)\chi(x) = \sum_{\chi\in\hat{G}} \chi(a^{-1}x)$ equals $|G|$ if $x = a$ and 0 if not. □

4.7. Definition. Let $\chi : \mathbf{Z}/m\mathbf{Z}^* \to \mathbf{C}^*$ be a character of $\mathbf{Z}/m\mathbf{Z}^*$. The *Dirichlet character* modulo m (also denoted by χ) is the map from $\mathbf{Z}$ to $\mathbf{C}$ defined by

$$\chi(n) = \begin{cases} \chi(n \bmod m) & \text{if } \gcd(m,n) = 1, \\ 0 & \text{if } \gcd(m,n) > 1. \end{cases}$$

Remark. We have the multiplicativity property: $\forall n, n' \in \mathbf{Z}$, $\chi(nn') = \chi(n)\chi(n')$, in other words, the function χ is completely multiplicative.

We will use these characters in the following way: we have the equality (at least formally)

$$\begin{aligned} \sum_{p\equiv a \bmod m} f(p) &= \frac{1}{\phi(m)} \sum_{\chi} \sum_{p} \overline{\chi}(a)\chi(p)f(p) \\ &= \frac{1}{\phi(m)} \sum_{p\nmid m} f(p) + \frac{1}{\phi(m)} \sum_{\chi\neq 1} \overline{\chi}(a) \left(\sum_{p} \chi(p)f(p) \right). \end{aligned}$$

We have already looked at sums like the first term $\sum_p f(p)$; to be able to deal with sums of the type $\sum_p \chi(p)f(p)$, we introduce the following series.

4.8. Definition. Let χ be a Dirichlet character modulo m. We define the Dirichlet "L"-series by the following series:

$$L(\chi, s) := \sum_{n=1}^{\infty} \chi(n)n^{-s}.$$

Remark. If χ_0 is the unitary character or principal character modulo m, we have $\chi_0(n) = 1$ or 0 depending on whether n is relatively prime to m or not. We can easily deduce from this that $L(\chi_0, s)$ is almost equal to the function $\zeta(s)$; to be more precise,

$$L(\chi_0, s) = \sum_{\gcd(n,m)=1} n^{-s} = \prod_{p \nmid m} (1 - p^{-s})^{-1} = \prod_{p \mid m} (1 - p^{-s}) \zeta(s).$$

4.9. Proposition. *The abscissa of convergence of the series $L(\chi, s)$ is $\sigma = 0$, except when χ is the unitary character, in which case $\sigma = 1$.*

Proof. We have seen from the previous remark that the series $L(\chi_0, s)$ where χ_0 is the unitary character has the same abscissa of convergence as the series which defines the zeta function, in other words 1. The terms in the series do not tend to 0 if $\mathrm{Re}(s) \leqslant 0$, hence the series cannot converge. If χ is a character modulo m which is not the unitary character, then $\sum_{n=r+1}^{r+m} \chi(n) = 0$, and hence

$$\left| \sum_{n \leqslant x} \chi(n) \right| = \left| \sum_{m \lfloor \frac{x}{m} \rfloor < n \leqslant x} \chi(n) \right| \leqslant m.$$

Furthermore, we have seen (cf. Remark 4-3.3) that the Dirichlet series therefore converges when $\mathrm{Re}(s) > 0$. □

4.10. Remark. The abscissa of absolute convergence is 1 and is strictly larger than the abscissa of convergence, which is 0 in this case. For any Dirichlet series, if we denote by σ_c its abscissa of convergence and σ_a its abscissa of absolute convergence, we can show that we always have the following inequality: $\sigma_c \leqslant \sigma_a \leqslant \sigma_c + 1$.

4.11. Theorem. *The generalized Euler formula holds:*

$$L(\chi, s) = \sum_{n=1}^{\infty} \frac{\chi(n)}{n^s} = \prod_p \left(1 - \frac{\chi(p)}{p^s}\right)^{-1} \quad \text{when } \mathrm{Re}(s) > 1. \tag{4.19}$$

Proof. When $\mathrm{Re}(s) > 0$, we can consider the following convergent geometric series: $(1 - \chi(p)p^{-s})^{-1} = \sum_{m=0}^{\infty} \chi(p)^m p^{-ms}$, and by taking the product over the prime numbers $p_1, \ldots, p_r$ which are $\leqslant T$, we therefore obtain

$$\begin{aligned}\prod_{p \leqslant T} (1 - \chi(p)p^{-s})^{-1} &= \prod_{p \leqslant T} \left(\sum_{m=0}^{\infty} \chi(p)^m p^{-ms} \right) \\ &= \sum \chi(p_1)^{m_1} \cdots \chi(p_r)^{m_r} (p_1^{m_1} \cdots p_r^{m_r})^{-s}\end{aligned}$$

$$= \sum_{n \in \mathcal{N}(T)} \chi(n) n^{-s},$$

where $\mathcal{N}(T)$ denotes the integers all of whose prime factors are $\leqslant T$. Thus whenever $\mathrm{Re}(s) > 1$, we have

$$\left| \sum_{n=1}^{\infty} \frac{\chi(n)}{n^s} - \prod_{p \leqslant T} \left(1 - \frac{\chi(p)}{p^s}\right)^{-1} \right| = \left| \sum_{n \notin \mathcal{N}(T)} \frac{\chi(n)}{n^s} \right|$$

$$\leqslant \sum_{n > T} \left| \frac{\chi(n)}{n^s} \right| \leqslant \sum_{n > T} \frac{1}{n^\sigma}.$$

The last sum is the tail-end of a real convergent series (when $\sigma := \mathrm{Re}(s) > 1$). It therefore tends to zero as T tends to infinity, which proves both the convergence of the product and the generalized Euler formula. □

4.12. Corollary. *When* $\mathrm{Re}(s) > 1$, *we have* $L(\chi, s) \neq 0$.

Proof. This is obvious since the Euler product is convergent and $1 - \chi(p)p^{-s} \neq 0$. □

4.13. Corollary. *We also have the following formulas.*

i)

$$\log L(\chi, s) = \sum_p \sum_{m \geqslant 1} \frac{\chi(p)^m}{m} p^{-ms}. \tag{4.20}$$

ii)

$$-\frac{L'(\chi, s)}{L(\chi, s)} = \sum_p \sum_{m \geqslant 1} \chi(p)^m \frac{\log p}{p^{ms}} = \sum_n \chi(n) \frac{\Lambda(n)}{n^s}. \tag{4.21}$$

Proof. We can use an argument similar to the one we used for the function $\zeta(s)$. □

Interlude (II). Let $K = \mathbf{Q}(\sqrt{d})$ where d is square-free and $\neq 1$. The decomposition law in $\mathcal{O}_K$ of primes of $\mathbf{Z}$ allows us to describe the contribution of p to the Dedekind function $\zeta_K(s)$:

$$\begin{cases} (1-p^{-s})^{-2} = (1-p^{-s})^{-1}(1-\left(\frac{d}{p}\right)p^{-s})^{-1} & \text{if } p \text{ is split in } K, \\ (1-p^{-2s})^{-1} = (1-p^{-s})^{-1}(1-\left(\frac{d}{p}\right)p^{-s})^{-1} & \text{if } p \text{ is inert in } K, \\ (1-p^{-s})^{-1} = (1-p^{-s})^{-1}(1-\left(\frac{d}{p}\right)p^{-s})^{-1} & \text{if } p \text{ is ramified in } K, \end{cases}$$

for odd p (there is a similar statement for $p = 2$). We therefore have that

$$\zeta_K(s) = \zeta(s) L(\chi_d, s),$$

where χ is the character defined by $\chi_d(p) = \left(\frac{d}{p}\right)$ for odd p and $\chi_d(2) = 1$ (resp. $\chi_d(2) = -1$, $\chi_d(2) = 0$) if $d \equiv 1 \bmod 8$ (resp. $d \equiv 5 \bmod 8$, $d \equiv 2$ or $3 \bmod 4$). It can be shown as an exercise that if $D := |d|$ for $d \equiv 1 \bmod 4$ (resp. $D := 4|d|$ for $d \equiv 2$ or $3 \bmod 4$), then χ_d is a character modulo D. We will prove below that $L(\chi_d, 1) \neq 0$, and therefore the function $\zeta_K(s)$ has, just like $\zeta(s)$, a pole of order 1 at $s = 1$, at which the residue equals $L(\chi_d, 1)$. One of the nicest results in analysis, the *class number formula* for a quadratic field, is given by:

$$\operatorname{Res}(\zeta_K, 1) = \begin{cases} \dfrac{2\pi h_K}{w\sqrt{D}} & \text{if } K \text{ is imaginary,} \\ \dfrac{2h_K \log \epsilon}{\sqrt{D}} & \text{if } K \text{ is real,} \end{cases}$$

where h_K is the class number, w the number of roots of unity (equal to 2 if $d < -4$, and equal to 4 or 6 if $d = -1$ or $d = -3$) and ϵ is the fundamental unit > 1 (the generator of $\mathcal{O}_K$ modulo ± 1). This formula, together with the explicit computation of $L(\chi, 1)$ (see Exercise 4-6.6) is very useful, namely for studying h_K.

For the proof of the theorem on arithmetic progressions, we will need the following key result.

4.14. Theorem. *Let χ be a Dirichlet character different from the unitary character. Then,*

$$L(\chi, 1) \neq 0.$$

4.15. Remark. If we knew that the Euler product converged at $s = 1$, we would immediately have this result since $1 - \chi(p)p^{-1} \neq 0$. Luckily we can show, by using the fact that $L(\chi, 1)$ is non-zero, that the Euler product at 1 converges (see Exercise 4-6.7).

Before proving the theorem, we will see how to deduce Dirichlet's theorem (4-1.4) from it.

Proof. For $\operatorname{Re}(s) > 1$, we write the formula as

$$\sum_{p \equiv a \bmod m} p^{-s} = \frac{1}{\phi(m)} \sum_{p \nmid m} p^{-s} + \frac{1}{\phi(m)} \sum_{\chi \neq 1} \overline{\chi}(a) \left(\sum_p \chi(p) p^{-s} \right).$$

The generalized Euler formula then yields for $\operatorname{Re}(s) > 1/2$,

$$\sum_p \chi(p) p^{-s} = \log L(\chi, s) + \text{a holomorphic function.}$$

The expression $\left|\sum_{p,m\geqslant 2}\frac{\chi(p^m)}{m}p^{-ms}\right|$ is bounded above by $\sum_{p,m\geqslant 2}p^{-m\sigma}/m$, which converges when $\sigma > 1/2$. Consequently, when χ is not the unitary character and knowing that $L(\chi,1)\neq 0$, we can deduce that $\sum_p \chi(p)p^{-s} = O(1)$ in a neighborhood of $s=1$. However, $\sum_p p^{-s} = -\log(s-1)+O(1)$. We can therefore conclude that

$$\sum_{p\equiv a \bmod m} p^{-s} = -\frac{\log(s-1)}{\phi(m)} + O(1),$$

which indeed proves the theorem on arithmetic progressions. □

4.16. Remark. If $\mathscr{Q}$ designates a subset of the set $\mathscr{P}$ of prime numbers, we can define various notions of density. The preceding proof suggests that we should introduce the notion of *analytic density*:

$$d_{\mathrm{an}}(\mathscr{Q}) := \lim_{s\to 1}\frac{\displaystyle\sum_{p\in\mathscr{Q}} p^{-s}}{\displaystyle\sum_{p\in\mathscr{P}} p^{-s}}.$$

Thus we have just shown that the analytic density of prime numbers congruent to a modulo m is $1/\phi(m)$. We could also define the "natural" density as

$$d(\mathscr{Q}) := \lim_{x\to\infty}\frac{\mathrm{card}\{p\in\mathscr{Q}\mid p\leqslant x\}}{\mathrm{card}\{p\in\mathscr{P}\mid p\leqslant x\}}.$$

It can be shown, but we will not do it, that the natural density of prime numbers congruent to a modulo m is $1/\phi(m)$.

To prove that $L(\chi,1)\neq 0$, we use the following lemma about Dirichlet series with positive real coefficients.

4.17. Lemma. *Let (a_n) be a sequence of positive real numbers. Suppose that the series $F(s)=\sum_{n=1}^{\infty}a_n n^{-s}$ converges for* $\mathrm{Re}(s)>\sigma_0$ *and that the function can be analytically continued in a neighborhood of σ_0. Then the abscissa of convergence of the series which defines $F(s)$ is strictly less than σ_0.*

Proof. Choose $r>0$ and $\sigma<\sigma_0<\sigma_1$ so that $\sigma\in D(\sigma_1,r)$, where this disk is contained in the domain of holomorphy of $F(s)$. The point σ_1 is in the half-plane of convergence, hence

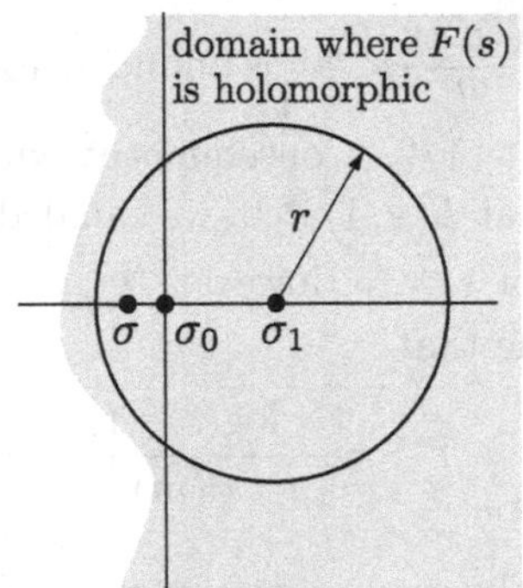

$$F^{(k)}(\sigma_1) = \sum_{n=1}^{\infty} a_n(-\log n)^k n^{-\sigma_1}.$$

By writing the expansion of F as a power series in the disk $D(\sigma_1, r)$ at the point σ, we obtain:

$$\begin{aligned}
F(\sigma) &= \sum_{k=0}^{\infty} \frac{F^{(k)}(\sigma_1)}{k!}(\sigma - \sigma_1)^k \\
&= \sum_{k=0}^{\infty} \frac{1}{k!} \sum_{n=1}^{\infty} a_n(-\log n)^k n^{-\sigma_1}(\sigma - \sigma_1)^k \\
&= \sum_{k=0}^{\infty} \frac{1}{k!} \sum_{n=1}^{\infty} a_n(\log n)^k n^{-\sigma_1}(\sigma_1 - \sigma)^k \\
&= \sum_{n=1}^{\infty} a_n n^{-\sigma_1} \sum_{k=0}^{\infty} \frac{1}{k!}(\log n)^k(\sigma_1 - \sigma)^k \\
&= \sum_{n=1}^{\infty} a_n n^{-\sigma_1} \exp\left(\log n(\sigma_1 - \sigma)\right) \\
&= \sum_{n=1}^{\infty} a_n n^{-\sigma_1} n^{\sigma_1 - \sigma} \\
&= \sum_{n=1}^{\infty} a_n n^{-\sigma},
\end{aligned}$$

where the rearrangement of the order of summation is justified by the fact that the terms are positive. This shows that the series converges at σ. □

4.18. Lemma. *Let $\hat{G}$ be the set of characters modulo m and p a prime number which does not divide m. We denote by f_p the order of $p \bmod m$*

and $g_p := \phi(m)/f_p$. *Then we have the identity:*

$$\prod_{\chi\in\hat{G}} (1-\chi(p)T) = \left(1-T^{f_p}\right)^{g_p}.$$

Proof. This follows from Lemma 4-4.3 on the values of characters at a point. □

4.19. Corollary. *The function* $F(s) := \prod_{\chi\in\hat{G}} L(\chi,s)$ *is a Dirichlet series with positive coefficients in the half-plane* $\mathrm{Re}(s) > 1$ *and has a simple pole at* $s = 1$.

Proof. To prove the first statement, we compute

$$\begin{aligned}\prod_{\chi} L(\chi,s) &= \prod_{p\nmid m}\prod_{\chi}\left(1-\frac{\chi(p)}{p^s}\right)\\ &= \prod_{p\nmid m}\left(1-\frac{1}{p^{sf_p}}\right)^{-g_p} = \prod_{p\nmid m}\left(\sum_{r=0}^{\infty} p^{-rf_ps}\right)^{g_p},\end{aligned}$$

which is clearly a Dirichlet series with positive coefficients. By furthermore noticing that $g_p \geqslant 1$ and $f_p \leqslant \phi(m)$, we have, for $\sigma\in\mathbf{R}$,

$$\prod_{\chi} L(\chi,\sigma) = \prod_{p\nmid m}\left(\sum_{r=0}^{\infty} p^{-rf_p\sigma}\right)^{g_p} \geqslant \prod_{p\nmid m}\left(1+p^{-\sigma\phi(m)}\right).$$

Thus the series and the product diverge for $\sigma = 1/\phi(m)$.

Furthermore, the function $L(\chi_0,s)$, like $\zeta(s)$, is meromorphic on $\mathrm{Re}(s) > 0$, with a unique simple pole at $s = 1$. The other $L(\chi,s)$ are holomorphic on $\mathrm{Re}(s) > 0$, hence the product of these functions is meromorphic on $\mathrm{Re}(s) > 0$, with a simple pole at $s = 1$ if $\prod_{\chi\neq\chi_0} L(\chi,1) \neq 0$ and no poles if one of the $L(\chi,1)$ is zero. We will now show that the latter case cannot happen. To see why this is true, if the product function were holomorphic up to $\mathrm{Re}(s) > 0$, the abscissa of convergence would therefore be $\leqslant 0$ in light of the lemma about Dirichlet series with positive coefficients (Lemma 4-4.17), which is a contradiction. □

We can deduce from the previous argument that $L(\chi,1)$ is non-zero for every χ different from the unitary character, and therefore we have indeed finished the proof of Dirichlet's theorem on arithmetic progressions.

4.20. Remark. We can show (up to factors corresponding to prime numbers p which divide m) that the product $\prod_{\chi} L(\chi,\sigma)$ is equal to the

Dedekind zeta function of the field $\mathbf{Q}(\exp(2\pi i/m))$, which explains in a more conceptual manner why the coefficients are positive.

5. The Prime Number Theorem

We will prove the following form of the prime number theorem.

5.1. Theorem. *The integral $\int_1^\infty (\theta(t) - t)t^{-2}dt$ is convergent.*

We will show that the convergence of this integral implies $\theta(x) \sim x$, and hence $\pi(x) \sim x/\log x$. To see this, suppose that $\limsup \theta(x)x^{-1} > 1$, then there exists $\epsilon > 0$ and x_n tending to infinity such that $\theta(x_n)x_n^{-1} \geqslant 1 + \epsilon$. For $t \in [x_n, (1+\epsilon/2)x_n]$, we therefore have

$$\frac{\theta(t) - t}{t^2} \geqslant \frac{\theta(x_n) - (1+\epsilon/2)x_n}{t^2} \geqslant \frac{\epsilon x_n/2}{t^2} \geqslant \frac{\epsilon x_n}{2(1+\epsilon/2)^2 x_n^2},$$

and consequently

$$\int_{x_n}^{(1+\epsilon/2)x_n} \frac{\theta(t) - t}{t^2} dt \geqslant \frac{\epsilon^2}{4(1+\epsilon/2)^2},$$

which contradicts the convergence of the integral. We can therefore conclude that $\limsup \theta(x)x^{-1} \leqslant 1$. A symmetric argument shows that $\liminf \theta(x)x^{-1} \geqslant 1$. It follows that $\lim \theta(x)x^{-1} = 1$.

To prove the theorem, we will use the following result from complex analysis (due to Newman, see [55, 81]) concerning the Laplace transform.

5.2. Theorem. ("The analytic theorem") *Let $h(t)$ be a bounded, piecewise continuous function. Then the integral*

$$F(s) = \int_0^{+\infty} h(u)e^{-su}du$$

is convergent and defines a holomorphic function on the half-plane $\mathrm{Re}(s) > 0$. *Suppose that this function can be analytically continued to a holomorphic function on the* closed *half-plane* $\mathrm{Re}(s) \geqslant 0$. *Then the integral converges for $s = 0$ and*

$$F(0) = \int_0^{+\infty} h(u)du.$$

Let us provisionally admit that this result is true and see how to apply it to the function

$$\begin{aligned} F(s) &= \int_1^{+\infty} \frac{\theta(t)-t}{t^{s+2}}\,dt \\ &= \int_0^{+\infty} \frac{\theta(e^u)-e^u}{e^{u(s+2)}} e^u du = \int_0^{+\infty} \left[\theta(e^u)e^{-u}-1\right] e^{-us}du. \end{aligned}$$

The function $h(u) := \theta(e^u)e^{-u} - 1$ is indeed bounded and piecewise continuous. If the analytic continuation hypothesis is satisfied, then we know that $F(0) = \int_0^{+\infty}(\theta(e^u)e^{-u}-1)du = \int_1^{+\infty} \dfrac{\theta(t)-t}{t^2}dt$ is indeed convergent.

We could transform the integral which defines $F(s)$ (for $\mathrm{Re}(s) > 0$) as follows:

$$\begin{aligned} F(s) &= \int_1^{+\infty} \frac{\theta(t)-t}{t^{s+2}}dt \\ &= \sum_{n=1}^{\infty} \int_n^{n+1} \theta(t)t^{-s-2}dt - \int_1^{+\infty} t^{-s-1}dt \\ &= \sum_{n=1}^{\infty} \theta(n) \frac{n^{-s-1}-(n+1)^{-s-1}}{s+1} - \frac{1}{s} \\ &= \frac{1}{s+1}\sum_{n=1}^{\infty} n^{-s-1}\left(\theta(n)-\theta(n-1)\right) - \frac{1}{s} \\ &= \frac{1}{s+1}\sum_p p^{-s-1}\log(p) - \frac{1}{s}. \end{aligned}$$

We have also seen that

$$-\frac{\zeta'(s)}{\zeta(s)} = \sum_{p,m\geqslant 1} \log(p)p^{-ms} = \sum_p \log(p)p^{-s} + \sum_{p,m\geqslant 2} \log(p)p^{-ms}.$$

The second term in the last expression is a convergent series and hence holomorphic for $\mathrm{Re}(s) > 1/2$. From this, we can deduce that $\sum_p \log(p)p^{-s} = -\dfrac{\zeta'(s)}{\zeta(s)} +$ a holomorphic function on $\mathrm{Re}(s) > 1/2$ and finally that

$$F(s) = -\frac{\zeta'(s+1)}{(s+1)\zeta(s+1)} - \frac{1}{s} + \text{ a holomorphic function on } \mathrm{Re}(s) > -1/2.$$

The key point in the proof is therefore the following result.

5.3. Theorem. (Hadamard, de la Vallée-Poussin) *The function* $\zeta(s)$ *does not have any zeros on the line* $\mathrm{Re}(s) = 1$.

Proof. We start with the formula

$$4\cos(x) + \cos(2x) + 3 = 2(1 + \cos(x))^2 \geqslant 0.$$

Recall that

$$\log \zeta(\sigma + it) = \sum_{p,m} \frac{p^{-m\sigma - mit}}{m}, \quad \text{and}$$

$$\log |\zeta(\sigma + it)| = \sum_{p,m} \frac{p^{-m\sigma}}{m} \cos(mt \log p).$$

This implies that

$$\log \left(|\zeta(\sigma + it)|^4 |\zeta(\sigma + 2it)| \zeta(\sigma)^3\right)$$
$$= \sum_{p,m \geqslant 1} \frac{p^{-m\sigma}}{m} (4\cos(mt\log p) + \cos(2mt \log p) + 3) \geqslant 0.$$

We can conclude from this, assuming $\sigma > 1$, that

$$|\zeta(\sigma + it)|^4 |\zeta(\sigma + 2it)| \zeta(\sigma)^3 \geqslant 1. \tag{4.22}$$

Now, if $\zeta(s)$ had a zero of order k at $1 + it$ and with order ℓ at $1 + 2it$, then $|\zeta(\sigma + it)| \sim a(\sigma - 1)^k$, $|\zeta(\sigma + 2it)| \sim b(\sigma - 1)^\ell$ and $\zeta(\sigma) \sim (\sigma - 1)^{-1}$ (where σ tends to 1 from above). The left-hand side of inequality (4.22) is therefore (asymptotically) equivalent to $c(\sigma - 1)^{4k+\ell-3}$, which implies that $4k + \ell - 3 \leqslant 0$, and hence $k = 0$. □

5.4. Corollary. *The function defined on* $\mathrm{Re}(s) > 1$ *by*

$$G(s) := -\frac{\zeta'(s)}{s\zeta(s)} - \frac{1}{s-1}$$

extends to a holomorphic function on $\mathrm{Re}(s) \geqslant 1$.

Proof. The previous theorem shows that the function $\zeta'(s)/\zeta(s)$ is holomorphic on the line $\mathrm{Re}(s) = 1$, except for $s = 1$. Consequently, the function $G(s)$ also is. To study $G(s)$ in a neighborhood of $s = 1$, we use the fact that $\zeta(s)$ has a simple pole at $s = 1$, and consequently $\zeta'(s)/\zeta(s) = -1/(s-1) + g(s)$, where $g(s)$ is holomorphic in a neighborhood of 1. Thus $G(s)$ is indeed holomorphic in a neighborhood of 1 and hence on the line $\mathrm{Re}(s) = 1$. □

Appendix. Proof of the "analytic theorem"

Recall the statement of the analytic result used in the proof of the prime number theorem.

5.5. Theorem. *If $h(t)$ is a bounded piecewise continuous function, then the integral (the Laplace transform of h)*

$$F(s) = \int_0^{+\infty} h(u)e^{-su}du$$

is convergent and defines a function which is holomorphic on the half-plane $\mathrm{Re}(s) > 0$. *Suppose that this function can be analytically continued to a holomorphic function on the closed half-plane* $\mathrm{Re}(s) \geqslant 0$. *Then the integral for $s = 0$ converges and*

$$F(0) = \int_0^{+\infty} h(u)du.$$

Proof. The first part is analogous to the theorem of convergence for Dirichlet series (see Exercise 4-6.2). We will therefore prove the second statement. For a (large) real number T, let $F_T(s) := \int_0^T h(t)e^{-st}dt$; these are functions which are holomorphic for all $s \in \mathbf{C}$. We now need to show that $\lim_{T\to\infty} F_T(0)$ exists and equals $F(0)$. To do this, we consider for some large R the contour $\gamma = \gamma(R, \delta)$ which bounds the region $S := \{s \in \mathbf{C} \mid \mathrm{Re}(z) \geqslant -\delta \text{ and } |s| \leqslant R\}$. Once we have fixed R, we can choose $\delta > 0$ sufficiently small so that $F(s)$ is analytic on this region.

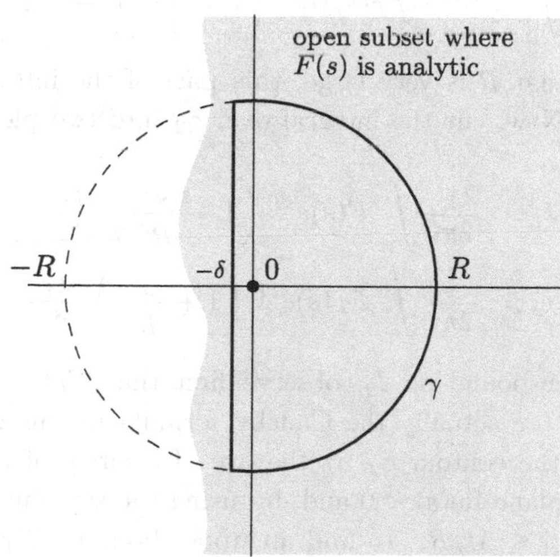

The trick lies in introducing the function

$$G_T(s) := (F(s) - F_T(s))\, e^{sT}\left(1 + \frac{s^2}{R^2}\right),$$

so that $G_T(0) = F(0) - F_T(0)$. Therefore, everything comes back to proving that $\lim_{T\to\infty} G_T(0) = 0$. To do this, we will use the residue theorem a first time, noticing that

$$G_T(0) = F(0) - F_T(0) = \frac{1}{2\pi i}\int_{\gamma} (F(s) - F_T(s))\, e^{sT}\left(1 + \frac{s^2}{R^2}\right)\frac{ds}{s}.$$

To find an upper bound on this integral, we cut the contour into two pieces: γ_1, which is the piece of γ which lives in the half-plane $\operatorname{Re}(s) > 0$, and γ_2, which lives in the half-plane $\operatorname{Re}(s) < 0$. We then carry out the following computation.

Let s be a number such that $|s| = R$ or $s = Re^{i\theta}$. Then we have

$$\left| e^{sT}\left(1 + \frac{s^2}{R^2}\right)\frac{1}{s}\right| = e^{\operatorname{Re}(s)T}\left|e^{-i\theta} + e^{i\theta}\right|\frac{1}{R} = e^{\operatorname{Re}(s)T}\,\frac{2\operatorname{Re}(s)}{R^2}.$$

We also have the upper bound

$$|F(s) - F_T(s)| = \left|\int_T^{\infty} h(t)e^{-st}dt\right| \leqslant M\int_T^{\infty}\left|e^{-st}\right| dt = \frac{Me^{-\operatorname{Re}(s)T}}{\operatorname{Re}(s)}.$$

This gives us

$$\left|\frac{1}{2\pi i}\int_{\gamma_1} (F(s) - F_T(s))\, e^{sT}\left(1 + \frac{s^2}{R^2}\right)\frac{ds}{s}\right| \leqslant \frac{M}{R}.$$

Thus assuming that R is very large, this part of the integral will be arbitrarily small. Now, cut the integral over γ_2 into two pieces, I_1 and I_2, where

$$I_1 := \frac{1}{2\pi i}\int_{\gamma_2} F(s)e^{sT}\left(1 + \frac{s^2}{R^2}\right)\frac{ds}{s},$$

$$I_2 := \frac{1}{2\pi i}\int_{\gamma_2} F_T(s)e^{sT}\left(1 + \frac{s^2}{R^2}\right)\frac{ds}{s}.$$

To find an upper bound on I_2, observe first that $F_T(s)$ is entire. The residue theorem (or actually the Cauchy formula in this case) allows us to then replace the contour γ_2 by the arc of a circle of radius R which lives in the half-plane $\operatorname{Re}(s) < 0$ and, by using the same upper bounds, to conclude that $|I_2| \leqslant M/R$. To find an upper bound of I_1, simply notice that the function $F(s)e^{sT}\left(1 + \frac{s^2}{R^2}\right)\frac{1}{s}$ converges to 0 when T tends to $+\infty$ and converges uniformly on every compact set contained in $\operatorname{Re}(s) < 0$. Consequently,

$$\lim_{T\to\infty}\frac{1}{2\pi i}\int_{\gamma_2} F(s)e^{sT}\left(1 + \frac{s^2}{R^2}\right)\frac{ds}{s} = 0.$$

By putting the three upper bounds together, we see that

$$|F_T(0) - F(0)| \leqslant \frac{2M}{R} + \epsilon(T)$$

where $\epsilon(T)$ tends to zero (in a way dependent on R). We needed to show that $\lim F_T(0) = F(0)$, which is now accomplished. □

Supplement. Analytic continuation and the functional equation

We will now outline the main steps of the proof of the following theorem due to Riemann.

5.6. Theorem. (The functional equation of the Riemann zeta function) *The function $\zeta(s)-1/(s-1)$ can be analytically continued to the whole complex plane. Furthermore, the function $\zeta(s)$ satisfies the functional equation given by*

$$\xi(s) = \xi(1-s), \tag{4.23}$$

where $\xi(s) := \pi^{-s/2}\Gamma(s/2)\zeta(s)$.

As a preliminary, we will recall the construction of the function $\Gamma(s)$ and the Poisson formula which gives the functional equation for the theta series.

5.7. Lemma. *The integral $\Gamma(s) := \int_0^\infty e^{-t}t^{s-1}dt$ defines a holomorphic function for $\mathrm{Re}(s) > 0$, which satisfies the functional equation $\Gamma(s+1) = s\Gamma(s)$. It can be continued to all of $\mathbf{C}$ as a meromorphic function with simple poles at $0, -1, -2, -3, \ldots$.*

Proof. Showing that the integral is convergent does not pose any problems. The functional equation can be obtained by integrating by parts. The functional equation also allows us to analytically continue by induction from $\mathrm{Re}(s) > -n$ to $\mathrm{Re}(s) > -n-1$ by using the fact that $\Gamma(s) = s^{-1}\Gamma(s+1)$. Finally, the expression

$$\Gamma(s) = \frac{1}{s(s+1)\cdots(s+n)}\Gamma(s+n+1)$$

makes it clear where the poles are. □

We can also prove that for all s, $\Gamma(s) \neq 0$ (see Exercise 4-6.19).

5.8. Lemma. (Poisson formula) *Let $f(x)$ be an integrable function over $\mathbf{R}$ (i.e., in $L^1(\mathbf{R})$). We define its Fourier transform by*

$$\hat{f}(y) := \int_{-\infty}^{+\infty} f(x)\exp(2\pi ixy)dx$$

and assume that the function $\sum_{n\in\mathbf{Z}} f(x+n)$ *is of bounded variation on* $[0,1]$ *and continuous. Then the following formula holds:*

$$\sum_{n\in\mathbf{Z}} f(n) = \sum_{m\in\mathbf{Z}} \hat{f}(m). \tag{4.24}$$

Proof. We introduce the function $G(x) := \sum_{n\in\mathbf{Z}} f(x+n)$ (the hypotheses guarantee the existence and continuity of such a function), which is clearly a periodic function. Dirichlet's theorem on Fourier series allows us to write its Fourier expansion as

$$G(x) = \sum_{m\in\mathbf{Z}} \hat{G}(m)\exp(2\pi imx),$$

where the Fourier coefficients can be calculated as follows:

$$\begin{aligned}\hat{G}(m) &:= \int_0^1 G(t)\exp(-2\pi imt)dt = \sum_{n\in\mathbf{Z}}\int_0^1 f(t+n)\exp(-2\pi imt)dt\\ &= \int_{-\infty}^{+\infty} f(x)\exp(-2\pi ixm)dx = \hat{f}(-m).\end{aligned}$$

This gives

$$\sum_{n\in\mathbf{Z}} f(x+n) = \sum_{m\in\mathbf{Z}} \hat{f}(m)\exp(-2\pi imx).$$

The Poisson formula follows from that by taking $x=0$. □

This formula is most often applied to a function f which is continuously differentiable and fast decreasing (i.e., $f(x) = O(|x|^{-M})$ for all M), and therefore the function G is itself continuously differentiable. This is the case when applying the formula to the following "theta" function.

5.9. Corollary. *The function*[2] $\theta(u) := \sum_{n\in\mathbf{Z}} \exp\left(-\pi un^2\right)$ *satisfies the functional equation for all* $u \in \mathbf{R}_+^*$ *given by:*

$$\theta(1/u) = \sqrt{u}\,\theta(u). \tag{4.25}$$

Proof. It suffices to apply the Poisson formula to the function $f(x) = \exp(-\pi ux^2)$ and to verify that $\hat{f}(y) = \exp(-\pi y^2/u)/\sqrt{u}$. □

Proof. (of Theorem 4-5.6) We start with the following computation (where

[2] We hope that the context will allow the reader to distinguish this function from the Tchebychev function $\theta(x) = \sum_{p\leqslant x}\log p$.

we introduce $t = \pi n^2 u$) which is valid for $\mathrm{Re}(s) > 1$.

$$\begin{aligned}\xi(s) = \pi^{-s/2}\Gamma(s/2)\zeta(s) &= \sum_{n\geqslant 1}\int_0^\infty e^{-t}t^{s/2}\pi^{-s/2}n^{-s}\frac{dt}{t}\\ &= \int_0^\infty \left\{\sum_{n\geqslant 1}\exp(-\pi u n^2)\right\} u^{s/2}\frac{du}{u}\\ &= \int_0^\infty \tilde\theta(u)\frac{u^{s/2}du}{u}\end{aligned}$$

where

$$\tilde\theta(u) := \sum_{n\geqslant 1}\exp\left(-\pi u n^2\right) = \frac{\theta(u)-1}{2}.$$

Let us point out that $\tilde\theta(u) = O(\exp(-\pi u))$ when u tends to infinity and that the functional equation of the function θ can be translated into

$$\tilde\theta\left(\frac{1}{u}\right) = \sqrt{u}\,\tilde\theta(u) + \frac{1}{2}\left(\sqrt{u}-1\right). \tag{4.26}$$

By using the simple computation $\int_1^\infty t^{-s} = 1/(s-1)$ and the functional equation of the theta function (4.25), we obtain

$$\begin{aligned}\xi(s) &= \int_0^1 \tilde\theta(u)\frac{u^{s/2}du}{u} + \int_1^\infty \tilde\theta(u)\frac{u^{s/2}du}{u}\\ &= \int_1^\infty \tilde\theta(1/u)\frac{u^{-s/2}du}{u} + \int_1^\infty \tilde\theta(u)\frac{u^{s/2}du}{u}\\ &= \int_1^\infty \left\{\sqrt{u}\tilde\theta(u) + \frac{1}{2}\left(\sqrt{u}-1\right)\right\}\frac{u^{-s/2}du}{u} + \int_1^\infty \tilde\theta(u)\frac{u^{s/2}du}{u}\\ &= \int_1^\infty \tilde\theta(u)\left\{u^{\frac{s}{2}} + u^{\frac{1-s}{2}}\right\}\frac{du}{u} + \frac{1}{s-1} - \frac{1}{s}.\end{aligned}$$

We have a priori obtained the desired expression only for $\mathrm{Re}(s) > 1$, but we can easily see that the integral defines an entire function since $\tilde\theta(u) = O(\exp(-\pi u))$ and since it is symmetric under the transformation $s \mapsto 1-s$. □

Supplement without proofs

1) To establish the prime number theorem, we could, in the place of the "analytic theorem", use Ikehara's theorem [40] (sometimes called the Ikehara-Wiener theorem), which is more powerful but also more tricky to prove. We will settle with stating the theorem. Its extension to the case of a multiple pole was proven by Delange [25].

5.10. Theorem. (Ikehara) *Let $A(t)$ be an increasing function such that the integral $F(s) = \int_1^{+\infty} A(t)t^{-s-1}dt$ is convergent whenever* $\mathrm{Re}(s) > 1$ *and can be analytically continued to the line* $\mathrm{Re}(s) = 1$ *except for a simple pole at $s = 1$ with residue λ (in other words, the function $F(s) - \lambda/(s-1)$ can be analytically continued to* $\mathrm{Re}(s) \geqslant 1$*). Then,*

$$A(x) = \lambda x + o(x).$$

If $\lambda = 0$ (in other words, if there is not a pole at $s = 1$), then $A(x) = o(x)$, if not $A(x) \sim \lambda x$. More generally (Delange), if $F(s)$ can be analytically continued to the line $\mathrm{Re}(s) = 1$ *with a pole of order t at $s = 1$ and principal term equal to $\lambda/(s-1)^t$, then $A(x) \sim \dfrac{\lambda}{(t-1)!} x(\log x)^{t-1}$.*

The prime number theorem follows from this theorem by using the fact that the hypotheses are satisfied when $A(x) = \psi(x)$, since

$$-\frac{\zeta'(s)}{\zeta(s)} = s\int_1^{+\infty} \psi(t)t^{-s-1}dt.$$

There are other paths or variations to arrive at the prime number theorem: see, for example, the proofs found in [18], [41] and [72]. Besides these, we would like to bring to your attention the proofs found in [4], [23], [51] and [53], which rely on "elementary" methods (not using a complex variable). The first elementary proof (1949) is due to Erdös and Selberg.

2) The result on the non-vanishing of the ζ function on the line $\mathrm{Re}(s) = 1$ could be considerably stronger, at least conjecturally. First of all, knowing that $\zeta(s)$ does not vanish in $\mathrm{Re}(s) \geqslant 1$, we can deduce from the functional equation that, in the closed half-plane $\mathrm{Re}(s) \leqslant 0$, the function $\zeta(s)$ vanishes uniquely at the points $s = -2, -4, -6, \ldots$, with order equal to one. To see why this is true, the function $\xi(s) = \pi^{-s/2}\Gamma(s/2)\zeta(s)$ does not vanish for $\mathrm{Re}(s) \geqslant 1$ (and has a simple pole at $s = 1$) hence, by the functional equation, it does not vanish for $\mathrm{Re}(s) \leqslant 0$ (and has a simple pole at $s = 0$). Furthermore, the function $\Gamma(s/2)$ never vanishes (see Exercise 4-6.19) and has a simple pole at $s = -2n$, for $n \in \mathbf{N}$ (see Lemma 4-5.7), hence $\zeta(s)$ should vanish at $s = -2n$, for $n \geqslant 1$. The question of describing the zeros in the *critical strip*, in other words in the strip $0 < \mathrm{Re}(s) < 1$, is much more delicate. The functional equation implies that the zeros are situated symmetrically with respect to the line $\mathrm{Re}(s) = \frac{1}{2}\cdot$

Riemann, in his extraordinary essay [60], suggested that the zeros are all situated on the line of symmetry.

"*Man Findet nun der That etwa so viel reelle Wurzeln innerhalb dieser Grenzen, und es ist sehr wahrscheinlich, dass alle Wurzeln reell sind. Hiervon wäre allerdings ein strenger Beweis zu wünschen; ich habe indess die Aufsuchung desselben nach einigen flüchtigen vergeblichen Versuchen vorläufig bei Seite gelassen, da er für den nächsten Zweck meiner Untersuchung entbehrlich schien.*"[3]

Keeping the previous notation, this can be formulated as follows.

5.11. Conjecture. "Riemann hypothesis"[4] *Let* $s \in \mathbf{C}$ *where* $\mathrm{Re}(s) > 1/2$, *then* $\zeta(s) \neq 0$. *In other words, if* $0 \leqslant \mathrm{Re}(s) \leqslant 1$ *and* $\zeta(s) = 0$, *then* $\mathrm{Re}(s) = \dfrac{1}{2}\cdot$

If this result proves to be true, it would follow that for every $\alpha > 1/2$, $\psi(x) = x + O(x^\alpha)$ and $\theta(x) = x + O(x^\alpha)$. By using the formula gotten above via Abel's formula, we can also deduce a much more precise equivalence for $\pi(x)$:

$$\pi(x) = \frac{\theta(x)}{\log(x)} + \int_2^x \frac{\theta(t)dt}{t(logt)^2}\cdot$$

This formula can be transformed into

$$\pi(x) = \int_2^x \frac{dt}{\log t} + 2\log 2 + \frac{\theta(x) - x}{\log(x)} + \int_2^x \frac{\theta(t) - t}{t(logt)^2}dt.$$

By introducing the "*logarithmic integral function*":

$$Li(x) := \int_2^x \frac{dt}{\log t},$$

we can see that the Riemann hypothesis implies that $\pi(x) = Li(x) + O(x^\alpha)$ for every $\alpha > 1/2$. By observing that

$$Li(x) = \frac{x}{\log x} + \frac{1}{2}\frac{x}{(\log x)^2} + O\left(\frac{x}{(\log x)^3}\right),$$

we can see that $Li(x)$ constitutes an estimate which is much more precise than $x/\log(x)$. Alas, the best proven result is far from our hopes, but we nonetheless know how to prove statements such as

$$\pi(x) = Li(x) + O\left(x \exp\left(-c\sqrt{\log x}\right)\right).$$

Because the zeta function is intimately linked to prime numbers, we can

[3]"*One finds, indeed, approximately so many real roots between these limits, and it is very likely that all of the roots are real. Certainly, a strict proof thereof needs to be done; I have however left aside the exploration of this question after some fleeting attempts in vain, since it seemed to be not essential for my current research objectives.*"

[4]The Riemann hypothesis is one of the major open problems in mathematics; the Clay Mathematics Institute also offers a million dollars for its solution.

reinterpret the functional equation and the Riemann hypothesis to be the expression of a higher order symmetry which brings a mysterious balance to the apparent chaos of the distribution of prime numbers. It provides in some sense an answer to Euler's thought with which we will close this chapter.[5]

"*Les mathématiciens ont tâché jusqu'ici en vain à découvrir un ordre quelconque dans la progression des nombres premiers, et on a lieu de croire que c'est un mystère auquel l'esprit humain ne saurait jamais pénétrer. Pour s'en convaincre, on n'a qu'à jeter les yeux sur les tables des nombres premiers, que quelques personnes se sont donné la peine de continuer au-delà de cent-mille: et on s'apercevra d'abord qu'il n'y règne aucun ordre ni règle.*"[6]

6. Exercises

6.1. Exercise. *Prove that Euclid's argument showing that the set of prime numbers is not finite implies that* $p_k \leqslant 2^{2^k}$. *Deduce from this the lower bound* $\pi(x) \geqslant \log\log x$ *for* $x \geqslant 2$.

6.2. Exercise. *Let* $h : [0, +\infty) \to \mathbf{C}$ *be a piecewise continuous function (or locally integrable). Prove that the Laplace transform* $F(s) := \int_0^{+\infty} h(u)e^{-us}du$ *is convergent in the half-plane* $\mathrm{Re}(s) > \sigma_0$ *and defines a holomorphic function there.*

Hint.– *You could use a procedure analogous to Proposition 4-3.1; more generally, you could extend the statement to functions of the type* $F(s) = \int_{\mathbf{R}_+} e^{-us}d\mu(u)$.

6.3. Exercise. *Recall that* $(f * g)(n) = \sum_{d|n} f(d)g(n/d)$. *Prove that if the two Dirichlet series* $F(s) = \sum_{n=1}^{\infty} f(n)n^{-s}$ *and* $G(s) = \sum_{n=1}^{\infty} g(n)n^{-s}$ *converge absolutely for* $\mathrm{Re}(s) > \sigma_0$, *then in the same half-plane, the following*

[5] Leonhard Euler: citation taken from his article *Découverte d'une loi tout extraordinaire des nombres par rapport à la somme de leurs diviseurs*, Bibliothèque impartiale 3, 1751, 10–31. The citation reproduced here can also be found in the reissue of the article from *Opera Posthuma* 1, 1862, 76–84 and is available on the website: http://math.dartmouth.edu/~euler.

[6] "*Mathematicians have tried, so far in vain, to discover some order in the progression of prime numbers, and we are led to believe that it is a mystery which the human mind will never know how to penetrate. To be convinced of this, we only have to have a look at the tables of prime numbers, which some people have taken the pain to continue to more than one-hundred thousand: and we realize right away that neither order nor rule prevails there.*"

equations hold:

$$F(s)G(s) = \left(\sum_{n=1}^{\infty} f(n)n^{-s}\right)\left(\sum_{n=1}^{\infty} g(n)n^{-s}\right) = \sum_{n=1}^{\infty}(f * g)(n)n^{-s}.$$

In particular, prove that for $\mathrm{Re}(s) > 1$, *we have*

$$\zeta(s)^{-1} = \sum_{n=1}^{\infty} \mu(n)n^{-s},$$

where μ *is the arithmetic Möbius function (i.e.,* $\mu(1) = 1$, $\mu(p_1 \cdots p_k) = (-1)^k$ *and* $\mu(n) = 0$ *if* n *has a square factor).*

6.4. Exercise. (Möbius inversion formula) *From among the arithmetic functions from* $\mathbf{N} \setminus \{0\}$ *to* $\mathbf{C}$, *we define the function* δ *by* $\delta(n) = 0$, *except* $\delta(1) = 1$ *and the function* $\mathbf{1}$ *by* $\mathbf{1}(n) = 1$, *for all* n. *Prove that* δ *is the identity element for the product* $*$ *and that* $\mu * \mathbf{1} = \delta$. *Deduce from this the Möbius inversion formula:*

$$g(n) = \sum_{d \mid n} f(d) \qquad \Leftrightarrow \qquad f(n) = \sum_{d \mid n} \mu(d) g(n/d).$$

6.5. Exercise. (Second Möbius inversion formula) *Let* F *and* G *be two functions of a positive real variable. Prove that if* $G(x) = \sum_{n \leqslant x} F\left(\frac{x}{n}\right)$, *then* $F(x) = \sum_{n \leqslant x} \mu(n) G\left(\frac{x}{n}\right)$.

6.6. Exercise. *Let* χ *be a nontrivial Dirichlet character modulo* N. *In this exercise, you are asked to give a* finite *explicit formula for* $L(\chi, 1)$.

a) Let $L(\theta) := \sum_{n \geqslant 1} \exp(in\theta) n^{-1}$. *By using the complex logarithm, prove that if* $\theta \in]0, 2\pi[$, *then*

$$L(\theta) = -\log(2\sin(\theta/2)) + i\left(\frac{\pi}{2} - \frac{\theta}{2}\right).$$

b) Prove that

$$\chi(a) = G(\chi)^{-1} \sum_{x \bmod N} \bar{\chi}(x) \exp\left(\frac{2\pi i a x}{N}\right).$$

c) Deduce from this that if χ *is even (i.e.,* $\chi(-1) = 1$*), then*

$$L(\chi, 1) = -G(\chi)^{-1} \sum_{u=1}^{N-1} \bar{\chi}(u) \log\sin\left(\frac{\pi u}{N}\right)$$

and if χ *is odd (i.e.,* $\chi(-1) = -1$*), then*

$$L(\chi, 1) = \frac{i\pi}{NG(\chi)} \sum_{u=1}^{N-1} \bar{\chi}(u)u.$$

d) Let χ *be a character modulo 4 such that* $\chi(-1) = -1$. *Verify that* $L(\chi, 1) = \pi/2\sqrt{2}$. *Let* χ' *be a character modulo 5 such that* $\chi(-1) = 1$, $\chi(2) = \chi(3) = -1$. *Verify that* $L(\chi', 1) = \log \eta/\sqrt{5}$ *where*

$$\eta = \frac{\sin(2\pi/5)\sin(3\pi/5)}{\sin(\pi/5)\sin(4\pi/5)} = \frac{1+\sqrt{5}}{2}.$$

We point out that η *is the fundamental unit of the field* $\mathbf{Q}(\sqrt{5})$ *and that this formula is a particular case of the class number formula.*

6.7. Exercise. *Let* χ *be a nontrivial Dirichlet modulo* N.

a) Prove that $\sum_{m \leqslant Y} \chi(m)m^{-1} = L(\chi, 1) + O(Y^{-1})$.

b) By using the formula $\log n = \sum_{m \mid n} \Lambda(m)$ *(which you should first verify), prove that*

$$\sum_{n \leqslant x} \frac{\chi(n) \log n}{n} = L(\chi, 1) \sum_{m \leqslant x} \frac{\chi(m)\Lambda(m)}{m} + O(1).$$

c) By using the fact that $L(\chi, 1) \neq 0$, *prove that* $\sum_{m \leqslant x} \dfrac{\chi(m)\Lambda(m)}{m}$ *is bounded when* x *tends to infinity, and deduce the convergence of the series* $\sum_p \dfrac{\chi(p)}{p}$ *from this.*

d) Finally, deduce that the Euler product $\prod_p (1 - \chi(p)p^{-1})^{-1}$ *is convergent and equals* $L(\chi, 1)$.

6.8. Exercise. *We denote by* $p_1, p_2, p_3, \ldots$ *the increasing sequence of prime numbers.*

a) Prove that

$$\sum_{n=1}^{N} p_n \sim \frac{N^2}{2} \log N.$$

b) Deduce from this a function equivalent to the sum $\sum_{p \leqslant X} p$, *when* X *tends to infinity.*

6.9. Exercise. *Let* d *be an odd integer. We define*

$$L_d(s) := \sum_{n=1}^{\infty} \left(\frac{n}{d}\right) n^{-s}.$$

a) Determine the abscissa of convergence of the series, and prove that the following equation holds:

$$L_d(s) = \prod_p \left(1 - \left(\frac{p}{d}\right) p^{-s}\right)^{-1}.$$

b) Prove that the function $\zeta(s)L_d(s)$ *can be written as a Dirichlet series* $\sum_{n=1}^{\infty} a_n n^{-s}$ *where* $a_n \geqslant 0$.

6.10. Exercise. *Prove the* Euler-Maclaurin formula, *which generalizes Abel's formula (Lemma 4-1.9):*

$$\sum_{a<n\leqslant b} f(n) = \int_a^b f(t)dt + \sum_{k=0}^{r} \frac{(-1)^{k+1} b_{k+1}}{(k+1)!} \left(f^{(k)}(b) - f^{(k)}(a)\right)$$
$$+ \frac{(-1)^r}{(r+1)!} \int_a^b B_{r+1}(t) f^{(r+1)}(t)dt,$$

where $b_k = B_k(0)$ *and the functions* B_k *are defined on* $t \in [0,1[$ *by* $B_0(t) = 1$, $B_k'(t) = kB_{k-1}(t)$ *and* $\int_0^1 B_k(t)dt = 0$, *and then extended by periodicity.*
Hint.– *The case* $k = 0$ *is Abel's formula, and for* $k > 0$, *proceed by integration by parts and induction.*

6.11. Exercise. *Let* $\gamma = \lim_{n\to\infty}(1 + \frac{1}{2} + \cdots + \frac{1}{n} - \log n)$ *be Euler's constant. Prove that*

$$\lim_{s\to 1} \left\{\zeta(s) - \frac{1}{s-1}\right\} = \gamma.$$

Hint.– *You could attempt a direct computation or compare the formula from Proposition 4-3.6, which gives a continuation of the* $\zeta(s)$ *function, to the expression for* γ *from application 1) of Lemma 4-1.9.*

6.12. Exercise. *Let* $f(n) := \operatorname{lcm}(1,2,3,\ldots,n)$. *Prove that the prime number theorem implies* $\log f(n) = n + o(n)$.

6.13. Exercise. *We define the following arithmetic function:*

$$\gamma(n) := \max_{m_1+\cdots+m_r=n} \operatorname{lcm}(m_1,\ldots,m_r),$$

where the m_i *are integers* $\geqslant 1$. *The integer* $\gamma(n)$ *represents the maximal order of an element in the permutation group on* n *letters. This is mainly why we are interested in the function* $n \mapsto \gamma(n)$. *In this exercise you are asked to prove that*

$$\lim_{n\to\infty} \frac{\log \gamma(n)}{\sqrt{n \log n}} = 1.$$

a) Show that you can write

$$\gamma(n) := \max_{p_1^{\alpha_1}+\cdots+p_s^{\alpha_s} \leqslant n} (p_1^{\alpha_1} \cdots p_s^{\alpha_s}),$$

where the p_i are distinct primes.

b) By using the inequality of arithmetic and geometric means:

$$\sqrt[m]{y_1 \cdots y_m} \leqslant \frac{y_1 + \cdots + y_m}{m},$$

prove that if $p_1^{\alpha_1} + \cdots + p_r^{\alpha_r} \leqslant n$, then $p_1^{\alpha_1} \cdots p_r^{\alpha_r} \leqslant (n/r)^r$.

c) By using the prime number theorem, prove that the sum of the r first prime numbers is asymptotically equivalent to $\frac{r^2}{2} \log r$. Deduce from this, referring back to the previous question, that $n \geqslant \frac{r^2}{2} \log r(1 + o(1))$, and hence $r \leqslant 2\sqrt{n/\log n}(1 + o(1))$.

d) By observing that the function $f(x) = (n/x)^x$ is increasing on the interval $[1, n/e]$, conclude that

$$\log \gamma(n) \leqslant \sqrt{n \log n}(1 + o(1)).$$

e) For a given (large) n, we can choose $r = r(n)$ to be the largest integer such that $p_1 + \cdots + p_r \leqslant n$, where (p_i) denotes the sequence of prime numbers ordered increasingly. Prove that r is asymptotically equivalent to $2\sqrt{n/\log n}$. Prove that $\log \gamma(n) \geqslant \theta(p_r)$, and finish the exercise by again using the prime number theorem.

6.14. Exercise. *Recall that an arithmetic function is multiplicative (resp. completely multiplicative) if $f(mn) = f(m)f(n)$ whenever $\gcd(m, n) = 1$ (resp. for all m, n).*

a) If f is a multiplicative arithmetic function, prove that

$$\sum_{n\geqslant 1} \frac{f(n)}{n^s} = \prod_p \sum_{m=0}^{\infty} \frac{f(p^m)}{p^{ms}}.$$

b) If f is a completely multiplicative arithmetic function, prove that

$$\sum_{n\geqslant 1} \frac{f(n)}{n^s} = \prod_p \left(1 - \frac{f(p)}{p^s}\right)^{-1}.$$

6.15. Exercise. *For $n \in \mathbf{N}^*$, we define the arithmetic function "number of integer divisors":*

$$\tau(n) := \sum_{d \mid n} 1 = \operatorname{card}\{(d,e) \in \mathbf{N}^2 \mid de = n\},$$

and you are asked in this exercise to study some of its properties.

a) Prove the identity

$$\sum_{n=1}^{\infty} \frac{\tau(n)}{n^s} = \zeta(s)^2.$$

b) Prove that if $n = p_1^{a_1} \cdots p_k^{a_k}$, then $\tau(n) = (a_1+1)\cdots(a_k+1)$, and deduce from this that $\liminf \tau(n) = 2$.

c) Prove that, on average, $\tau(n)$ equals $\log n$ in the following sense:

$$\sum_{n \leqslant X} \tau(n) = \sum_{d \leqslant X} \left\lfloor \frac{X}{d} \right\rfloor \sim X \log X.$$

d) We set $P(x) := \prod_{p \leqslant x} p$. By using the prime number theorem, prove that:

$$\lim_{x \to \infty} \frac{\log \tau\left(P(x)\right) \log\log P(x)}{\log P(x)} = \log 2.$$

You are now asked to show that

$$\alpha := \limsup_{n \to \infty} \frac{\log \tau\left(n\right) \log\log n}{\log n} = \log 2. \qquad (*)$$

To do this, if $n = p_1^{a_1} \ldots p_k^{a_k} \in \mathbf{N}^*$, then we divide $\tau(n) = D_1 D_2$ into two pieces in the following manner. We choose a real number $M \geqslant 2$ which depends on n, and we set $I_1 := \{1 \leqslant i \leqslant r \mid p_i \leqslant M\}$, $I_2 := \{1 \leqslant i \leqslant r \mid p_i > M\}$, $D_1 := \prod_{i \in I_1}(a_i + 1)$ and $D_2 := \prod_{i \in I_2}(a_i+1)$. (Let us point out that if $I_i = \emptyset$, then $D_i = 1$.)

e) Prove that $D_2 \leqslant 2^{\sum_{i \in I_2} a_i} \leqslant 2^{\log n / \log M}$.

f) Prove that there exists $c > 0$ (independent of n and M) such that $D_1 \leqslant \exp\left(cM \log\log n / \log M\right)$.

Hint.– *You could prove and use that $a_i \leqslant \log n / \log 2$ and $\operatorname{card}(I_1) \leqslant \pi(M)$.*

g) By choosing $M := \log n / (\log\log n)^2$ in the preceding questions, find an upper bound of $\tau(n)$, and then conclude that equation $(*)$ holds.

6.16. Exercise. *Let $k \geqslant 1$ be an integer. We keep the notation $\tau(n)$ for the function defined in the previous exercise.*

a) Prove the identity

$$\sum_{m=0}^{\infty} \tau(p^m)^k T^m = \sum_{m=0}^{\infty} (m+1)^k T^m = \frac{P_k(T)}{(1-T)^{k+1}},$$

where P_k is a polynomial of degree $k-1$ defined by $P_1(T) = 1$ and $P_{k+1}(T) = P_k(T)(1+kT) + P_k'(T)T(1-T)$.

b) Prove that the polynomial P_k can be written as $P_k(T) = 1 + (2^k - k - 1)T + \cdots + T^{k-1}$; deduce from this that the Euler product

$$G_k(s) := \prod_p \left(1 - p^{-s}\right)^{2^k - k - 1} P_k(p^{-s})$$

defines a holomorphic function in the half-plane $\mathrm{Re}(s) > 1/2$, *and verify that the following identity is true:*

$$\sum_{n=1}^{\infty} \tau(n)^k n^{-s} = \zeta(s)^{2^k} G_k(s).$$

c) By using Ikehara's theorem (Theorem 4-5.10), deduce from the previous question that the following estimate holds:

$$\sum_{n \leqslant x} \tau(n)^k \sim \lambda_k x (\log x)^{2^k - 1},$$

where $\lambda_k := G_k(1)/(2^k - 1)!$.

6.17. Exercise. *We introduce the following arithmetic function:*

$$\tau_k(n) := \mathrm{card}\left\{(n_1, \ldots, n_k) \in \mathbf{N}^k \mid n = n_1 \cdots n_k\right\}.$$

1) Prove that τ_k is multiplicative and that

$$\tau_k(p^m) = \frac{(m+k-1)\cdots(m+1)}{(k-1)!} = \binom{m+k-1}{k-1}.$$

2) Prove that $\tau_{k+1}(n) = \sum_{d \mid n} \tau_k(d)$, and deduce from this the equality

$$\sum_{n=1}^{\infty} \frac{\tau_k(n)}{n^s} = \zeta(s)^k.$$

3) By using the generalized (by Delange) theorem of Ikehara, prove that

$$\sum_{n \leqslant x} \tau_k(n) \sim \frac{x}{(k-1)!} (\log x)^{k-1}.$$

4) Observe that $\tau_k(p) = k$, and, by imitating the steps in Exercise 4-6.15, *prove that*

$$\limsup_{n\to\infty} \frac{\log \tau_k(n) \log\log n}{\log n} = \log k. \qquad (**)$$

Hint.– *You could proceed in a similar manner and replace the inequality* $m+1 \leqslant 2^m$ *by* $\dfrac{(m+k-1)\cdots(m+1)}{(k-1)!} \leqslant k^m$ *then find a combinatorial interpretation of the inequality.*

6.18. Exercise. *Recall that* $\phi(n) = \operatorname{card}(\mathbf{Z}/n\mathbf{Z})^*$ *denotes the Euler totient and that* $\phi(n) = n\prod_{p|n}\left(1-\frac{1}{p}\right)$.

1) For every $n \geqslant 2$, *check that* $\phi(n) \leqslant n-1$.

2) Let $P(x) := \prod_{p\leqslant x}\left(1-\frac{1}{p}\right)$; *by comparing* $\log P(x)$ *to* $\sum_{p\leqslant x} p^{-1}$, *prove that there exists a constant* $C_0 > 0$ *such that*

$$P(x) \sim \frac{C_0}{\log x}.$$

3) Let $N := \prod_{p\leqslant x} p$. *By using the prime number theorem and the previous question, prove that:*

$$\phi(N) \sim \frac{C_0 N}{\log\log N}.$$

4) Let $p_1 < p_2 < p_3 < \ldots$ *be the increasing sequence of prime numbers. For* $n \geqslant 2$, *we denote by* $\omega(n)$ *the arithmetic function which denotes the number of distinct prime numbers which divide* n. *Prove that* $p_1\cdots p_{\omega(n)} \leqslant n$, *and deduce that there exists a constant* $c > 0$ *such that*

$$\omega(n) \leqslant \frac{c\log n}{\log\log n}.$$

5) Now let $n \geqslant 2$. *Prove that*

$$\prod_{k=1}^{\omega(n)}\left(1-\frac{1}{p_k}\right) \leqslant \frac{\phi(n)}{n},$$

and deduce from this an estimate of the form

$$\forall n \geqslant 2, \qquad \frac{C_0 n}{\log\log n}(1+\epsilon(n)) \leqslant \phi(n),$$

where $\lim \epsilon(n) = 0$.[7]

[7] One could show that in fact $C_0 = e^{-\gamma}$ where γ is Euler's constant.

6.19. Exercise. *1) Prove that the sequence* $g_n(z) := \dfrac{z(z+1)\cdots(z+n)n^{-z}}{n!}$ *converges uniformly on every compact set in the complex plane and therefore defines an entire function* $G(z) := \lim_n g_n(z)$*, which only has simple zeros at* $z = 0, -1, -2, \ldots, -n,$ *etc.*

2) Verify that G *satisfies the formulas* $G(z+1) = G(z)/z$ *and*

$$G(z)G(1-z) = z\prod_{n=1}^{\infty}\left(1 - \frac{z^2}{n^2}\right) = \frac{\sin(\pi z)}{\pi}.$$

Hint.– *The second equality is well-known and can be shown by comparing the zeros of the two functions.*

3) Deduce from this that $U(z) := \Gamma(z)G(z)$ *is periodic with period 1 and satisfies* $U(z)U(z-1) = 1$.

Hint.– *You could use the "reflection formula" given by*

$$\Gamma(z)\Gamma(1-z) = \frac{\pi}{\sin(\pi z)},$$

which is usually proven using the formula

$$\Gamma(x)\Gamma(y) = \Gamma(x+y)\int_0^1 (1-t)^{x-1}t^{y-1}dt.$$

You could either prove this formula or consult a real and complex variable analysis text.

4) Deduce from this that $U(z)$*, and consequently* $\Gamma(z)$*, does not vanish anywhere in the complex plane.*

Let us point out that we could compute even further and prove that $U \equiv 1$ *and* $G(z) = \Gamma(z)^{-1}$*, which proves that*

$$\Gamma(z) = \lim_{n\to\infty} \frac{n!}{z(z+1)\cdots(z+n)} n^z.$$

6.20. Exercise. *Recall that the Möbius function* $\mu(n)$ *can be defined by the formula*

$$\zeta(s)^{-1} = \sum_{n=1}^{\infty} \frac{\mu(n)}{n^s} \quad \textit{for } \mathrm{Re}(s) > 1,$$

and let $M(x) := \sum_{n\leqslant x}\mu(n)$*. Observe that* $|M(x)| \leqslant x$ *for* $x > 0$.

a) Verify that the function defined on the open half-plane $\mathrm{Re}(s) > 1$ *by the Dirichlet series* $\sum_{n=1}^{\infty} \dfrac{\mu(n)}{n^s}$ *can be analytically continued to an open set which contains the* closed *half-plane* $\mathrm{Re}(s) \geqslant 1$*. Prove that this function vanishes at* $s = 1$.

b) Prove that the following formula is valid for all $s \in \mathbf{C}$*:*

$$\sum_{n=1}^{N} \frac{\mu(n)}{n^s} = s \int_1^N M(t) t^{-1-s} dt + M(N) N^{-s}.$$

c) Deduce from this that if $\mathrm{Re}(s) > 1$*, then*

$$\zeta(s)^{-1} = s \int_1^\infty M(t) t^{-1-s} dt,$$

and conclude that the integral $\int_1^\infty \frac{M(t)dt}{t^2}$ *is convergent and zero.*

Hint.– *You could use a), the fact that* $M(t)t^{-1}$ *is bounded and Newman's "analytic theorem" to prove that the integral is equal to value of* $\zeta(s)^{-1}$ *at* $s = 1$.

d) Prove that

$$\lim_{x \to \infty} \left\{ \sum_{n \leqslant x} \frac{\mu(n)}{n} - \frac{M(x)}{x} \right\} = 0.$$

e) We would like to show that the prime number theorem implies that $M(x) := \sum_{n \leqslant x} \mu(x) = o(x)$*. Let* $H(x) := \sum_{n \leqslant x} \mu(n) \log n$*. Prove that*

$$H(x) = M(x) \log x - \int_0^x \frac{M(t)}{t} dt$$

then that $H(x) = \sum_{n \leqslant x} \mu(n) \psi(x/n)$*, and conclude, by using the prime number theorem (i.e.,* $\psi(x) \sim x$*), that*

$$M(x) := \sum_{n \leqslant x} \mu(x) = o(x).$$

f) Deduce from this the value of the sum

$$\sum_{n=1}^{\infty} \frac{\mu(n)}{n}.$$

6.21. Exercise. *In this exercise, we denote by* $\pi(X)$ *the number of prime numbers smaller than* X*. We will use the following form of the prime number theorem:*

$$\pi(X) = \frac{X}{\log X} + O\left(\frac{X}{(\log X)^2}\right).$$

a) Prove that the following two estimates hold whenever $\alpha > -1$*):*

$$\int_2^X \frac{t^\alpha}{(\log t)^2} dt = O\left(\frac{X^{\alpha+1}}{(\log X)^2}\right),$$

$$\int_2^X \frac{t^\alpha}{\log t} dt = \frac{X^{\alpha+1}}{(\alpha+1)\log X} + O\left(\frac{X^{\alpha+1}}{(\log X)^2}\right).$$

b) By using Abel's summation formula, prove that if f is a continuously differentiable function, then

$$\sum_{p \leqslant X} f(p) = \pi(X) f(X) - \int_2^X \pi(t) f'(t) dt.$$

c) Prove, still assuming that $\alpha > -1$, the following generalization of the prime number theorem:

$$\sum_{p \leqslant X} p^\alpha = \frac{X^{\alpha+1}}{(\alpha+1)\log X} + O\left(\frac{X^{\alpha+1}}{(\log X)^2}\right).$$

Chapter 5

Elliptic Curves

"Mais où sont les neiges d'antan?"
François Villon

An elliptic curve can be defined as a smooth projective curve of degree 3 in the projective plane with a distinguished point chosen on it. The set of points on the curve can thus be endowed with a natural additive group structure. The most concrete description of an elliptic curve comes from its affine equation, written as

$$y^2 = x^3 + ax + b, \quad \textit{with } 4a^3 + 27b^2 \neq 0.$$

The theory of elliptic curves is a marvelous mixture of elementary mathematics and profound, advanced mathematics, a mixture which moreover lies on the crossroads of multiple theories: arithmetic, algebraic geometry, group representations, complex analysis, etc. Here, we will provide an introduction to the subject and prove the main Diophantine theorems: the group of rational points is finitely generated *(the Mordell-Weil theorem)* and the set of integral points is finite *(Siegel's theorem). Finally, we will evoke the famous theorem of Wiles—whose proof resulted in the proof of Fermat's last theorem—and the Birch & Swinnerton-Dyer conjecture.*

1. Group Law on a Cubic

Here, the word "cubic" designates an algebraic curve C in the projective plane $\mathbf{P}^2$ defined by a homogeneous equation, $F(X,Y,Z) = 0$, of degree 3. The curve is *smooth* if it has a tangent line at each point, i.e., if $\left(\frac{\partial F}{\partial X}, \frac{\partial F}{\partial Y}, \frac{\partial F}{\partial Z}\right) \neq (0,0,0)$ (see Appendix B for an introduction to projective geometry). If $F \in K[X,Y,Z]$, recall that we denote by

M. Hindry, *Arithmetics*, Universitext,
DOI 10.1007/978-1-4471-2131-2_5,

$C(K)$ the set of rational points on K, in other words, the set $\{(x, y, z) \in \mathbf{P}^2(K) \mid F(x, y, z) = 0\}$.

1.1. Definition. Let C be a smooth cubic. If P and Q are distinct points on C, the line joining P and Q cuts the cubic at three points, P, Q and a third point R (possibly equal to P or Q) that is denoted by $R = P \circ Q$. If $P = Q$, we do the same operation with the line tangent to the curve C at P. We define the law $+$ by choosing a distinguished point called the "origin" $O \in C$ and setting $O' = O \circ O$, then

$$P + Q := O \circ (P \circ Q) \qquad \text{and} \qquad -P := O' \circ P.$$

The procedure which defines this addition law is called the "*chord-tangent*" method.

1.2. Theorem. *The law defined by the chord-tangent method on a smooth cubic is a commutative group law, where the identity element is given by the distinguished point O. If $O \in C(K)$, then $C(K)$ is an abelian group.*

Since $P \circ Q = Q \circ P$, the law $+$ is obviously commutative. Indeed, we know that $O + P = P$, since O, P and $O \circ P$ are colinear. If $Q = -P$, then Q, O' and P are colinear, hence $O' = Q \circ P$ and $O \circ O' = O$, and therefore $O = Q + P$. The only tricky point is the associativity. We will use the following two classical lemmas to prove this.

1.3. Lemma. *Let $P_1, \ldots, P_8$ be eight distinct points in $\mathbf{P}^2$. Assume that no subset of four of them is ever colinear and no subset of seven of them ever appears on the same conic. Then the vector space of homogeneous polynomials of degree 3 which vanish at $P_1, \ldots, P_8$ is of dimension 2.*

Proof. Let n be the dimension that we are looking for. No matter how the eight points P_i are positioned, we know that $n \geqslant 10 - 8 = 2$. Without loss of generality, if P_1, P_2, P_3 are colinear, then we can choose P_9 on the same line, whose equation is given by $L = 0$. Every cubic F which vanishes at $P_1, \ldots, P_9$ is therefore of the form LQ, where Q vanishes at $P_4, \ldots, P_8$. But by Lemma B-1.18, given five points such that any four of which are not colinear, there is only one conic which contains all five, say $Q_0 = 0$, and F is a multiple of LQ_0. The dimension n_0 of the space of these cubics is therefore equal to 1. Thus $n \leqslant n_0 + 1 = 2$. Now suppose that $P_1, \ldots, P_6$ lie on a conic $Q = 0$ and choose P_9 on this conic. Every cubic F vanishing at $P_1, \ldots, P_9$ is therefore of the form LQ, where $L = 0$ is the equation of the line (P_7, P_8). The dimension n_0 of the space of these cubics is therefore equal to 1. Thus, $n \leqslant n_0 + 1 = 2$. In the general case (no three-tuple of points lie on the same line and no six-tuple of points lie on the same conic), we introduce P_9 and P_{10} which lie on the line (P_1, P_2) with equation $L = 0$. If $n \geqslant 3$, there would exist a nontrivial cubic $F = 0$ passing through $P_1, \ldots, P_{10}$, but then $F = LQ$ and the conic with equation $Q = 0$ would pass through $P_3, \ldots, P_8$. □

1.4. Lemma. *Let $P_1, \ldots, P_9$ be the intersection points of two cubics C_1 and C_2, one of which is irreducible. Suppose that $P_1, \ldots, P_8$ are distinct. If a cubic C passes through $P_1, \ldots, P_8$, then it also passes through P_9.*

Proof. If, for example, C_1 is irreducible, then it contains neither four colinear points, nor 7 points on the same conic. By the previous lemma, the vector space of cubics vanishing at $P_1, \ldots, P_8$ has dimension 2 and is therefore generated by the equations of C_1 and C_2. □

Proof. (of Theorem 5-1.2) Let P, Q and R be three distinct points on the cubic C. The line $L_1 = (P, Q)$ intersects C at P, Q and T; the line $L_2 = (T, O)$ intersects C at T, O and T'; the line $L_3 = (R, T')$ intersects C at R, T' and U; and finally, the line $L_4 = (U, O)$ intersects C at U, O and U', so that $(P+Q)+R = U'$. Moreover, the line $M_1 = (Q, R)$ intersects C at Q, R and S; the line $M_2 = (S, O)$ intersects C at S, O and S'; the line $M_3 = (P, S')$ intersects C at P, S' and V; and finally, the line $M_4 = (V, O)$ intersects C at V, O and V', so that $(Q + R) + P = V'$. We want to show that $U' = V'$, which is equivalent to $U = V$. To do this, consider the cubic $C_1 = L_1 + M_2 + L_3$ and $C_2 = M_1 + L_2 + M_3$. Then,

$$C \cap C_1 = \{P, Q, R, O, T, T', S, S', U\} \quad \text{and}$$
$$C \cap C_2 = \{P, Q, R, O, T, T', S, S', V\}.$$

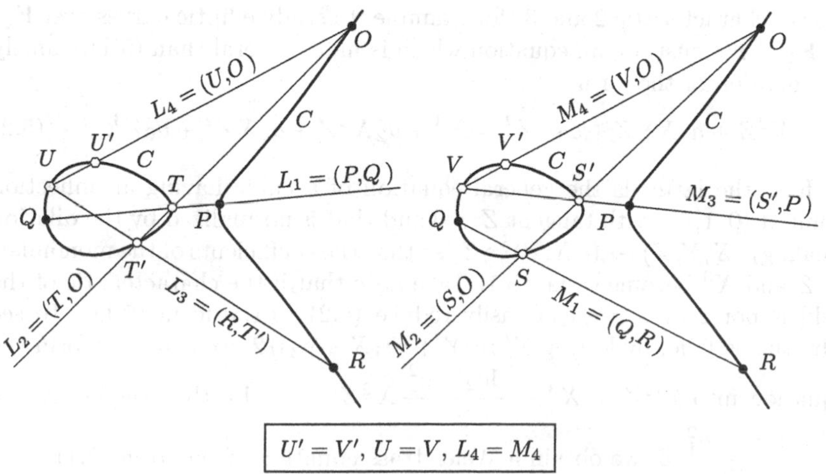

$U' = V'$, $U = V$, $L_4 = M_4$

If the points P, Q, R, O, T, T', S, S' are distinct, we can conclude, by Lemma 5-1.4 applied to C and C_1, that $U = V$. This is the case in general, and we can conclude that the equality $(P+Q)+R = (Q+R)+P$ is always true by invoking a continuity argument (either for the usual topology if we are working over $\mathbf{R}$ or $\mathbf{C}$ or the Zariski topology in the general case—see Appendix B, Lemma B-1.9). □

We would like to point out that this construction only uses the simplest two cases of Bézout's theorem on the intersection with a line or a conic (see Lemmas B-1.13 and B-1.14). Finally, we have the following operations over a curve: "translation by Q", defined by $P \mapsto P + Q$, and "multiplication by N", defined by $P \mapsto P + \cdots + P$ (N times) and denoted by $[N]$.

We will now explain this group law with a simpler model, known as the "Weierstrass" model.

1.5. Definition. A Weierstrass cubic is a curve given by a plane cubic equation of the form

$$Y^2Z = X^3 + aXZ^2 + bZ^3, \tag{5.1}$$

where $\Delta := 4a^3 + 27b^2 \neq 0$.

1.6. Remarks. The condition $\Delta \neq 0$ precisely means that the curve does not have a singular point. The curve defined by (5.1) has an obvious point, $O := (0, 1, 0)$, that we will take to be the origin and which is an inflection point, i.e., the tangent $Z = 0$ intersects the curve only at this point with multiplicity 3. It can be shown that every smooth cubic which has a rational point over K is isomorphic to a Weierstrass cubic, at least whenever K does not have characteristic 2 or 3. If we wish to include the case of characteristic 2 and 3 (for example, to study elliptic curves over $\mathbf{F}_{2^f}$ or $\mathbf{F}_{3^f}$), we must use an equation which is more general than (5.1), namely an equation of the form

$$Y^2Z + a_1XYZ + a_3YZ^2 = X^3 + a_2X^2Z + a_4XZ^2 + a_6Z^3. \tag{5.2}$$

In fact, the latter is the general equation of a cubic having an inflection point at $(0, 1, 0)$, with tangent $Z = 0$ and that is normalized by the dilation (scaling) $(X, Y, Z) \mapsto (\alpha X, \beta Y, \gamma Z)$ so that the coefficients of the monomials Y^2Z and X^3 become equal to 1. Take note that if the characteristic of the field is not 2 or 3, we can easily reduce (5.2) to the form (5.1). To see why this is true, by letting $Y' := Y + (a_1X + a_3)/2$, we can transform the equation into $Y'^2Z = X^3 + \dfrac{4a_2 + a_1^2}{4}X^2Z + \ldots$; by then setting $X' := X + \dfrac{4a_2 + a_1^2}{12}Z$, we obtain a Weierstrass equation of the form (5.1).

Coming back to the Weierstrass model, (5.1), we will often work in the affine coordinates $x := X/Z$ and $y := Y/Z$ by considering the point O as "the point at infinity". The affine equation is therefore of the type announced in the introduction:

$$y^2 = x^3 + ax + b. \tag{5.3}$$

A possible singular point would satisfy $2y = 3x^2 + a = 0$, hence $y = 0$ and x is a double root of the equation $x^3 + ax + b = 0$, whose discriminant is precisely $4a^3 + 27b^2$. Moreover, if α is a root of $x^3 + ax + b$, then the point $P := (\alpha, 0)$ is a point of order 2; thus there are three points of order 2.

1.7. Proposition. (Explicit group law) *If $P_1 = (x_1, y_1)$ and $P_2 = (x_2, y_2)$ are points on the curve whose equation is given by (5.3), then*

$$[-1](P_1) = (x_1, -y_1). \tag{5.4}$$

If $P_2 = [-1](P_1)$ (i.e., if $x_1 = x_2$ and $y_2 = -y_1$), then $P_1 + P_2 = O$. If $P_2 = P_1$, we set $\lambda = \dfrac{3x_1^2 + a}{2y_1}$ and $\mu = y_1 - \lambda x_1$, and if $P_2 \neq \pm P_1$ (i.e., if $x_2 \neq x_1$), we set $\lambda = \dfrac{y_1 - y_2}{x_1 - x_2}$ and $\mu = y_1 - \lambda x_1$. Then,

$$P_1 + P_2 = (\lambda^2 - x_1 - x_2, -\lambda^3 + \lambda(x_1 + x_2) - \mu). \tag{5.5}$$

Proof. Let $y = \lambda x + \mu$ be the equation of the line (P_1, P_2) (resp. of the tangent to the curve at P_1) if $P_1 \neq P_2$ (resp. if $P_1 = P_2$). Then λ and μ are given as in the statement. If $P_3 = (x_3, y_3)$ is the third point of intersection, then $P_1 + P_2 = (x_3, -y_3)$. To compute the intersection points of the line and the curve, we make a substitution for y to obtain the equation

$$x^3 + ax + b - (\lambda x + \mu)^2 = x^3 - \lambda^2 x^2 + (a - 2\lambda\mu)x + (b - \mu^2) = 0,$$

of which we know two roots: x_1 and x_2. From this, we obtain $x_1 + x_2 + x_3 = \lambda^2$ and $y_3 = \lambda x_3 + \mu$ as in the statement of the proposition. □

To verify continuity of the addition law in the Zariski topological sense (Definition B-1.7), observe that

$$\frac{y_1 - y_2}{x_1 - x_2} = \frac{x_1^2 + x_1 x_2 + x_2^2 + a}{y_1 + y_2}. \tag{5.6}$$

For future use, we also have the following formulas (that can be checked by direct computation):

$$x(P+Q) + x(P-Q) = \frac{2(x(P) + x(Q))(a + x(P)x(Q)) + 4b}{(x(P) - x(Q))^2}, \tag{5.7}$$

$$x(P+Q)x(P-Q) = \frac{(x(P)x(Q)-a)^2 - 4b(x(P)+x(Q))}{(x(P)-x(Q))^2}, \tag{5.8}$$

$$x(2P) = \frac{x(P)^4 - 2ax(P)^2 - 8bx(P) + a^2}{4\left(x(P)^3 + ax(P) + b\right)}. \tag{5.9}$$

2. Heights

We will introduce a precise notion of "size" or "arithmetic complexity" for the algebraic points in a projective space, which will be christened "height". The first version is sometimes called the Weil height *and the refined version, over elliptic curves, the* Néron-Tate height, *which we will prove has a quadratic nature.*

2.1. Weil Heights

We will start by defining the height of a point in a projective space first with rational coordinates, then with algebraic coordinates. From this, we will deduce the notion of the height of an algebraic number.

2.1.1. Definition. If P is a point in $\mathbf{P}^n(\mathbf{Q})$, we can choose projective coordinates for it, $(x_0, \ldots, x_n)$, where $x_i \in \mathbf{Z}$ and $\gcd(x_0, \ldots, x_n) = 1$. We thus define the *height* (resp. the *logarithmic height*) of P by

$$H(P) := \max(|x_0|, \ldots, |x_n|) \qquad (\text{resp. } h(P) := \log\max(|x_0|, \ldots, |x_n|)).$$

This very simple and natural definition does not translate very easily into algebraic coordinates, and it is technically more convenient to reinterpret height in terms of the set of absolute values of a field.

2.1.2. Definition. An *absolute value* v over a field K is a map $x \mapsto |x|_v$ from K to $\mathbf{R}_+$ such that for every $x, y \in K$,

i) $|x|_v = 0$ if and only if $x = 0$;
ii) $|xy|_v = |x|_v|y|_v$;
iii) there exists a constant $C_v > 0$ such that $|x+y|_v \leqslant C_v \max\{|x|_v, |y|_v\}$.

If v satisfies the more precise inequality $|x+y|_v \leqslant \max(|x|_v, |y|_v)$ (i.e., we can take C_v to be 1), v is said to be *ultrametric.*

2.1.3. Example. The standard absolute values over the field $K = \mathbf{Q}$ are the usual absolute value (denoted $|x|$ or $|x|_\infty$) and the p-adic absolute values (denoted $|x|_p$). For every prime number p, the p-adic absolute value is defined for $x = \pm p_1^{e_1} \cdots p_r^{e_r}$, where $e_i = \operatorname{ord}_{p_i}(x) \in \mathbf{Z}$, by

$$|x|_p := p^{-\operatorname{ord}_p(x)}.$$

The p-adic absolute values are ultrametric. We denote by $M_{\mathbf{Q}}$ the set of absolute values, which we will also call *places* of the field $\mathbf{Q}$.

2.1.4. Remark. If $|\cdot|$ is an absolute value, then $|\cdot|^\alpha$ is another one (for every $\alpha > 0$). If a map $|\cdot|_v$ satisfies *i)* and *ii)* and the triangle inequality *iii)'* given by $|x+y|_v \leqslant |x| + |y|_v$, then it satisfies *iii)* with $C_v = 2$ and is hence an absolute value. Conversely, we will leave it as an exercise to prove that an absolute value for which we can take the constant in inequality *iii)* to be 2 satisfies the triangle inequality. The reason that we take *iii)* in the definition is that the condition is stable when we replace $|\cdot|$ by $|\cdot|^\alpha$, which is not the case for the triangle inequality.

2.1.5. Theorem. (Product formula for $\mathbf{Q}$) *Let $x \in \mathbf{Q}^*$. Then,*

$$\prod_{v \in M_{\mathbf{Q}}} |x|_v = 1. \tag{5.10}$$

Proof. We write $x = \pm p_1^{e_1} \cdots p_r^{e_r}$, where $e_i = \operatorname{ord}_{p_i}(x) \in \mathbf{Z}$. For $1 \leqslant i \leqslant r$, we have $|x|_{p_i} = p_i^{-e_i}$. If p does not appear in x, then $|x|_p = 1$ and the usual absolute value equals $|x|_\infty = p_1^{e_1} \cdots p_r^{e_r}$. The formula follows from this. □

2.1.6. Corollary. *Let $P \in \mathbf{P}^n(\mathbf{Q})$ and $(x_0, \ldots, x_n)$ be (any) projective coordinates of P. Then,*

$$H(P) = \prod_{v \in M_{\mathbf{Q}}} \max\left(|x_0|_v, \ldots, |x_n|_v\right). \tag{5.11}$$

Proof. By the product formula, we know that the right-hand side is independent of the projective coordinates. Therefore, if we choose $x_i \in \mathbf{Z}$ to be relatively prime, we have, for each prime number p, $\max\left(|x_0|_p, \ldots, |x_n|_p\right) = 1$, and the right-hand side will indeed be equal to $\max\left(|x_0|_\infty, \ldots, |x_n|_\infty\right)$, in other words to $H(P)$. □

In order to generalize heights to algebraic coordinates, we will define standard absolute values over a number field K.

2.1.7. Example. Let K be a number field with r_1 real embeddings and r_2 pairs of complex embeddings, so that $n := [K : \mathbf{Q}] = r_2 + 2r_2$. Every embedding $\sigma : K \hookrightarrow \mathbf{R}$ or $\mathbf{C}$ produces, by composition with the modulus, an absolute value. If the embedding is complex, then σ and its conjugate produce the same absolute value. This then gives us r_1+r_2 absolute values:

$$|x|_\sigma := \begin{cases} |\sigma(x)| & \text{if } \sigma \text{ is real,} \\ |\sigma(x)|^2 & \text{if } \sigma \text{ is complex.} \end{cases}$$

Now if p is a prime number which factors into $p\mathscr{O}_K = \mathfrak{p}_1^{e_1}\cdots\mathfrak{p}_g^{e_g}$, with $\mathrm{N}\,\mathfrak{p}_i = p^{f_i}$ and $\sum_{i=1}^{g} e_i f_i = n$, then for every prime ideal $\mathfrak{p}$, we can define the absolute value

$$|x|_\mathfrak{p} := \mathrm{N}\,\mathfrak{p}^{-\operatorname{ord}_\mathfrak{p}(x)}.$$

We denote by M_K the set of these absolute values and by $M_{K,\infty}$ the subset of Archimedean absolute values.

It follows from these choices that for $x \in K$, we have

$$\prod_{\mathfrak{p}\,|\,p} |x|_\mathfrak{p} = \left|\mathrm{N}_\mathbf{Q}^K(x)\right|_p \qquad \text{and} \qquad \prod_{v\in M_{K,\infty}} |x|_v = \left|\mathrm{N}_\mathbf{Q}^K(x)\right|_\infty. \tag{5.12}$$

To understand this statement, if $x\mathscr{O}_K = \prod_\mathfrak{p} \mathfrak{p}^{\operatorname{ord}_\mathfrak{p}(x)}$, then we can write

$$\mathrm{N}_\mathbf{Q}^K(x) = \pm\,\mathrm{N}\,(x\mathscr{O}_K) = \pm\prod_\mathfrak{p} \mathrm{N}\,\mathfrak{p}^{\operatorname{ord}_\mathfrak{p}(x)} = \pm\prod_p p^{\sum_{\mathfrak{p}\,|\,p} f_\mathfrak{p} \operatorname{ord}_\mathfrak{p}(x)}.$$

Hence we have

$$\left|\mathrm{N}_\mathbf{Q}^K(x)\right|_p = p^{-\sum_{\mathfrak{p}\,|\,p} f_\mathfrak{p} \operatorname{ord}_\mathfrak{p}(x)} = \prod_{\mathfrak{p}\,|\,p} \mathrm{N}\,\mathfrak{p}^{-\operatorname{ord}_\mathfrak{p}(x)} = \prod_{\mathfrak{p}\,|\,p} |x|_\mathfrak{p}.$$

For the Archimedean places, we have

$$\left|\mathrm{N}_\mathbf{Q}^K(x)\right| = \left|\prod_{\sigma:K\hookrightarrow\mathbf{C}} \sigma(x)\right| = \prod_{i=1}^{r_1} |\sigma_i(x)| \prod_{j=r_1+1}^{r_1+r_2} \left|\sigma_j(x)\overline{\sigma_j(x)}\right| = \prod_{v\in M_{K,\infty}} |x|_v\,.$$

This gives us the following formula, which is analogous to Theorem 5-2.1.5.

2.1.8. Theorem. (Product formula for K) *Let $x \in K^*$. Then,*

$$\prod_{v\in M_K} |x|_v = 1. \tag{5.13}$$

Proof. We can regroup the places of K into packets over a certain place of $\mathbf{Q}$, and we can use the previous formula:

$$\prod_{w\in M_K} |x|_w = \prod_{v\in M_\mathbf{Q}} \prod_{w\,|\,v} |x|_w = \prod_{v\in M_\mathbf{Q}} \left|\mathrm{N}_\mathbf{Q}^K x\right|_v = 1. \qquad \square$$

2.1.9. Definition. Let $P \in \mathbf{P}^n(K)$, and let $(x_0,\ldots,x_n)$ be (any) projective coordinates of P. The *height* relative to the K is the number

$$H_K(P) = \prod_{v \in M_K} \max\left(|x_0|_v, \dots, |x_n|_v\right). \tag{5.14}$$

The height of a point considered in different algebraic extensions varies in a simple way.

2.1.10. Lemma. *If K' is a finite extension of K and $P \in \mathbf{P}^n(K)$, then*

$$H_{K'}(P) = H_K(P)^{[K':K]}. \tag{5.15}$$

Proof. Let $(x_0, \dots, x_n)$ be projective coordinates in P. We can assume that $x_i \in K$. If v is a place of K and w ranges over the places of K' over K, we clearly have

$$\prod_{w \mid v} \max_i |x_i|_w = \prod_{w \mid v} \max_i |x_i|_v^{e_w f_w} = \max_i |x_i|_v^{[K':K]}.$$

We thus have

$$\begin{aligned} H_{K'}(P) &= \prod_{w \in M_{K'}} \max_i |x_i|_w = \prod_{v \in M_K} \prod_{w \mid v} \max_i |x_i|_w \\ &= \prod_{v \in M_K} \max_i |x_i|_v^{[K':K]} = H_K(P)^{[K':K]}. \end{aligned}$$

□

This lemma allows us to define the absolute height, which is defined on the set of points with coordinates in $\bar{\mathbf{Q}}$, the algebraic closure of $\mathbf{Q}$.

2.1.11. Definition. We define $H : \mathbf{P}^n(\bar{\mathbf{Q}}) \to \mathbf{R}$ as follows: if $P \in \mathbf{P}^n(K)$, we let

$$H(P) := H_K(P)^{1/[K:\mathbf{Q}]}.$$

If $\alpha \in K$, we define the *height* of α (relative to the field K) as the height of the point $(1, \alpha) \in \mathbf{P}^1(K)$.

To establish a connection between the height of an algebraic number and its minimal polynomial, we will use the following classical lemma (see Lemma 2-6.2.3), which is valid for $\mathbf{Z}[X]$).

2.1.12. Lemma. (Gauss's lemma) *Let $P, Q \in K[X]$, where K is a number field. We denote by $||P||_v$ the sup-norm of the coefficients of P for an ultrametric absolute value v. Then,*

$$||PQ||_v = ||P||_v \, ||Q||_v. \tag{5.16}$$

Proof. By localizing (i.e., replacing $\mathscr{O}_K$ by $\mathscr{O} := \{x \in K \mid \operatorname{ord}_v x \geqslant 0\}$), we can reduce to the case of showing that if $||P||_v = ||Q||_v = 1$, then $||PQ||_v = 1$. Thus if π is a generator of the maximal ideal in $\mathcal{O}$ associated to the absolute value, then we can write $P = \pi^m P^*$ and $Q = \pi^n Q^*$, where $||P^*||_v = ||Q^*||_v = 1$. Now, $||P||_v = 1$ means that $P \in \mathcal{O}[X]$ and is non-zero in $\mathcal{O}/\pi\mathcal{O}[X]$. Since the latter ring is integral, the product PQ stays non-zero in $\mathcal{O}/\pi\mathcal{O}[X]$, and thus $||PQ||_v = 1$. □

2.1.13. Lemma. *Let α be an algebraic number and $K = \mathbf{Q}(\alpha)$. Let the minimal polynomial of α in $\mathbf{Z}[X]$ be written in the form*

$$P(X) = a_0(X - \alpha_1)\cdots(X - \alpha_d) = a_0 X^d + \ldots$$

Then,

$$H_K(\alpha) = |a_0| \prod_{i=1}^{d} \max\{1, |\alpha_i|\}. \tag{5.17}$$

Proof. Consider the field $L := \mathbf{Q}(\alpha_1, \ldots, \alpha_d)$. Then

$$H_L(\alpha) = H_K(\alpha)^{[L:K]} = \prod_{\mathfrak{q}\in M_L} \max(1, |\alpha|_{\mathfrak{q}}) \prod_{w\in M_{L,\infty}} \max(1, |\alpha|_w).$$

First of all, we have

$$\prod_{w\in M_{L,\infty}} \max(1, |\alpha|_w) = \prod_{v\in M_{K,\infty}} \max(1, |\alpha|_v)^{[L:K]} = \left(\prod_{i=1}^{d} \max\{1, |\alpha_i|\}\right)^{[L:K]}.$$

Gauss's lemma applied to P and its factorization shows that, for $\mathfrak{q} \in M_L$, we have

$$1 = ||P||_{\mathfrak{q}} = |a_0|_{\mathfrak{q}} \prod_{i=1}^{d} \max(1, |\alpha_i|_{\mathfrak{q}}).$$

By taking the product over $\mathfrak{q}$ and invoking the product formula for L (applied to a_0), we obtain

$$1 = \prod_{\mathfrak{q}\in M_L} |a_0|_{\mathfrak{q}} \prod_{i=1}^{d} \prod_{\mathfrak{q}\in M_L} \max(1, |\alpha_i|_{\mathfrak{q}}) = |a_0|^{-[L:\mathbf{Q}]} \left(\prod_{\mathfrak{q}\in M_L} \max(1, |\alpha|_{\mathfrak{q}})\right)^d.$$

Combining these results gives

$$H_K(\alpha)^{[L:K]} = |a_0|^{[L:\mathbf{Q}]/d} \prod_{i=1}^{d} \max(1, |\alpha_i|)^{[L:K]},$$

which yields the desired equality by taking the $[L : K]$th roots. □

The main merit of the height function introduced in this section is the following finiteness theorem, whose first part is due to Northcott and the second to Kronecker.

2.1.14. Theorem. (Northcott, Kronecker) *Let $d \geqslant 1$ and $X > 0$. Then the set $S(n,d,X) = \{P \in \mathbf{P}^n(\bar{\mathbf{Q}}) \mid [\mathbf{Q}(P) : \mathbf{Q}] \leqslant d,\ H(P) \leqslant X\}$ is finite. Furthermore, we have $H(P) > 1$, except if the point P has projective coordinates all equal to zero or a root of unity.*

Proof. Let $P = (x_0, \ldots, x_n) \in \mathbf{P}^n(\bar{\mathbf{Q}})$. Up to permuting the coordinates, we can assume that $x_0 \neq 0$. Then we can write $P = (1, \alpha_1, \ldots, \alpha_n)$, where the α_i are algebraic. It is trivially true that $H(\alpha_i) \leqslant H(P)$ and $[\mathbf{Q}(\alpha_i) : \mathbf{Q}] \leqslant [\mathbf{Q}(P) : \mathbf{Q}]$. It therefore suffices to prove that the set of algebraic numbers $\{\alpha \in \bar{\mathbf{Q}} \mid [\mathbf{Q}(\alpha) : \mathbf{Q}] \leqslant d,\ H(\alpha) \leqslant X\}$ is finite. A bound on the degree and the height gives, by Lemma 5-2.1.13, a bound on the coefficients of the minimal polynomial of α, which proves the finiteness. For the second assertion, we can again only consider points $P = (1, \alpha_1, \ldots, \alpha_n)$ where the α_i are algebraic. If $H(P) = 1$, then $|\alpha_i|_v \leqslant 1$ (for all i all v), hence this stays true for α^m. Thus the set of points $(1, \alpha_1^m, \ldots, \alpha_n^m)$ is finite, which implies that every α_i is zero or a root of unity. □

2.1.15. Lemma. *If α and β are two algebraic numbers, then*

$$\frac{1}{2} H(\alpha)H(\beta) \leqslant H(1, \alpha+\beta, \alpha\beta) \leqslant 2H(\alpha)H(\beta). \tag{5.18}$$

Proof. We can reason according to the cases $|\alpha|_v \leqslant |\beta|_v \leqslant 1$, $|\alpha|_v \leqslant 1 \leqslant |\beta|_v$ and $1 \leqslant |\alpha|_v \leqslant |\beta|_v$. Whenever v is an ultrametric absolute value, the following equality can be checked directly:

$$\max(1, |\alpha+\beta|_v, |\alpha\beta|_v) = \max(1, |\alpha|_v) \max(1, |\beta|_v).$$

For an Archimedean absolute value satisfying the triangle inequality, we obtain the bounds

$$\begin{aligned} \frac{1}{2} \max(1, |\alpha|_v) \max(1, |\beta|_v) &\leqslant \max(1, |\alpha+\beta|_v|, |\alpha\beta|_v) \\ &\leqslant 2 \max(1, |\alpha|_v) \max(1, |\beta|_v). \end{aligned}$$

The proof of the lemma follows from taking the product of these inequalities. □

2.1.16. Theorem. *Let $P_0, \ldots, P_m$ be a family of homogeneous polynomials of degree d in $x = (x_0, \ldots, x_n)$. Let Z be the location of the common zeros of the P_i and $\Phi : \mathbf{P}^n \setminus Z \to \mathbf{P}^m$ the map defined by $\Phi(x) = (P_0(x), \ldots, P_m(x))$.*

i) There exists a constant $C_1 = C_1(\Phi)$ *such that for* $x \in (\mathbf{P}^n \setminus Z)(\bar{\mathbf{Q}})$ *we have*

$$H(\Phi(x)) \leqslant C_1 H(x)^d. \tag{5.19}$$

ii) Let V *be a closed subvariety of* $\mathbf{P}^n$ *such that* $V \cap Z = \emptyset$. *Then there exist two constants* $C_1 = C_1(\Phi)$ *and* $C_2 = C_2(\Phi)$ *such that for* $x \in V(\bar{\mathbf{Q}})$,

$$C_2 H(x)^d \leqslant H(\Phi(x)) \leqslant C_1 H(x)^d. \tag{5.20}$$

Proof. The first inequality can be deduced by repeatedly applying the triangle inequality (usual and ultrametric). Let $x = (x_0, \ldots, x_n)$ and $x^i := x_0^{i_0} \cdots x_n^{i_n}$ and we call K a field of rationality of x. We write $P_i = \sum_j a_j^{(i)} x^j$ and denote by $N = \binom{n+d}{d}$ the number of monomials of degree d. Finally, let N_v be a constant such that for all $x_1, \ldots, x_N \in K$ we have:

$$|x_1 + \cdots + x_N|_v \leqslant N_v \max(|x_1|_v, \ldots, |x_N|_v).$$

Observe that we can take $N_v = 1$ for the ultrametric places and $N_v = N$ for the Archimedean places. We can therefore write, for every place v of K,

$$|P_i(x)|_v \leqslant N_v \max_j \left|a_j^{(i)}\right|_v \max_i |x_i|_v^d.$$

By setting $A_v = \max_{i,j} \left|a_j^{(i)}\right|_v$, we see that $A_v = 1$, except for a finite number of places. This gives us

$$H_K(\Phi(x)) = \prod_v \max_i |P_i(x)|_v \leqslant \prod_v N_v A_v \max_i |x_i|_v^d = \left(\prod_v N_v A_v\right) H_K(x)^d,$$

and, by taking the $[K:\mathbf{Q}]$th roots, we obtain the first inequality with $C := (\prod_v N_v A_v)^{1/[K:\mathbf{Q}]}$. For the second inequality, we rely on the Hilbert's Nullstellensatz (see Theorem B-2.1), which says, in light of the given hypotheses, that if $Q_1 = \cdots = Q_r = 0$ is a system of equations of V, there exist polynomials $A_i^{(j)}$ and $B_i^{(j)}$ and an integer $M \geqslant 1$ such that

$$X_j^M = \sum_{i=0}^m A_i^{(j)} P_i + \sum_{i=1}^r B_i^{(j)} Q_i.$$

Observe also that we can assume that the $A_i^{(j)}$ are homogeneous of degree $M - d$ and with coefficients in K. If we apply this to a point $x \in V$, we obtain

$$x_j^M = \sum_{i=0}^m A_i^{(j)}(x) P_i(x).$$

By applying the triangle inequality as before, we obtain

$$|x_j|_v^M \leqslant (m+1)_v \max_i \left|A_i^{(j)}(x)\right|_v \max_i |P_i(x)|_v \leqslant A'_v \max_i |x_i|_v^{M-d} \max_i |P_i(x)|_v,$$

with $A'_v = 1$, except for a finite number of places. This gives us

$$\max_j |x_j|_v^d \leqslant A'_v \max_i |P_i(x)|_v .$$

By taking the product over the places v and the $[K : \mathbf{Q}]$th root, we obtain the desired result. □

Notation. We set $h_K = \log H_K$ and $h = \log H$, and we call it the *logarithmic height.* With this convention, the conclusion of the inequalities in part *ii*) from the previous theorem can be rewritten as

$$h(\Phi(x)) = dh(x) + O(1).$$

We will now return to the study of elliptic curves and define a Weil height.

2.1.17. Definition. Let $E \subset \mathbf{P}^2$ be an elliptic curve given by a Weierstrass equation $Y^2Z = X^3 + aXZ^2 + bZ^3$. For $P \in E(\bar{\mathbf{Q}})$, we define the *height*[1] of P by

$$h(P) = \begin{cases} h(x(P)) & \text{if } P \neq 0_E, \\ 0 & \text{if } P = 0_E. \end{cases}$$

2.1.18. Theorem. *There exists a constant (dependent on E) such that the height over E satisfies*

$$-C_1 \leqslant h([2](P)) - 4h(P) \leqslant C_1. \tag{5.21}$$

Proof. We can ignore the case where $P = 0$ or is 2-torsion. By invoking the duplication formula (5.9), we see that if we set

$$\Phi(T, X) := (4T(X^3 + aXT^2 + bT^3, X^4 - 2aX^2T^2 - 8bXT^3 + a^2T^4),$$

then $\Phi(1, x(P)) = (1, x(2P))$. On the other hand, the polynomials $x^3 + ax + b$ and $x^4 - 2ax^2 - 8bx + a^2$ are relatively prime under the condition that $\Delta_0 = 4a^3 + 27b^2$ is non-zero. To see why this is true, a direct computation or applying the Euclidean algorithm yields the identity

$$(3x^2+4a)(x^4-2ax^2-8bx+a^2)-(3x^3-5ax-27b)(x^3+ax+b) = 4a^3+27b^2. \tag{5.22}$$

By applying Theorem 5-2.1.16 to $\Phi : \mathbf{P}^1 \to \mathbf{P}^1$ with $d = 4$, we have

$$\begin{aligned} h([2](P)) &= h(1, x(2P)) = h(\Phi(1, x(P))) \\ &= 4h(1, x(P)) + O(1) = 4h(P) + O(1). \end{aligned}$$

□

[1] We are talking about a logarithmic height; furthermore, for reasons which are unimportant in this context, this height is equal to 2 times the height commonly used.

2.1.19. Theorem. *The height over E is symmetric (i.e., $h(-P) = h(P)$) and almost satisfies the parallelogram law, in other words:*

$$h(P+Q) + h(P-Q) = 2h(P) + 2h(Q) + O(1). \tag{5.23}$$

Proof. The formula is trivially true when P or Q is zero; it is also true if $Q = \pm P$ by the previous theorem. We can therefore assume that $P, Q \in E \setminus \{0_E\}$ and $Q \neq \pm P$. Let $x_1 = x(P)$, $x_2 = x(Q)$, $x_3 = x(P+Q)$ and $x_4 = x(P-Q)$, and also $x_1 + x_2 = u$, $x_1 x_2 = v$. The formulas (5.7) and (5.8) can be rewritten as

$$\begin{cases} x_3 + x_4 & = \dfrac{2u(a+v) + 4b}{u^2 - 4v}, \\ x_3 x_4 & = \dfrac{(v-a)^2 - 4bu}{u^2 - 4v}. \end{cases}$$

Thus if we introduce the map from $\mathbf{P}^2$ to $\mathbf{P}^2$ given by

$$\Phi(T, U, V) := (U^2 - 4TV, 2U(aT + V) + 4bT^2, (aT - V)^2 - 4bTU),$$

then the three polynomials do not have any common zeros in $\mathbf{P}^2$ (the verification of this below uses the condition that $4a^3 + 27b^2 \neq 0$). By the second part of Theorem 5-2.1.16, we thus obtain

$$h(\Phi(T, U, V)) = 2h(T, U, V) + O(1).$$

Furthermore, if we let $\psi : (E \setminus \{0_E\})^2 \to \mathbf{P}^2$ be defined by $\psi(P, Q) = (1, x(P) + x(Q), x(P)x(Q))$ and $\mu(P, Q) = (P + Q, P - Q)$, we see that

$$h(\psi(P, Q)) = h(x(P)) + h(x(Q)) + O(1)$$

by Lemma 5-2.1.15 and also, by using formulas (5.7) and (5.8), that

$$\psi \circ \mu = \Phi \circ \psi.$$

This implies that

$$\begin{aligned} h(P+Q) + h(P-Q) &= h(x(P+Q)) + h(x(P-Q)) \\ &= h(1, x(P+Q) \\ &\qquad + x(P-Q), x(P+Q)x(P-Q)) + O(1) \\ &= h(\psi \circ \mu(P, Q)) + O(1) \\ &= h(\Phi(\psi(P, Q))) + O(1) \\ &= 2h(\psi(P, Q)) + O(1) \\ &= 2h(P) + 2h(Q) + O(1). \end{aligned}$$

To complete the proof, we will check that if $\Phi(T, U, V) = (0, 0, 0)$, then $T = U = V = 0$. This is immediate if $T = 0$. If $T \neq 0$, we set $x = U/2T$ and

$w = V/T$; we thus obtain $x^2 - w = 2x(a+w)+4b = 0$ and $(w-a)^2 - 8bx = 0$. By eliminating w, we find $x^4 - 2ax^2 - 8bx + a^2 = x^3 + ax + b = 0$, which is impossible according to the identity (5.22). □

2.2. Néron-Tate Heights

If C is a curve embedded in the projective space $\mathbf{P}^n$, we can define the height of a point on C as the height of the point in the projective space that contains it. The inconvenience of this definition is its nonintrinsic character. We will now offer a modification of this height which will get rid of this inconvenience.

2.2.1. Lemma. *Let S be a set and $d > 1$. If $h : S \to \mathbf{R}$ and $f : S \to S$ satisfy $|h \circ f - dh| \leqslant C$, then for all $x \in S$, the sequence $(d^{-n}h(f^n(x)))$ is a convergent sequence, which we will denote by the limit $\hat{h}_f(x)$. Furthermore, for every $x \in S$,*

$$\left|h(x) - \hat{h}_f(x)\right| \leqslant \frac{C}{d-1}, \tag{5.24}$$

$$\hat{h}_f(f(x)) = d\hat{h}_f(x). \tag{5.25}$$

Proof. By writing the inequality in the statement of the lemma at the point $f^{k-1}(x)$ and dividing by d^k, we obtain

$$-\frac{C}{d^k} \leqslant d^{-k}h(f^k(x)) - d^{-k+1}h(f^{k-1}(x)) \leqslant \frac{C}{d^k}.$$

By summing these inequalities from $n+1$ to m (with $n < m$), we can conclude that

$$-\frac{C}{d^n(d-1)} \leqslant d^{-m}h(f^m(x)) - d^{-n}h(f^n(x)) \leqslant \frac{C}{d^n(d-1)}.$$

Thus $d^{-n}h(f^n(x))$ is the general term in a Cauchy sequence, which we will denote by the limit $\hat{h}_f(x)$. By letting m tend to infinity, we thus obtain

$$-\frac{C}{d^n(d-1)} \leqslant \hat{h}_f(x) - d^{-n}h(f^n(x)) \leqslant \frac{C}{d^n(d-1)}.$$

In particular, $\frac{C}{d-1} \leqslant \hat{h}_f(x) - h(x) \leqslant \frac{C}{d-1}$. Finally,

$$\hat{h}_f(f(x)) = \lim_{n\to\infty} d^{-n}h(f^{n+1}(x)) = d \lim_{n\to\infty} d^{-n-1}h(f^{n+1}(x)) = d\hat{h}_f(x). \quad \square$$

By applying this lemma to the Weil height of an elliptic curve and to the morphism $[2] : E \to E$ (with $d = 4$), we obtain the following theorem.

2.2.2. Theorem. (Néron-Tate) *Let E be an elliptic curve defined over a number field K. We define a height, called the "canonical" or "Néron-Tate" height, by the formula*

$$\hat{h}(P) = \lim_{n\to\infty} \frac{h(x(2^n P))}{4^n}. \tag{5.26}$$

This height over E satisfies $\hat{h}(P) = h(P) + O(1)$ and also the parallelogram law:

$$\hat{h}(P+Q) + \hat{h}(P-Q) = 2\hat{h}(P) + 2\hat{h}(Q). \tag{5.27}$$

It is therefore quadratic. In particular, $\hat{h}(mP) = m^2\hat{h}(P)$. Finally, $\hat{h}(P) = 0$ if and only if P is a torsion point.

Proof. We can apply Lemma 5-2.2.1 to the height $h(P) = h(x(P))$ and to the map $P \mapsto [2](P)$ with $d = 4$. The inequality in Theorem 5-2.1.19 applied to the points $P' = [2^n](P)$ and $Q' = [2^n](Q)$ gives

$$-C \leqslant h([2^n](P+Q)) + h([2^n](P-Q)) - 2h([2^n](P]) - 2h([2^n](Q) \leqslant C.$$

By dividing by 4^n and letting n tend to infinity, we obtain the desired formula. Thus $\hat{h}$ is quadratic and in particular satisfies $\hat{h}(mP) = m^2\hat{h}(P)$. If $mP = 0$, we can immediately deduce that $\hat{h}(P) = 0$. Conversely, if $\hat{h}(P) = 0$, then for all $m \in \mathbf{Z}$, we have $\hat{h}(mP) = 0$. Therefore, the set $\{mP \mid m \in \mathbf{Z}\}$ is of bounded height and is hence finite, which implies that P is torsion. □

2.2.3. Corollary. *If an elliptic curve E is defined over a number field K, the torsion subgroup $E(K)_{\mathrm{tor}}$ is finite and the group $E(K)/E(K)_{\mathrm{tor}}$ is free abelian.*

By skipping ahead to a theorem which we will prove in the following section (the group $E(K)$ is finitely generated), we can try to specify the size of the generators of $E(K)$ in the following manner. Theorem 5-2.2.2 can be interpreted as saying that the quadratic form on the lattice $E(K)/E(K)_{\mathrm{torsion}}$ is nondegenerate. We can be even more precise and prove the following theorem.

2.2.4. Theorem. *The real quadratic form $E(K) \otimes \mathbf{R} \to \mathbf{R}$ induced by $\hat{h}$ is positive-definite.*

We should point out that the fact that a quadratic form $Q(x)$ satisfies $Q(x) > 0$ for all $x \in \mathbf{Q}^n \setminus \{0\}$ does not imply that it is positive-*definite* (consider $Q(x_1, x_2) = (x_1 + x_2\sqrt{2})^2$).

Proof. Let Q be the quadratic form on $\mathbf{R}^r$ gotten from $\hat{h}$ by tensoring with $\mathbf{R}$. It is clearly positive. If it were degenerate, it could then be written, after a base change, as $Q(x_1,\ldots,x_r) = x_1^2 + \cdots + x_s^2$ where $s < r$. The sets $\{x \in \mathbf{R}^r \,; Q(x) \leqslant \epsilon\}$ would be, for every $\epsilon > 0$, cylinders with infinite volume and would therefore contain, by Minkowski's theorem (3-5.3), a non-zero point in every lattice. This would contradict the fact that the set $\{P \in E(K) \mid \hat{h}(P) \leqslant \epsilon\}$ is reduced, for small enough ϵ, to the torsion subgroup. □

There is also a *scalar product* on $E(K) \otimes \mathbf{R}$ defined by

$$\langle P, Q\rangle := \frac{1}{2}\left(\hat{h}(P+Q) - \hat{h}(P) - \hat{h}(Q)\right).$$

2.2.5. Definition. Let $P_1,\ldots,P_r$ be a basis for $E(K)$ modulo the finite torsion subgroup F. We define the *regulator* of E by

$$\operatorname{Reg}(E/K) := \det\left(\langle P_i, P_j\rangle\right)_{1\leqslant i,j\leqslant r},$$

and we define the minimal height of a point of infinite order by

$$\hat{h}_{\min}(E/K) := \min_{P\in E(K)\setminus F} \hat{h}(P).$$

These two quantities are exactly the necessary quantities needed to bound the height of possible generators of the Mordell-Weil group $E(K)$, in light of the following result coming from the geometry of numbers and due to Hermite (see Exercise 3-6.6).

2.2.6. Proposition. *There exist constants C_r such that for every lattice L in $\mathbf{R}^r$, equipped with the Euclidean norm, there exists a basis for L, $e_1,\ldots,e_r$, such that*

$$\det(L) \leqslant \|e_1\|\cdots\|e_r\| \leqslant C_r \det(L).$$

3. The Mordell-Weil Theorem

The goal of this section is to prove the following theorem.

3.1. Theorem. (Mordell-Weil) *Let E be an elliptic curve defined over a number field K (for example $K = \mathbf{Q}$). Then the group $E(K)$ is finitely generated.*

We could of course reinterpret this theorem by saying that all of the rational points on the curve can be found starting with a finite set of points and applying the chord-tangent method to these. An important intermediate step is the following result.

3.2. Theorem. ("Weak" Mordell-Weil) *Let E be an elliptic curve defined over a number field K (for example $K = \mathbf{Q}$). Then the group $E(K)/2E(K)$ is finite.*

Actually, the "weak" Mordell-Weil theorem, together with the theory of heights from the previous section implies Theorem 5-3.1, thanks to the following descent lemma.

3.3. Lemma. *Let G be an abelian group endowed with a quadratic form $q : G \to \mathbf{R}$. Suppose that the sets $\{x \in G \mid q(x) \leqslant X\}$ are finite for all $X \in \mathbf{R}$ and that the quotient $G/2G$ is finite. Then the group G is finitely generated. More precisely, if S is a set of representations modulo $2G$ and if $C := \max_{x \in S} q(x)$, then $\{x \in G \mid q(x) \leqslant C\}$ generates G.*

Proof. Let us first point out that the hypotheses imply that $q(x) \geqslant 0$: if there existed x_0 where $q(x_0) < 0$, we would have, by homogeneity, infinitely many elements where $q(x) < 0$. Let $|x| := \sqrt{q(x)}$, so that $|mx| = m|x|$ and $|x + y| \leqslant |x| + |y|$ (for $x, y \in G$ and $m \in \mathbf{N}$). Let $x \in G$ where $q(x) > C$. We can define a sequence (x_n) of points in G as follows: start with $x_0 = x$, then write $x_0 = y_1 + 2x_1$ where $y_1 \in S$ and $x_1 \in G$, then $x_1 = y_2 + 2y_2$, etc. Observe that

$$|x_1| = \frac{|x_0 - y_1|}{2} \leqslant \frac{|x_0| + |y_1|}{2} \leqslant \frac{|x_0| + \sqrt{C}}{2} < |x_0|.$$

We can iterate this procedure and obtain a sequence which satisfies

$$|x_n| < |x_{n-1}| < \cdots < |x_1| < |x_0|,$$

as long as $|x_n| > \sqrt{C}$. The finiteness hypothesis implies that, after a finite number steps, we will have $|x_n| \leqslant \sqrt{C}$. The point $x = x_0$ can be expressed as a linear combination of the y_i and the x_n, which are all in the finite set $\{y \in G \mid q(y) \leqslant C\}$, so this set indeed generates G. □

We will now lay out a plan for the proof of Theorem 5-3.2. To make things simpler, we will assume that the equation of the curve is given by:

$$y^2 = f(x) = x^3 + ax + b = (x - \alpha_1)(x - \alpha_2)(x - \alpha_3),$$

in other words, we will assume that the roots of f are rational over K. In particular, the 2-torsion points, $P_i = (\alpha_i, 0)$, are in $E(K)$. This does not interfere with the generality of the proof of the Mordell-Weil theorem since we can always replace K by the extension $K(\alpha_1, \alpha_2, \alpha_3)$. However, from an algorithmic point of view, it is better to work in K. At the end of the section, we will indicate how the proof should be modified to this effect.

3.4. Definition. We define the map $\psi = (\psi_1, \psi_2, \psi_3)$ from $E(K)$ to $\left(K^*/K^{*2}\right)^3$ by the following formulas:

$$\psi_i(P) = \begin{cases} x(P) - \alpha_i & \text{if } P \neq P_i, 0_E, \\ (\alpha_i - \alpha_j)(\alpha_i - \alpha_k) & \text{if } P = P_i, \\ 1 & \text{if } P = 0_E. \end{cases}$$

3.5. Remark. In the definition of the *homomorphism* ψ, the formula for $P = P_i = (\alpha_i, 0)$ is natural since $(x - \alpha_i)K^{*2} = (x - \alpha_j)(x - \alpha_j)K^{*2}$. Another possible definition would be to take $\psi_i(P_i) = f'(\alpha_i) \bmod K^{*2}$.

By proving the following three lemmas, we will have finished the proof of Theorem 5-3.2 since we can deduce from them that $E(K)/2E(K) \cong \psi(E(K))$.

3.6. Lemma. *The map $\psi : E(K) \to \left(K^*/K^{*2}\right)^3$ is a homomorphism.*

3.7. Lemma. *The kernel of the map ψ is equal to $2E(K)$.*

3.8. Lemma. *The image $\psi(E(K))$ in $\left(K^*/K^{*2}\right)^3$ is finite.*

Proof. (of Lemma 5-3.6) If P, Q and R are three points on the curve E, the equality $P+Q+R = 0_E$ is equivalent to saying that P, Q and R are colinear. Therefore, let $y = \lambda x + \mu$ be the equation of the line D which intersects E at P, Q and R. First assume that $\{P, Q, R\} \cap \{0_E, P_1, P_2, P_3\} = \emptyset$. The equation $f(x) - (\lambda x + \mu)^2 = 0$ therefore has $x(P)$, $x(Q)$ and $x(R)$ as roots. If we set $x' = x - \alpha_i$, then

$$f(x' + \alpha_i) - (\lambda x' + \lambda\alpha_i + \mu)^2 = 0$$

has $x(P) - \alpha_i$, $x(Q) - \alpha_i$ and $x(R) - \alpha_i$ as solutions, and since $f(\alpha_i) = 0$, the constant term is $-(\lambda\alpha_i + \mu)^2$. This yields

$$(x(P) - \alpha_i)(x(Q) - \alpha_i)(x(R) - \alpha_i) = (\lambda\alpha_i + \mu)^2,$$

and thus $\psi_i(P)\psi_i(Q)\psi_i(R) = 1$. The equality $R = P + Q$ implies that P, Q and $-R$ are colinear, in other words $\psi_i(P)\psi_i(Q)\psi_i(-R) = 1$; since $\psi_i(-R) = \psi_i(R) = \psi_i(R)^{-1}$, we indeed obtain $\psi_i(R) = \psi_i(P)\psi_i(Q)$. This finishes the proof if $\{P, Q, R\} \cap \{0_E, P_1, P_2, P_3\} = \emptyset$. If $R = 0_E$, the relation becomes obvious. If not, observe that $(x(P) - \alpha_1)(x(P) - \alpha_2)(x(P) - \alpha_3) =$

$y(P)^2$; we can check case by case that the relation $\psi_i(P+Q) = \psi_i(P)\psi_i(Q)$ is always true. □

Proof. (of Lemma 5-3.7) It is clear that $2E(K) \subset \operatorname{Ker}\psi$, since the exponent of K^*/K^{*2} is 2. We need to show that $\cap_i \operatorname{Ker}\psi_i \subset 2E(K)$. Assume then that

$$\text{for } i = 1, 2, 3,\ \exists z_i \in K^* \text{ such that } x(P) - \alpha_i = z_i^2. \tag{5.28}$$

We will solve the Vandermonde linear system of equations $u + v\alpha_i + w\alpha_i^2 = z_i$. From the equations $(u + v\alpha_i + w\alpha_i^2)^2 = x - \alpha_i$, we obtain the system

$$\begin{cases} u^2 - 2vwb - x = 0, \\ 2uv - 2vwa - bw^2 + 1 = 0, \\ v^2 + 2uw - aw^2 = 0, \end{cases} \tag{5.29}$$

which in particular yields $v^3 + vw^2a + bw^3 - w = 0$ and also (by noticing that w must be non-zero because if not, then $v = 0$ and $1 = 0$!) the following equation:

$$\left(\frac{v}{w}\right)^3 + a\left(\frac{v}{w}\right) + b = \left(\frac{1}{w}\right)^2.$$

Therefore, $Q := \left(\frac{v}{w}, \frac{1}{w}\right) \in E(K)$. A direct computation using the duplication formula (5.9) and the relations (5.29) therefore gives us $P = 2Q$. This is because

$$\begin{aligned} x(2Q) &= \frac{\left(\frac{v}{w}\right)^4 - 2a\left(\frac{v}{w}\right)^2 - 8b\left(\frac{v}{w}\right) + a^2}{4\left(\left(\frac{v}{w}\right)^3 + a\left(\frac{v}{w}\right) + b\right)} \\ &= \frac{v^4 - 2av^2w^2 - 8bvw^3 + a^2w^4}{4w^2} \\ &= \frac{(aw^2 - 2uw)^2}{4w^2} + \frac{1}{4}(-2av^2 - 8bvw + aw^2) \\ &= u^2 - 2vwb - \frac{a}{2}(v^2 - aw^2 + 2uw) \\ &= x. \end{aligned}$$

□

Proof. (of Lemma 5-3.8) Choose a finite set S of places of the field K such that

i) the element $2\Delta_E = 2(4a^3 + 27b^2)$ is an S-unit,
ii) the ring $\mathcal{O}_{K,S}$ is principal.

This is possible because of Remark 3-5.14. Condition i) implies that $\alpha_i - \alpha_j \in \mathcal{O}_{K,S}^*$ since $\Delta = ((\alpha_1 - \alpha_2)(\alpha_1 - \alpha_3)(\alpha_2 - \alpha_3))^2$. We can now write $x = A/B$ and $y = C/D$ where $A, B, C, D \in \mathcal{O}_{K,S}$ and $\gcd(A, B) = \gcd(C, D) = 1$ (in the ring $\mathcal{O}_{K,S}$). The equation $y^2 = (x - \alpha_1)(x - \alpha_2)(x -$

$\alpha_3)$ can be transformed into $C^2B^3 = D^2(A-\alpha_1 B)(A-\alpha_2 B)(A-\alpha_3 B)$. Since D is relatively prime to C, we know that D^2 divides B^3, and since B is relatively prime to A, we know that B^3 divides D^2. Hence up to modifying B and D by a unit, we can assume that $B^3 = D^2$, $B = E^2$ and $D = E^3$, which yields

$$(x,y) = \left(\frac{A}{E^2}, \frac{C}{E^3}\right) \qquad \text{and} \qquad C^2 = (A-\alpha_1 E^2)(A-\alpha_2 E^2)(A-\alpha_3 E^2).$$

If p (a prime in $\mathcal{O}_{K,S}$) divides $(A-\alpha_1 E^2)$ and $(A-\alpha_2 E^2)$, then it also divides $(\alpha_1-\alpha_2)E^2$ and $(\alpha_1-\alpha_2)A$, hence $(\alpha_1-\alpha_2)$, which is invertible. The factors are relatively prime and are therefore squares, up to a unit. Thus we obtain

$$x(P) - \alpha_i = \frac{A-\alpha_i E^2}{E^2} = \epsilon_i t_i^2,$$

where $\epsilon_i \in \mathcal{O}_{K,S}^*$. As a corollary to the generalized unit theorem, we have that $\mathcal{O}_{K,S}^*/\mathcal{O}_{K,S}^{*2}$ is finite, and we can therefore choose the ϵ_i from a finite set. Thus, $\psi(P) = (\epsilon_1, \epsilon_2, \epsilon_3)$ takes a finite number of possible values in $(K^*/K^{*2})^3$. □

3.9. Remark. To make the proof of the Mordell-Weil theorem effective computationally, it suffices to find representatives of $E(K)/2E(K)$. The proof indicates that it thus suffices to decide, for $(\epsilon_1,\epsilon_2,\epsilon_3) \in (\mathcal{O}_{K,S}^*/\mathcal{O}_{K,S}^{*2})^3$, if the curve defined by the equations $A-\alpha_i E^2 = \epsilon_i Z_i^2$ has a rational point and then compute it. Unfortunately, no such algorithm is currently known.

We will finish this section by briefly indicating the modifications necessary for working with a curve $y^2 = f(x)$ without leaving the field K of coefficients of the polynomial f. We introduce the ring $A := K[X]/(f(X))$. By letting α be the image of X, we set $\psi(P) = x(P) - \alpha$ with values in $G := A^*/A^{*2}$ if $x(P)$ is not a root of $f(X)$. For the particular case of 2-torsion points, we proceed as in Definition 5-3.4.

4. Siegel's Theorem

We are now interested in integer solutions. The main result is the following.

4.1. Theorem. (Siegel) *Let C be an affine curve given by the equation*

$$y^2 = f(x) = x^3 + ax + b,$$

where $a, b \in \mathcal{O}_K$ and $\Delta := 4a^3 + 27b^2 \neq 0$. Then the set of points $P = (x,y)$ on the curve, where $x, y \in \mathcal{O}_K$, is finite.

4.2. Remark. The smoothness condition is necessary because, for example, the curve $y^2 = x^3$ has every pair (t^2, t^3) as a solution (where $t \in \mathcal{O}_K$), while the curve $y^2 = x^3 - x^2$ has every pair (t^2+1, t^3+t) as a solution (where $t \in \mathcal{O}_K$). We could deduce from the previous theorem (but will not prove it) an apparently more general theorem, namely the following.

4.3. Theorem. (Siegel) *Let C be an affine curve given by the equation*

$$ax^3 + bx^2y + cxy^2 + dy^3 + ex^2 + fxy + gy^2 + hx + iy + j = 0$$

such that the corresponding projective curve is smooth. Then the set of points $P = (x, y)$ on the curve, with $x, y \in \mathcal{O}_{K,S}$, is finite.

We will now deduce Siegel's theorem (Theorem 5-4.1) from the following result, also due to Siegel.

4.4. Theorem. (The S-unit equation) *Let K be a number field and S a finite set of places. The set of pairs of S-units $(x, y) \in (\mathcal{O}_{K,S}^*)^2$ which satisfy*

$$x + y = 1 \tag{5.30}$$

is finite.

Proof. (Theorem 5-4.4 implies Theorem 5-4.1) We can, if we want to, expand the set S and the field K. We will therefore assume that $\mathcal{O}_{K,S}$ is principal, $\Delta \in \mathcal{O}_{K,S}^*$ and $f(x) = (x-\alpha_1)(x-\alpha_2)(x-\alpha_3)$. Then let $(x, y) \in (\mathcal{O}_{K,S})^2$ be an integer solution. As in the proof of the Mordell-Weil theorem, we deduce from this the factorization:

$$x - \alpha_i = b_i z_i^2,$$

where b_i are representatives of $\mathcal{O}_{K,S}^*/\mathcal{O}_{K,S}^{*2}$. We will introduce the algebraic numbers $\beta_i = \sqrt{b_i}$, which are in a finite extension K' of K. From $x - \alpha_i = (\beta_i z_i)^2$, we can deduce the relations

$$\alpha_i - \alpha_j = (\beta_i z_i - \beta_j z_j)(\beta_i z_i + \beta_j z_j) \in \mathcal{O}_{K,S}^*.$$

We will now make use of the "Siegel identities":

$$\frac{\beta_i z_i \pm \beta_j z_j}{\beta_i z_i - \beta_k z_k} \mp \frac{\beta_j z_j \pm \beta_k z_k}{\beta_i z_i - \beta_k z_k} = 1. \tag{5.31}$$

We know from Theorem 5-4.4 that the set of values taken by $\epsilon := \dfrac{\beta_i z_i \pm \beta_j z_j}{\beta_i z_i - \beta_k z_k}$ is finite. It easily follows that there are only a finite number of values $\beta_i z_i$ and likewise of values for x and hence for y. □

4.5. Remark. Observe that the reduction of Theorem 5-4.1 to Theorem 5-4.4 is computationally effective in the following sense: if we had an algorithm for calculating solutions of the equation of S-units, we would have an algorithm for calculating the set of integer solutions to $y^2 = f(x)$.

We essentially know two proofs of Theorem 5-4.4: one due to Siegel, based on a rational approximation theorem, and one due to Baker, based on his theorem of linear forms of logarithms. The disadvantage of Siegel's proof is that it does not explicitly determine the finite set of solutions. This is nevertheless the one that we will present. For a sketch of Baker's proof, see the Chap. 6, Sect. 6-4.

Reduction of Theorem 5-4.1 to Theorem 5-4.4. Let $m \geqslant 2$. We know from the generalized unit theorem that the group $\mathcal{O}_{K,S}^*/\mathcal{O}_{K,S}^{*m}$ is finite. In other words, there exists a finite set of S-units ϵ_i such that all S-units can be written as $x = \epsilon_i z^m$ with $z \in \mathcal{O}_{K,S}^*$. Solutions (x, y) of the S-unit equation thus provide (a finite number of) solutions to one of the following equations:

$$\epsilon_1 z_1^m + \epsilon_2 z_2^m = 1, \tag{5.32}$$

and it suffices to prove that the latter only have a finite number of solutions $(z_1, z_2) \in (\mathcal{O}_{K,S}^*)^2$ or likewise in $(\mathcal{O}_{K,S})^2$.

4.6. Proposition. *Let $a, b \in \mathcal{O}_{K,S}$ and $m \geqslant 3$. The set of S-integral solutions of the equation*

$$ax^m + by^m = 1$$

is finite.

Proof. We will give the proof for the case which has the simplest notation, the case $\mathcal{O}_{K,S} = \mathbf{Z}$. Let $\alpha = \sqrt[m]{-\frac{b}{a}}$, and let $[\mathbf{Q}(\alpha) : \mathbf{Q}] = d \leqslant m$. We factor $aX^m + b = (X - \alpha)F(X)$ and thus obtain

$$\left(\frac{x}{y}\right)^m + \frac{b}{a} = \left(\frac{x}{y} - \alpha\right) a^{-1} F\left(\frac{x}{y}\right) = \frac{1}{ay^m}.$$

Observe that the ratio x/y must be close to one of the roots, for example to α. Since it must lie at a distance which is bounded below from the other roots (those of F), we get an inequality of the form:

$$\left|\frac{x}{y} - \alpha\right| \leqslant \frac{C_1}{|y|^m}, \tag{5.33}$$

where the constant C_1 only depends on α. To finish the proof, it suffices to have a Diophantine approximation statement of the type

$$\forall \frac{x}{y} \in \mathbf{Q}, \quad \frac{C_2}{|y|^\delta} \leqslant \left|\frac{x}{y} - \alpha\right|, \tag{5.34}$$

where C_2 is dependent on α and δ, and of course $\delta < m$. In fact, by combining inequalities (5.33) and (5.34), we obtain

$$|y| \leqslant \left(\frac{C_1}{C_2}\right)^{\frac{1}{m-\delta}}.$$

A statement of the same type as inequality (5.34) is provided by Roth's theorem, which allows us to choose any $\delta > 2$ and hence a $\delta < 3 \leqslant m$. An older result of Thue allows us to choose every $\delta > 1 + \frac{[\mathbf{Q}(\alpha):\mathbf{Q}]}{2}$, which finishes the proof (observe that if $m > 2$, then $1 + \frac{m}{2} < m$). The proof of Thue's theorem is given in the following chapter. Let us nonetheless point out that the proof of Roth's theorem, like that of Thue's theorem, is not constructive in the sense that it does not allow us to calculate the constant $C_2 = C_2(\alpha, \delta)$. A more constructive method was developed in the 1960's by Baker and will be briefly discussed in the next chapter. □

5. Elliptic Curves over the Complex Numbers

In this section, we will describe the connection to the classical theory of elliptic functions, thus justifying the name "elliptic curves", elliptic functions taking their name from the fact that they intervene in the calculation of the length of an arc of an ellipse.

We will need to following classical result on complex variables.

5.1. Theorem. (Liouville) *If a function is entire (i.e., holomorphic on all of $\mathbf{C}$) and bounded, then it is constant.*

We will now consider $\Omega := \mathbf{Z}\omega_1 \oplus \mathbf{Z}\omega_2$, a lattice in $\mathbf{C}$ and study the Ω-periodic functions, i.e., such that $f(z+\omega) = f(z)$ for $\omega \in \Omega$. Liouville's theorem indicates right away that the only entire functions which are Ω-periodic are constant functions, a fact which justifies the following definition.

5.2. Definition. An *elliptic* function is a meromorphic function on $\mathbf{C}$ which is Ω-periodic for some lattice Ω.

Let us point out that the set of Ω-elliptic functions forms a field, which is denoted by $\mathscr{M}(\Omega)$. This field contains the constants, i.e., the field $\mathbf{C}$, and is stable under derivation. We can see right away that this field is not reduced to only constants.

5.3. Definition. Let $\Omega := \mathbf{Z}\omega_1 \oplus \mathbf{Z}\omega_2$ be a lattice in $\mathbf{C}$. We define the *Weierstrass function* associated to Ω by the formula

$$\wp(z) = \wp(z;\Omega) = \frac{1}{z^2} + \sum_{\omega\in\Omega}{}' \left(\frac{1}{(z-\omega)^2} - \frac{1}{\omega^2} \right), \tag{5.35}$$

where $\sum'$ signifies that we leave out $\omega = 0$ in the sum.

The Weierstrass function allows us to give a complete description of elliptic functions and to establish a connection to elliptic curves.

5.4. Theorem. *The Weierstrass function $\wp$ is an elliptic function. The field of Ω-elliptic functions is generated by $\wp$ and its derivative $\wp'$, i.e., $\mathscr{M}(\Omega) = \mathbf{C}(\wp, \wp')$. Furthermore, these two elliptic functions satisfy the following algebraic relation:*

$$\wp'(z)^2 = 4\wp(z)^3 - g_2\wp(z) - g_3, \tag{5.36}$$

where the constants g_2 and g_3 are defined by

$$g_2 = g_2(\Omega) = 60 \sum_{\omega\in\Omega}{}' \frac{1}{\omega^4} \qquad \text{and} \qquad g_3 = g_3(\Omega) = 140 \sum_{\omega\in\Omega}{}' \frac{1}{\omega^6}.$$

Finally, we have $g_2^3 - 27g_3^2 \neq 0$.

Proof. (Sketch) The defining series of the derivative,

$$\wp'(z) = -2 \sum_{\omega\in\Omega} \frac{1}{(z-\omega)^3},$$

is absolutely convergent and uniformly convergent on every compact set which avoids Ω: it therefore defines a holomorphic function on $\mathbf{C} \setminus \Omega$, which is clearly Ω-periodic and odd. Furthermore, $\wp'$ has a pole of order 3 at every point of Ω, and thus $\wp' \in \mathscr{M}(\Omega)$. The defining series of $\wp$ shows that it is even and meromorphic with a double pole at every $\omega \in \Omega$. The periodicity of $\wp'$ implies that $\wp(z+\omega) = \wp(z) + C_\omega$. Let ω be one of the generators of Ω such that $\omega/2 \notin \Omega$. By taking $z := -\omega/2$, we obtain $\wp(-\omega/2) = \wp(\omega/2) = \wp(-\omega/2) + C_\omega$, hence $C_\omega = 0$. Thus we also know that $\wp \in \mathscr{M}(\Omega)$. In order to prove that $\mathscr{M}(\Omega) = \mathbf{C}(\wp, \wp')$, we can decompose a function in $\mathscr{M}(\Omega)$ into an even + an odd function and have thus reduced to showing that a function f which is Ω-elliptic and even is in $\mathbf{C}(\wp)$. To do this, we prove that its poles and zeros are symmetric under the map $z \mapsto -z$ and of even order in the periods and half-periods. Thus a function f has the same zeros and poles as a function of the type $\prod_i(\wp(z) - \wp(u_i))^{m_i}$: the two functions coincide up to a constant. In order to prove the relation of algebraic dependence, we compute the Taylor expansion of $\wp(z)$ (or rather of $\wp(z) - z^{-2}$) at $z = 0$:

$$\wp(z) = \frac{1}{z^2} + \sum_{n=1}^{\infty} a_n z^n, \qquad \text{where} \quad a_n = (n+1)\sum_{\omega\in\Omega}{}' \frac{1}{\omega^{n+2}}. \tag{5.37}$$

The calculation of the Taylor expansion (only the polar part and the constant term) of the function $\psi(z) = \wp'(z)^2 - 4\wp(z)^3 + g_2\wp(z) + g_3$ shows that it is holomorphic and zero at 0. Liouville's theorem thus implies that the function $\psi(z)$ is identically zero.

Finally, the equality $g_2^3 - 27g_3^2 = 0$ is equivalent to the fact that $4x^3 - g_2x - g_3 = 0$ has a double root, say h. If that were the case, then we would have an equation of the form $\left(\frac{\wp'(z)}{2(\wp(z)-h)}\right)^2 = \wp(z) + 2h$, which is a contradiction considering the zeros. □

5.5. Corollary. *Let Ω be a lattice in $\mathbf{C}$. The map $z \mapsto (\wp(z), \wp'(z), 1)$ extended by $\omega \mapsto (0,1,0)$ defines a biholomorphic map from $\mathbf{C}/\Omega$ to the projective cubic with points (X,Y,T) (in $\mathbf{P}^2$) given by the equation*

$$TY^2 = 4X^3 - g_2XT^2 - g_3T^3.$$

Furthermore, the map is an isomorphism of groups.

Proof. The first assertion follows essentially from the previous theorem. The second assertion can be proven by comparing the algebraic addition law to the following addition formula on the Weierstrass function:

$$\wp(u+v) = -\wp(u) - \wp(v) + \frac{1}{4}\left(\frac{\wp'(u) - \wp'(v)}{\wp(u) - \wp(v)}\right)^2. \tag{5.38}$$

For a fixed v, the poles of the left-hand side are double poles at every $u \in -v + \Omega$. The right hand side actually has the same poles because $\dfrac{\wp'(u) - \wp'(v)}{\wp(u) - \wp(v)}$ has a simple pole for $u \in \Omega$, which is compensated by the term $-\wp(u)$ and since $\wp(u) - \wp(v) = 0$ if and only if $u \pm v \in \Omega$, but $\wp'(u) - \wp'(v)$ vanishes for $u \in v + \Omega$. Once we have checked the equality of terms corresponding to the poles, formula (5.38) follows. □

Conversely, we can show, given $g_2, g_3 \in \mathbf{C}$ which satisfy $\Delta := g_2^3 - 27g_3^2 \neq 0$, that there exists a lattice Ω such that $g_2 = g_2(\Omega)$ and $g_3 = g_3(\Omega)$. Thus, over the field of complex numbers, we can consider an elliptic curve as a complex torus, i.e., a quotient $\mathbf{C}/\Omega$. This point of view clearly shows various properties of elliptic curves or of families of elliptic curves. The following two propositions illustrate this principle.

5.6. Proposition. *Let $E = \mathbf{C}/\Omega$ be an elliptic curve. Then*

$$\operatorname{Ker}[N]_E = \frac{1}{N}\Omega/\Omega \cong (\mathbf{Z}/N\mathbf{Z})^2. \tag{5.39}$$

Proof. To prove this, the map $[N]_E : \mathbf{C}/\Omega \to \mathbf{C}/\Omega$ is induced by the multiplication by N in $\mathbf{C}$, hence $\operatorname{Ker}[N]_E = \{z \in \mathbf{C} \mid Nz \in \Omega\}/\Omega$. Since $\Omega \cong \mathbf{Z}^2$, the proposition is clearly true. □

5.7. Remark. We can observe that the torsion points allow us to partially reconstruct the lattice Ω. To be more precise, we easily see that for every prime number ℓ, we have[2]

$$\varprojlim_n \operatorname{Ker}[\ell^n] = \varprojlim_n \Omega/\ell^n\Omega \cong \left(\varprojlim_n \mathbf{Z}/\ell^n\mathbf{Z}\right)^2.$$

We thus introduce the ring $\mathbf{Z}_\ell := \varprojlim_n \mathbf{Z}/\ell^n\mathbf{Z}$, and we see that

$$\varprojlim_n \operatorname{Ker}[\ell^n] = \Omega \otimes \mathbf{Z}_\ell.$$

This remark might appear to be pedantic when we are working with elliptic curves over $\mathbf{C}$, but it becomes fundamental when we want to work over other fields (for example a finite field) since the left-hand side is still meaningful, whereas the right-hand side (say Ω) does not exist anymore. The following definition will clarify things.

5.8. Definition. Let E be an elliptic curve defined over a field K and ℓ a prime number other than the characteristic of K. The ℓ-adic *Tate module* of an elliptic curve is defined to be

$$T_\ell(E) := \varprojlim_n E[\ell^n].$$

(Here, $E[\ell^n]$ is the subgroup of points with coordinates in the algebraic closure, $\bar{K}$, which are killed by ℓ^n.)

5.9. Remark. We point out that if $u \in \mathbf{C}^*$, then $\mathbf{C}/\Omega$ is isomorphic to

[2] Recall that if $(\phi_n : G_{n+1} \to G_n)$ is a sequence of homomorphisms, the projective limit, $\varprojlim_n G_n$, is defined to be the set of sequences $(x_n)_{n\geqslant 1}$, where $x_n \in G_n$, which satisfy $\phi_n(x_{n+1}) = x_n$. It is obviously a group.

$\mathbf{C}/u\Omega$ (by multiplication by u). Moreover, we can easily check that

$$u^2\wp(uz, u\Omega) = \wp(z, \Omega) \quad \text{and} \quad u^3\wp'(uz, u\Omega) = \wp'(z, \Omega).$$

Thus we can always, up to isomorphism, replace the lattice $\mathbf{Z}\omega_1 \oplus \mathbf{Z}\omega_2$ by the similar lattice $\mathbf{Z} \oplus \mathbf{Z}\tau$, where we set $\tau := \frac{\omega_2}{\omega_1}\cdot$ Up to exchanging ω_1 and ω_2, we can also assume that $\mathrm{Im}(\tau) > 0$. Every elliptic curve (over $\mathbf{C}$) is thus isomorphic to a torus $\mathbf{C}/(\mathbf{Z} \oplus \mathbf{Z}\tau)$, where τ is in the *Poincaré half-plane*,

$$\mathscr{H} := \{\tau \in \mathbf{C} \mid \mathrm{Im}(\tau) > 0\}.$$

The following result specifies when two such curves are isomorphic.

5.10. Proposition. *Two tori* $E_\tau = \mathbf{C}/(\mathbf{Z} \oplus \mathbf{Z}\tau)$ *and* $E_{\tau'} = \mathbf{C}/(\mathbf{Z} \oplus \mathbf{Z}\tau')$ *are isomorphic if and only if there exists* $\begin{pmatrix} a & b \\ c & d \end{pmatrix} \in \mathrm{SL}(2, \mathbf{Z})$ *such that*

$$\tau' = \frac{a\tau + b}{c\tau + d}\cdot$$

In particular, we can identify the space of isomorphism classes of complex elliptic curves with the space $\mathrm{SL}(2, \mathbf{Z})\backslash\mathscr{H}$.

Proof. A homomorphism $\phi : \mathbf{C}/(\mathbf{Z} \oplus \mathbf{Z}\tau') \to \mathbf{C}/(\mathbf{Z} \oplus \mathbf{Z}\tau)$ comes from a homomorphism from $\mathbf{C}$ to $\mathbf{C}$, i.e., by multiplication by $\alpha \in \mathbf{C}$ such that $\alpha(\mathbf{Z} \oplus \mathbf{Z}\tau') \subset \mathbf{Z} \oplus \mathbf{Z}\tau$. In particular, $\alpha = c\tau + d$ (where $c, d \in \mathbf{Z}$) and $\alpha\tau' = a\tau + b$ (where $a, b \in \mathbf{Z}$). Therefore, $\tau' = (a\tau + b)/(c\tau + d)$. The fact that ϕ is an isomorphism translates into $\begin{pmatrix} a & b \\ c & d \end{pmatrix} \in \mathrm{GL}(2, \mathbf{Z})$, and since $\mathrm{Im}(\tau') = \det\begin{pmatrix} a & b \\ c & d \end{pmatrix}\mathrm{Im}(\tau)/|c\tau + d|^2$, we see that the matrix is indeed in $\mathrm{SL}(2, \mathbf{Z})$. □

5.11. Proposition. *Let* $E = \mathbf{C}/\Omega$ *be an elliptic curve, where* $\Omega = \mathbf{Z} + \mathbf{Z}\tau$. *Then*

$$\mathrm{End}(E) = \{\alpha \in \mathbf{C} \mid \alpha\Omega \subset \Omega\} = \begin{cases} \mathbf{Z} & \text{if } [\mathbf{Q}(\tau) : \mathbf{Q}] > 2, \\ \mathbf{Z} + \mathbf{Z}A\tau & \text{if } [\mathbf{Q}(\tau) : \mathbf{Q}] = 2, \end{cases} \tag{5.40}$$

where, in the second case, the integer A *is the leading coefficient of the minimal equation* $A\tau^2 + B\tau + C = 0$.

In the case where $\mathrm{End}(E)$ is a subring of finite index of the ring of integers of an imaginary quadratic field, we say that E has "*complex multiplication*", or (in algebra) is of "*CM-type*".

Proof. In light of the previous discussion, an endomorphism is given by multiplication by $\alpha = c\tau + d$ and corresponds to a matrix $\begin{pmatrix} a & b \\ c & d \end{pmatrix}$ with integer coefficients such that $\tau = (a\tau+b)/(c\tau+d)$, in other words $c\tau^2+(d-a)\tau - b = 0$. If $[\mathbf{Q}(\tau) : \mathbf{Q}] > 2$, the only possibility is to have $c = b = 0$ and $a = d$, in other words, multiplication by d; if τ is quadratic and satisfies the minimal equation $A\tau^2 + B\tau + C = 0$, we can conclude that $c = mA$, $d - a = mB$ and $-b = mC$, hence $\alpha = mA\tau + d \in \mathbf{Z} + \mathbf{Z}A\tau$. □

5.12. Remark. If we consider $\mathrm{End}(E)$ as a subring of $\mathbf{C}$, we can easily verify the following formula:

$$\deg(\alpha) := \mathrm{card}\,\mathrm{Ker}(\alpha) = \mathrm{N}(\alpha) = \alpha\bar{\alpha}. \tag{5.41}$$

In particular, the map $\deg : \mathrm{End}(E) :\to \mathbf{Z}$ is quadratic. Furthermore, the ring $\mathrm{End}(E)$ acts naturally on the Tate module

$$T_\ell(E) := \lim_{\leftarrow} E[\ell^n].$$

6. Elliptic Curves over a Finite Field

In this section, we will translate some of results from the previous section into results on fields of characteristic p, notably to finite fields. See Silverman's book [70] for the complete proofs. Elliptic curves over finite fields are especially useful in cryptography—see for example the text [15].

Let E be a projective plane curve, as in (5.2), given by the equation

$$Y^2Z + a_1XYZ + a_3YZ^2 = X^3 + a_2X^2Z + a_4XZ^2 + a_6Z^3,$$

where $a_i \in \mathbf{F}_q$. The group of rational points $E(\mathbf{F}_q)$ is obviously finite, and in particular, all of the points are torsion. Any such curve also has the maps given by "multiplication by an integer n", but it has a remarkable endomorphism, specific to the characteristic p, as well.

6.1. Definition. The "Frobenius" endomorphism on $E/\mathbf{F}_q$ is defined by the formula

$$\Phi_q(x, y) = (x^q, y^q).$$

Let us point out that if $f(x, y) = 0$ (where $f(X, Y) \in \mathbf{F}_q[X, Y]$), then $f(x^q, y^q) = (f(x, y))^q = 0$, and hence Φ_q is indeed an endomorphism (it is clear that it will also respect the addition law).

We will admit the following proposition (see for example [70]), which is analogous to Proposition 5-5.6.

6.2. Proposition. *Let E be an elliptic curve defined over a finite field $\mathbf{F}_q$ and N an integer $\geqslant 2$. We denote by* $\mathrm{Ker}[N]_E = \{P \in E(\bar{\mathbf{F}}_q) \mid NP = 0_E\}$. *If N is not divisible by the characteristic of the field, then*

$$\mathrm{Ker}[N]_E \cong (\mathbf{Z}/N\mathbf{Z})^2 . \tag{5.42}$$

6.3. Remarks. In particular, for a prime number ℓ different from the characteristic of $\mathbf{F}_q$, we can define the Tate module as we did over the field of complex numbers:

$$T_\ell(E) := \varprojlim_n \mathrm{Ker}[\ell^n]_E .$$

We again have that $T_\ell(E) \cong (\mathbf{Z}_\ell)^2$ (as a $\mathbf{Z}_\ell$-module). Note however that we do not have a natural lattice $\Omega \cong \mathbf{Z}^2$ such that $T_\ell(E) \cong \Omega \otimes \mathbf{Z}_\ell$.

The statement of the proposition does not stay true for $N = p^m$, where p is the characteristic of $\mathbf{F}_q$. It can be shown that in this case either $\mathrm{Ker}[p^m]_E \cong \mathbf{Z}/p^m\mathbf{Z}$ (the "*ordinary*" case) or $\mathrm{Ker}[p^m]_E = \{0_E\}$ (the "*supersingular*" case).

The key result concerning the number of rational points over $\mathbf{F}_q$ is the following.

6.4. Theorem. (Hasse) *Let E be an elliptic curve defined over $\mathbf{F}_q$. Then,*

$$|\mathrm{card}\, E(\mathbf{F}_q) - q - 1| \leqslant 2\sqrt{q}. \tag{5.43}$$

More precisely, there exists an imaginary quadratic integer α which satisfies $\alpha\bar{\alpha} = q$ such that

$$\mathrm{card}\, E(\mathbf{F}_{q^m}) = q^m + 1 - \alpha^m - \bar{\alpha}^m. \tag{5.44}$$

Proof. (Partial) The set of rational points is also the set of fixed points of the Frobenius endomorphism Φ_q. We will assume that the degree of the endomorphism is given, as with complex numbers, by a quadratic function and thus, in particular, that

$$\deg(n\Phi_q + m) = P(n,m) = an^2 + 2bmn + cm^2.$$

Then we have, on the one hand, $\mathrm{card}\, E(\mathbf{F}_q) = P(1,-1) = a + c - 2b$ and, on the other hand, $c = P(0,1) = 1$ and $a = P(1,0) = q$. Finally, since the polynomial $P(n,m)$ is positive-valued, we have that $b^2 - ac \leqslant 0$ and therefore $|b| \leqslant \sqrt{q}$, which proves the inequality given in the statement.

For the formula which comes next, we use an analogy to the complex case, where the endomorphisms satisfy a quadratic relation. The Frobenius Φ_q can also be seen as an endomorphism of Tate modules, whose eigenvalues α, $\bar{\alpha}$ satisfy $\alpha\bar{\alpha} = q$. The eigenvalues of Φ_q^m are therefore α^m and $\bar{\alpha}^m$.

Since $\operatorname{card} E(\mathbf{F}_{q^m}) = \deg\left(\Phi_q^m - 1\right)$, the relation that we want to prove can be written as $\Phi_q^2 - (\alpha + \bar{\alpha})\Phi_q + q = 0$. □

7. The L-function of an Elliptic Curve

This section, which does not contain any proofs, is a stroll in the direction of the work of Wiles [80]—elliptic curves, modular forms and the great Fermat's last theorem—and the famous Birch & Swinnerton-Dyer conjecture. The stroll continues at the end of the following chapter. Two good references to continue along this path are [27] and [37].

Let E be an elliptic curve over $\mathbf{Q}$. Suppose, as above, that it is a projective plane curve given by the equation

$$y^2 + a_1xy + a_3y = x^3 + a_2x^2 + a_4x + a_6,$$

this time with $a_i \in \mathbf{Z}$ (cf. formula (5.2)). We can see fairly easily that the only coordinate changes that preserve the form of the generalized Weierstrass equation are of the type:

$$x = u^2x' + r, \qquad y = u^3y' + u^2sx' + t.$$

We can define the discriminant of the model in the following ad hoc manner. If the model can be written $y^2 = x^3 + Ax + B$, we set $\Delta := -16\Delta(A, B) = -16(4A^3 + 27B^2)$; in general, we can always transform the equation into a simpler model (A, B), and we will set $\Delta' = u^{-12}\Delta$. A Weierstrass equation is said to be *minimal* and Δ_E is its discriminant if the discriminant of every other equation with integer coefficients is of the form $\Delta = u^{12}\Delta_E$ where $u \in \mathbf{Z}$.

To simplify things, we will continue the discussion by staying in the field of rationals. Reduction modulo p has at most one singular point of cusp type, $y^2 = x^3$, or node type, $y^2 + axy + bx^2 = x^3$, where the polynomial $y^2 + axy + bx^2 = (y - \alpha x)(y - \alpha' x)$ is either irreducible (if $\alpha \in \mathbf{F}_{p^2} \setminus \mathbf{F}_p$) or not (if $\alpha \in \mathbf{F}_p$) over $\mathbf{F}_p$. Geometrically, in the second case, the cone tangent to the singular point is composed of two lines $y = \alpha x$ and $y = \alpha' x$, which could be either rational over $\mathbf{F}_p$ or defined over $\mathbf{F}_{p^2}$ and conjugate.

7.1. Definition. Let p be a prime number and $E/\mathbf{Q}$ have a minimal model

$$y^2 + a_1xy + a_3y = x^3 + a_2x^2 + a_4x + a_6.$$

The curve E is said to have *good reduction* at p if this model stays smooth modulo p (i.e., if p does not divide Δ_E). A curve E is said to have *additive reduction* at p if this model is singular modulo p and if the singularity has a unique tangent. The curve is said to have *multiplicative reduction* or *semi-stable reduction* at p if this model is singular modulo p with two distinct tangents; if the two tangents are defined over $\mathbf{F}_p$ (resp. are not defined over $\mathbf{F}_p$), we say that the reduction is *split* (resp. *non-split*) multiplicative. A fairly simple-to-calculate criterion is the following. If we write a minimal Weierstrass model, except maybe for 2 and 3, in the following way: $y^2 = x^3 - 27c_4x - 54c_6$, then we have additive reduction if p divides c_4 and Δ_E, and we have multiplicative reduction if p divides Δ_E but does not divide c_4. The model is minimal (except maybe for 2 and 3) under the condition that for every prime number p, $p^4 \not| c_4$ or $p^6 \not| c_6$. Finally, we define the invariant j of the elliptic curve E by

$$j := c_4^3/\Delta. \tag{5.45}$$

7.2. Remark. The adjectives "additive" and "multiplicative " come from the following observation, whose verification is left to the reader. If a Weierstrass cubic E is singular (necessarily at a unique point P_0), then the chord-tangent method defines a group law on $E \setminus \{P_0\}$. This group is isomorphic to the additive group if the singular point is a cusp point and isomorphic to the multiplicative group if the tangents are distinct. To be more precise, $E(K) \setminus \{P_0\}$ is isomorphic to $(K, +)$ if the reduction is additive, to $(K^*, \times)$ if the reduction is split multiplicative, and finally to $(K_1, \times)$ if the reduction is non-split multiplicative, where $K_1 = \operatorname{Ker}\left\{\mathrm{N}_K^{K'} : K'^* \to K^*\right\}$ and K' is a quadratic extension. Observe that $j = j_E$ is indeed independent of the model, since, by a change in coordinates, we have $c_4 = u^4c_4'$ and $\Delta = u^{12}\Delta'$. Finally, we can easily verify the following formula:

$$c_4^3 - c_6^2 = 1728\Delta. \tag{5.46}$$

Now we can resume the practical calculation on the generalized Weierstrass model (5.2) by the following list of formulas (due to Tate):

$$\begin{aligned}
b_2 &= a_1^2 + 4a_2 \\
b_4 &= 2a_4 + a_1a_3 \\
b_6 &= a_3^2 + 4a_6 \\
b_8 &= a_1^2a_6 + 4a_2a_6 - a_1a_3a_4 + a_2a_3^2 - a_4^2 \\
c_4 &= b_2^2 - 24b_4 \\
c_6 &= -b_2^3 + 36b_2b_4 - 216b_6 \\
\Delta &= -b_2^2b_8 - 8b_4^3 - 27b_6^2 + 9b_2b_4b_6.
\end{aligned}$$

7.3. Definition. The *conductor* of $E/\mathbf{Q}$ is defined by $N_E := \prod_p p^{n(E,p)}$, where

$$n(E,p) = \begin{cases} 0 & \text{if } E \text{ has good reduction at } p, \\ 1 & \text{if } E \text{ has multiplicative reduction at } p, \\ 2+\delta_{E,p} & \text{if } E \text{ has additive reduction at } p, \end{cases}$$

where $\delta_{E,p} = 0$ if $p \geqslant 5$ and $\delta_{E,2} \leqslant 8$, $\delta_{E,3} \leqslant 5$. (The precise values of $\delta_{E,2}$ and $\delta_{E,3}$ are described in Appendix C.)

7.4. Definition. Let E be an elliptic curve defined over $\mathbf{Q}$. If E has good reduction at p, we set

$$a_p := p + 1 - \operatorname{card} E(\mathbf{F}_p).$$

We then define the function $L(E,s)$ and its local factors by:

$$L_p(E,s) = \begin{cases} \left(1 - a_p p^{-s} + p^{1-2s}\right)^{-1} & \text{if } p \text{ does not divide } \Delta_E, \\ (1-p^{-s})^{-1} & \text{if } E \text{ has split multiplicative reduction,} \\ (1+p^{-s})^{-1} & \text{if } E \text{ has non-split multiplicative reduction,} \\ 1 & \text{if } E \text{ has additive reduction,} \end{cases} \tag{5.47}$$

$$L(E,s) = \sum_{n=1}^{\infty} a_n n^{-s} = \prod_p L_p(E,s). \tag{5.48}$$

7.5. Proposition. *The Dirichlet series and the Euler product defining the function $L(E,s)$ are absolutely convergent for* $\operatorname{Re}(s) > 3/2$.

Proof. This immediately follows from Hasse's theorem (Theorem 5-6.4). □

7.6. Theorem. (Wiles [80]) *The function $L(E,s)$ can be analytically continued to an entire function which satisfies the following functional equation, where we let $\Lambda(E,s) := N_E^{s/2}(2\pi)^{-s}\Gamma(s)L(E,s)$,*

$$\Lambda(E,s) = \pm\Lambda(E,2-s). \tag{5.49}$$

Observe the obvious analogy to functional equation of Riemann zeta function. The theorem of Wiles is actually more precise and "explains" the functional equation of the Dirichlet series $L(E,s) = \sum_{n=1}^{\infty} a_n n^{-s}$ by the fact that associated function (for z in the Poincaré half-plane), $f_E(z) = \sum_{n=1}^{\infty} a_n \exp(2\pi i n z)$, is *modular* with level N_E and weight 2 and satisfies the functional equation

$$f_E\left(-\frac{1}{N_E z}\right) = \pm N_E z^2 f_E(z). \tag{5.50}$$

A formal computation, analogous to the one done to prove the functional equation of the $\zeta(s)$ function, shows that the (5.50) for f_E implies the functional equation for $L(E,s)$ (see Chap. 6, the last section for more details).

The connection between Wiles's theorem and Fermat's last "theorem" is the following. Starting with a hypothetical solution to Fermat's equation

$$a^\ell + b^\ell + c^\ell = 0,$$

we construct an elliptic curve, called the Frey or Hellegouarch curve:

$$y^2 = x(x - a^\ell)(x + b^\ell).$$

By examining this hypothetical curve, we notice that it has (or would have) some remarkable properties. For example, $\Delta_E = 2^{-8}(abc)^{2\ell}$ and $N_E = \prod_{p \mid abc} p$; this implies that the associated Galois representation (see Appendix C) has very little ramification. The associated modular form (from Wiles's theorem) would have even more extraordinary properties: by appealing to a theorem due to Ribet [59], we could lower its level N_E down to 2. However, there does not exist such a non-zero form with level 2, which finishes the proof of Fermat's last theorem! We will add a couple more elements to this subject in Chap. 6. You could also consult the texts of Hellegouarch [37] and Diamond and Shurman [27].

To state the following conjecture, we remind you of the existence of a bilinear form $\langle \cdot, \cdot \rangle$ coming from the Néron-Tate height. We will also need to define the real *period* of E:

$$\Omega_E := \int_{E(\mathbf{R})} \frac{dx}{2y + a_1 x + a_3}. \tag{5.51}$$

7.7. Conjecture. (Birch & Swinnerton-Dyer [14])

I) The order of vanishing of the function $L(E,s)$ at $s = 1$ is equal to the rank $r = \operatorname{rank} E(\mathbf{Q})$ of the Mordell-Weil group.

II) Let $P_1, \ldots, P_r$ be a basis for $E(\mathbf{Q})$ modulo torsion and Ω_E the period of E. Then the leading coefficient of $L(E,s)$ at $s = 1$ is given by

$$\lim_{s \to 1} \frac{L(E,s)}{(s-1)^r} = u\Omega_E \det\left(\langle P_i, P_j \rangle\right) \tag{5.52}$$

where $u \in \mathbf{Q}^*$.

7.8. Remarks. In the complete formulation[3], the number u is explicitly defined. In fact,

$$u = M \prod_{p \mid \Delta_E} c_p \left|E(\mathbf{Q})_{\mathrm{tor}}\right|^{-2},$$

where $c_p \geqslant 1$ is an integer which depends on the bad reduction at p and M is the cardinality of the "Tate-Shafarevich group". This cardinality should be finite (it has only been shown in certain cases) and should be a perfect square (which is true if the group in question is finite).

We point out that the conjectural formula is strongly analogous to the formula which gives the residue at $s = 1$ of the Dedekind zeta function in a number field K (see formula (6.34)). The Tate-Shafarevich group corresponds to the class group of ideals in K, the torsion group corresponds to the group of roots of unity, the Néron-Tate regulator corresponds to the regulator R_K of the units of K, etc.

The first observation made by Birch & Swinnerton-Dyer is that, at least formally, $L(1) = \prod_p \frac{p}{N_p}$ where $N_p := \operatorname{card} E(\mathbf{F}_p)$ (more precisely, the number of nonsingular points in the reduction modulo p of E). Now, N_p is approximately equal to p with a variation of at most $2\sqrt{p}$. We denote this by $N_p = p + \delta(p)\sqrt{p}$, and thus we have (still formally) $L(1) = \prod_p (1 + \delta(p)p^{-1/2})^{-1}$. If $E(\mathbf{Q})$ is finite, we can imagine that N_p oscillates regularly and hence that $\delta(p)$ has the tendency to be well-distributed in the interval $[-2, 2]$, which would make the product converge. If now $E(\mathbf{Q})$ is infinite, we might think that we would find more points modulo p and thus that $\delta(p)$ would have the tendency to be positive, which would imply the divergence of the product, more precisely, would force $L(1)$ to be zero.

We can think of the conjecture as a sophisticated version of the local/global principle. To see this, the function $L(E, s)$ is constructed using fairly simple information of a local type, essentially the number of points modulo p, and it allows us, thanks to analytic continuation, to recover the rank of the group $E(\mathbf{Q})$.

Finally, the sign of the functional equation of $L(E, s)$ determines the parity of the order of the zero of $L(E, s)$ at $s = 1$. Thus, conjecturally, the sign of the functional equation determines the parity of the rank of the group $E(\mathbf{Q})$. This weakened version is called the *parity conjecture*.

[3]The Birch & Swinnerton-Dyer conjecture is one of the Millennium Prize Problems; the Clay Mathematics Institute offers a million dollars for its solution.

Chapter 6

Developments and Open Problems

> *"Une pierre*
> *deux maisons*
> *trois ruines*
> *quatre fossoyeurs*
> *un jardin*
> *des fleurs*
>
> *un raton laveur"*
>
> JACQUES PRÉVERT

The tone and the level of the prerequisites of this final chapter differ from the previous chapters. Here we will present a panorama—necessarily partial and one-sided—of some important research areas in number theory. In particular, every section contains at least one open problem. This last chapter also includes many statements whose proofs surpass the level of this book but which also provide an opportunity to combine and expand on the mathematics introduced in the preceding chapters. The chosen themes—the number of solutions of equations over a finite field, algebraic geometry, p-adic numbers, Diophantine approximation, the a, b, c conjecture and generalizations of zeta and L-series—have all been introduced, either implicitly or explicitly, in the previous chapters. We will freely use themes from algebraic geometry and Galois theory, described respectively in Appendices B and C.

M. Hindry, *Arithmetics*, Universitext,
DOI 10.1007/978-1-4471-2131-2_6,

1. The Number of Solutions of Equations over Finite Fields

In order to deepen your understanding of this section, you could first consult Weil's original article [78] and Appendix C of [35].

The successes brought about by the introduction of the Riemann ζ function, then the Dedekind ζ_K function, naturally lead to the study of following generalization. We consider a finitely generated ring A over $\mathbf{Z}$ or $\mathbf{F}_p$, in other words $A := \mathbf{Z}[t_1, \ldots, t_n] = \mathbf{Z}[X_1, \ldots, X_n]/I$ or $A := \mathbf{F}_p[t_1, \ldots, t_n] = \mathbf{F}_p[X_1, \ldots, X_n]/I$. It is easy to see that if $\mathfrak{p}$ is a *maximal* ideal in A, then $A/\mathfrak{p}$ is a finite field. We denote by $\mathrm{N}\,\mathfrak{p} = \mathrm{card}(A/\mathfrak{p})$. Letting $\mathscr{M}_A$ be the set of maximal ideals in A, we can therefore set

$$\zeta_A(s) := \prod_{\mathfrak{p} \in \mathscr{M}_A} \left(1 - \mathrm{N}\,\mathfrak{p}^{-s}\right)^{-1}.$$

If $A = \mathbf{Z}$, we recover the Riemann ζ function, and if $A = \mathscr{O}_K$ (for a number field K), we recover the Dedekind ζ_K function. Furthermore, in the case where $\mathbf{Z} \subset A$, every maximal ideal $\mathfrak{p}$ contains exactly one prime number since $\mathfrak{p} \cap \mathbf{Z}$ is a non-zero prime ideal. If we denote by $\mathscr{M}_{A,p}$ the maximal ideals which contain p (this set is in bijection with the maximal ideals of A/pA), then $\mathscr{M}_A = \cup_p \mathscr{M}_{A,p}$, and we can write:

$$\zeta_A(s) = \prod_p \prod_{\mathfrak{p} \in \mathscr{M}_{A,p}} \left(1 - \mathrm{N}\,\mathfrak{p}^{-s}\right)^{-1} = \prod_p \zeta_{A/pA}(s).$$

We can therefore, at least momentarily, concentrate on the case $A = \mathbf{F}_p[t_1, \ldots, t_n] = \mathbf{F}_p[X_1, \ldots, X_n]/I$ (we will come back to the case of varieties defined over $\mathbf{Q}$ or $\mathbf{Z}$ in the last section of this chapter). Let V denote the affine variety defined by the ideal I in $\mathbf{A}^n$. The maximal ideals of $\bar{\mathbf{F}}_p[t_1, \ldots, t_n]$ correspond to points of $V(\bar{\mathbf{F}}_p)$, and the maximal ideals of $A = \mathbf{F}_p[X_1, \ldots, X_n]/I$ correspond to conjugacy classes under $\mathrm{Gal}(\bar{\mathbf{F}}_p/\mathbf{F}_p)$ in $V(\bar{\mathbf{F}}_p)$. We denote by $|V|$ the set of these classes[1]. If $x \in V(\bar{\mathbf{F}}_p)$ and if $\mathfrak{p}$ is the corresponding maximal ideal in A, then $\mathrm{N}\,\mathfrak{p} = p^{\deg(x)}$ where $\deg(x) := [\mathbf{F}_p(x) : \mathbf{F}_p]$. Since a point in $V(\mathbf{F}_{p^m})$ has a field of definition equal to $\mathbf{F}_{p^d}$ where d divides m, we see that

$$\mathrm{card}\, V(\mathbf{F}_{p^m}) = \sum_{\substack{x \in |V| \\ \deg(x) \mid m}} \deg(x).$$

We thus obtain a second expression for $\zeta_A(s)$:

[1] In the language of Grothendieck schemes, we are talking about *closed* points of the scheme $V = \mathrm{spec}(A)$.

$$\zeta_A(s) = \prod_{x \in |V|} \left(1 - p^{-\deg(x)s}\right)^{-1} = \exp\left(\sum_{m=1}^{\infty} \operatorname{card} V(\mathbf{F}_{p^m}) \frac{p^{-ms}}{m}\right). \quad (6.1)$$

By setting $T = p^{-s}$, this leads to the following definition (where now V is not necessarily affine).

1.1. Definition. Let V be an algebraic variety defined over $\mathbf{F}_q$. Its *zeta function* is given by the formal series with integer coefficients:

$$Z(V/\mathbf{F}_q;T) = \prod_{x \in |V|} \left(1 - T^{\deg(x)}\right)^{-1} = \exp\left(\sum_{m=1}^{\infty} \operatorname{card} V(\mathbf{F}_{q^m}) \frac{T^m}{m}\right). \quad (6.2)$$

In fact, if we write $Z(V/\mathbf{F}_q;T) = \sum_{m \geqslant 0} a_m T^m$, we can see that a_m is the number of formal linear combinations[2] $m_1x_1 + \cdots + m_rx_r$, where $m_i \in \mathbf{N}$ and $\sum_{i=1}^{r} m_i \deg(x_i) = m$.

1.2. Examples. Let us compute this series for some varieties.

- The calculation for the affine space $\mathbf{A}^n$ of dimension n is simple:

$$Z(\mathbf{A}^n/\mathbf{F}_q, T) = \exp\left(\sum_{m=1}^{\infty} q^{mn} \frac{T^m}{m}\right) = (1 - q^nT)^{-1}. \quad (6.3)$$

Since $\mathbf{P}^n = \mathbf{A}^n \sqcup \mathbf{A}^{n-1} \sqcup \cdots \sqcup \mathbf{A}^1 \sqcup \mathbf{A}^0$, we have

$$Z(\mathbf{P}^n/\mathbf{F}_q, T) = \prod_{j=0}^{n} \left(1 - q^jT\right)^{-1}. \quad (6.4)$$

- If $V = G(n,k)$ is the Grassmannian which parametrizes the vector subspaces of dimension k in $\mathbf{A}^n$, we find positive integers B_{2i} such that

$$\operatorname{card} G(n,k)(\mathbf{F}_q) = \prod_{j=0}^{k-1} \frac{q^{n-j} - 1}{q^{j+1} - 1} = \sum_{i=0}^{k(n-k)} B_{2i} q^i \quad (6.5)$$

$$Z(G(n,k)/\mathbf{F}_q, T) = \prod_{j=0}^{k(n-k)} \left(1 - q^jT\right)^{-B_{2j}}. \quad (6.6)$$

For example, for $V = G(4,2)$ (the space of lines in $\mathbf{P}^3$), we find

[2] In more scholarly terms, we are referring to effective cycles of dimension zero and degree m.

$\operatorname{card} V(\mathbf{F}_q) = q^4 + q^3 + 2q^2 + q + 1$, and hence

$$Z(V/\mathbf{F}_q, T) = \frac{1}{(1-T)(1-qT)(1-q^2T)^2(1-q^3T)(1-q^4T)}.$$

- If E is an elliptic curve over $\mathbf{F}_q$, we can deduce from Hasse's theorem (Theorem 5-6.4) that there exists $\alpha \in \mathbf{C}$ with $|\alpha| = \sqrt{q}$ such that

$$Z(E/\mathbf{F}_q, T) = \exp\left(\sum_{m=1}^{\infty} (q^m + 1 - \alpha^m - \bar{\alpha}^m)\frac{T^m}{m}\right)$$
$$= \frac{(1-\alpha T)(1-\bar{\alpha}T)}{(1-T)(1-qT)}. \tag{6.7}$$

  We therefore realize that $Z(E/\mathbf{F}_q, T) = (1 - aT + qT^2)(1-T)^{-1}(1-qT)^{-1}$ where $a = q+1-|E(\mathbf{F}_q)|$. Hence, in this case, knowing $|E(\mathbf{F}_q)|$ is equivalent to knowing $Z(E/\mathbf{F}_q, T)$.

- Let V be a nondegenerate quadric in $\mathbf{P}^n$. Theorem 1-5.5, together with the Davenport-Hasse relation (see Theorem 1-5.15 and Exercise 1-6.25), gives

$$Z(V/\mathbf{F}_q, T) = \begin{cases} Z(\mathbf{P}^{n-1}/\mathbf{F}_q, T) & \text{if } n \text{ is even,} \\ Z(\mathbf{P}^{n-1}/\mathbf{F}_q, T)(1 - \epsilon q^{\frac{n-1}{2}} T)^{-1} & \text{if } n \text{ is odd,} \end{cases}$$

  where $\epsilon = \left(\frac{-1}{p}\right)^{\frac{n-1}{2}} \left(\frac{D}{p}\right)$ and D is the discriminant of the quadratic form.

- Let V be the smooth intersection of two quadrics $Q_1 = a_0x_0^2 + \cdots + a_nx_n^2 = 0$ and $Q_2 = b_0x_0^2 + \cdots + b_nx_n^2 = 0$ in $\mathbf{P}^n$ where n is even. The computations done in Exercise 1-6.24, together with the Davenport-Hasse relation, yield

$$Z(V/\mathbf{F}_p, T) = Z(\mathbf{P}^{n-2}/\mathbf{F}_p, T) \prod_{i=0}^{n} \left(1 - \eta_i p^{n/2} T\right)^{-1},$$

  where $\eta_i = \left(\frac{-1}{p}\right)^{\frac{n}{2}} \left(\frac{D_i}{p}\right)$ and $D_i := \prod_{j \neq i}(b_ia_j - a_ib_j)$.

- Let V be a Fermat hypersurface given by the equation $a_0x_0^d + \cdots + a_nx_n^d$. Theorem 1-5.13, together with the Davenport-Hasse relation, yields

$$Z(V/\mathbf{F}_q, T) = Z(\mathbf{P}^{n-1}/\mathbf{F}_q, T)P(T)^{(-1)^n}$$

  where

$$P(T) := \prod_{(\chi_0,\dots,\chi_n)\in S} \left(1-(-1)^{n-1}q^{-1}\bar{\chi}_0(a_0)\cdots\bar{\chi}_n(a_n)G(\chi_0)\cdots G(\chi_n)T\right).$$

Here, S designates the set of $(n+1)$-tuples of characters different from the unitary character and such that the χ_j^d, as well as $\chi_0\cdots\chi_n$, are equal to the unitary character. In particular, we have $B_{n-1}(d) := \deg(P(T)) = ((d-1)^{n+1} + (-1)^{n+1}(d-1))d^{-1}$ and the equality

$$\left|q^{-1}\bar{\chi}_0(a_0)\cdots\bar{\chi}_n(a_n)G(\chi_0)\cdots G(\chi_n)\right| = q^{\frac{n-1}{2}}.$$

This thus yields the estimate

$$\left|\operatorname{card} V(\mathbf{F}_q) - \operatorname{card} \mathbf{P}^{n-1}(\mathbf{F}_q)\right| \leqslant B_{n-1}(d)q^{\frac{n-1}{2}}. \qquad (6.8)$$

The following theorem (conjectured by André Weil [78]) lets us extend the previous inequality (6.8) to every *smooth* hypersurface of degree d in $\mathbf{P}^n$. We can verify its truthfulness in each of the previous examples, and its proof, which largely surpasses the level of this text, motivated specialists in algebraic geometry for twenty years. His work on developing the theory of algebraic geometry earned Grothendieck the Fields Medal in 1966. Deligne earned the same distinction in 1978 for completing this theory.

1.3. Theorem. (Weil conjectures, Grothendieck's and Deligne's theorems) *Let V be a smooth projective variety of dimension r.*

1) (Rationality) *The function $Z(V/\mathbf{F}_q;T)$ is a rational function in the indeterminate T.*

2) (Functional equation) *There exists an integer $\chi(V)$ and a sign $\epsilon = \pm 1$ such that*

$$Z\left(V/\mathbf{F}_q; \frac{1}{q^rT}\right) = \epsilon q^{\frac{r\chi(V)}{2}} T^{\chi(V)} Z\left(V/\mathbf{F}_q;T\right). \qquad (6.9)$$

3) (Riemann hypothesis) *There exist polynomials $P_i(T) \in \mathbf{Z}[T]$ such that*

$$Z\left(V/\mathbf{F}_q;T\right) = \frac{P_1(T)\cdots P_{2r-1}(T)}{P_0(T)\cdots P_{2r}(T)} = \prod_{i=0}^{2r} P_i(T)^{(-1)^{i+1}},$$

and $P_i(T) = \prod_{j=1}^{B_i}(1-\alpha_{i,j}T)$ with $|\alpha_{i,j}| = q^{i/2}$.

4) Suppose that V is the reduction modulo a prime ideal of a variety $\mathscr{V}$ defined over a number field. Then the numbers B_i are the topological Betti numbers of the variety $\mathscr{V}$.

1.4. Remarks. In particular, by *3)* we have

$$|V(\mathbf{F}_{q^m})| = q^{mr} + \sum_{i=0}^{2r-1} (-1)^i \sum_{j=1}^{B_i} \alpha_{i,j}^m \,.$$

The meaning of the last statement is the following. Let $\mathscr{V}$ be a smooth projective variety defined over a number field K and having good reduction at a prime ideal $\mathfrak{p}$ of $\mathscr{O}_K$ such that $\mathscr{O}_K/\mathfrak{p} \cong \mathbf{F}_q$ (see Appendix B). We can therefore consider the complex variety $\mathscr{V}(\mathbf{C})$, to which we can associate the Betti numbers $B_i := \dim H^i(\mathscr{V}(\mathbf{C}), \mathbf{Q})$. We can also consider the reduced variety modulo $\mathfrak{p}$ denoted $V/\mathbf{F}_q$. Statement *4)* therefore says that the numbers $B_i = \deg(P_i(T))$, given by the first part of the theorem, are exactly the Betti numbers of $\mathscr{V}(\mathbf{C})$.

With this interpretation, we can see that $\chi = \sum_{i=0}^{2n}(-1)^i B_i$, and it is therefore legitimate to call the latter ($\in \mathbf{Z}$) the *Euler-Poincaré characteristic* of the variety V.

The reason why Statement *3)* of Theorem 6-1.3 is called the Riemann hypothesis is an analogy. To be more specific, if we go back to the function initially associated to $V/\mathbf{F}_q$, it can be written

$$\zeta_V(s) = Z(V/\mathbf{F}_q, q^{-s}) = \prod_{i=0}^{2r} P_i(q^{-s})^{(-1)^{i+1}},$$

and the Riemann hypothesis—Statement *3)* of the theorem—can be translated into the assertion that the zeros (for odd i) or poles (for even i) are situated on the lines $\mathrm{Re}(s) = \dfrac{i}{2}$.

1.5. Remark. One of the most modern applications of Theorem 6-1.3 is the upper bound on the sum of exponentials, of which we studied a typical example given by Gauss sums. We could, for example, prove using these techniques the following estimate due to Deligne, where we assume that $F(x) = F_d(x) + F_{d-1}(x) + \cdots + F_1(x)$ with F_i homogeneous of degree i and F_d smooth, i.e., the hypersurface that it defines in $\mathbf{P}^{n-1}$ is smooth, and also that p does not divide d:

$$\left| \sum_{x \in (\mathbf{F}_p)^n} \exp\left(\frac{2\pi i F(x)}{p}\right) \right| \leqslant B_{n,d}\, p^{n/2}. \tag{6.10}$$

Before that, Weil proved these assertions over curves and deduced from them an upper bound for Kloosterman sums over integers modulo a prime p (see Exercise 1-6.18 for sums modulo any integers) which does not divide ab:

$$\left| \sum_{x \in \mathbf{F}_p^*} \exp\left(\frac{2\pi i(ax + bx^{-1})}{p}\right) \right| \leqslant 2\sqrt{p}. \tag{6.11}$$

The smoothness constraints are essential if we want the estimates to be fine. For example, the sum $\sum_{x,y,z\in\mathbf{F}_p}\exp\left(\frac{2\pi ixyz}{p}\right)$ equals $2p^2-p$, which is much larger than $p^{3/2}$. Nonetheless, Theorem 6-1.3 does not outweigh the following more robust (and older) result:

1.6. Theorem. (Lang-Weil [47]) *There exists a constant $C = C(n,r,d)$ such that for every closed (irreducible) subvariety V of $\mathbf{P}^n$ of dimension r and degree d, we have the following estimate:*

$$|\operatorname{card} V(\mathbf{F}_q) - \operatorname{card}\mathbf{P}^r(\mathbf{F}_q)| \leqslant (d-1)(d-2)q^{r-1/2}+C(n,r,d)q^{r-1}. \quad (6.12)$$

1.7. Remarks. 1) Let $P_i(T) = P_i(V/\mathbf{F}_q,T) = \prod_{j=1}^{B_i}(1-\alpha_{i,j}T)$ be one of the polynomials associated to V by Theorem 6-1.3. Let $m_i = m_i(V)$ be the multiplicity with which $\alpha_{i,j} = q^{i/2}$. Since $\bar{\alpha}_{i,j} = q^i/\alpha_{i,j}$, we have

$$P_i(T) = \prod_{j=1}^{B_i}\left(1-\frac{q^i}{\alpha_{i,j}}T\right) = \left(\prod_{j=1}^{B_i}(-\alpha_{i,j})\right)T^{B_i}P_i\left(\frac{1}{q^iT}\right).$$

Furthermore, we can easily see that $\prod_{j=1}^{B_i}(-\alpha_{i,j}) = (-1)^{m_i}q^{iB_i/2}$, which gives us a functional equation for $P_i(T)$ of the form:

$$P_i(T) = (-1)^{m_i}q^{\frac{iB_i}{2}}T^{B_i}P_i\left(\frac{1}{q^iT}\right). \quad (6.13)$$

2) Moreover, by identifying the factors according to the absolute value of their roots in the functional equation, we obtain

$$P_{2r-i}(T) = P_i(q^{r-i}T), \quad (6.14)$$

from which we can immediately deduce that $m_{2r-i} = m_i$ and $B_{2r-i} = B_i$. This yields the equality $\chi' := \sum_{i=0}^{2r}(-1)^i iB_i = r\sum_{i=0}^{2r}(-1)^iB_i = r\chi$.

3) By referring to formulas (6.13) and (6.14), we recover the functional equation of the function $Z(V,T)$ using the supplementary information that $\epsilon = (-1)^{m_r}$.

4) By using the fact that a symplectic isometry has determinant 1, we can prove that if $r := \dim(V)$ is odd, for example for an algebraic curve, then the sign of the functional equation of $Z(V/\mathbf{F}_q,T)$ is $+1$. Whenever r is even, the sign can be positive or negative and remains more or less mysterious. To conjecturally describe its behavior, Tate suggested to compare it to the rank of the group $\operatorname{Num}^i(V)$ of cycles of codimension i modulo numerical equivalence (see Appendix B, Definition B-2.9).

1.8. Conjecture. (Tate) *The multiplicity $m_{2i}(V)$ is equal to the dimension of the subspace generated by the algebraic subvarieties of codimension i modulo the numerical equivalence relation.*

In particular, the conjecture indicates that the sign of the functional equation should be $\epsilon = (-1)^{\operatorname{rank} \operatorname{Num}^i(V)}$. Tate proved that it is always the case that $\operatorname{rank} \operatorname{Num}^i(V) \leqslant m_{2i}(V)$, but we only know how to prove equality in certain cases, among which are the examples outlined previously in this section.

2. Diophantine Equations and Algebraic Geometry

In this section, we will consider the question—raised essentially by Serge Lang, [45] and [46]—of establishing a correspondence or dictionary between arithmetic and geometric properties of algebraic varieties. More concretely, if we are given a system of equations with integer coefficients

$$V \;:\; f_1(X_1,\ldots,X_n) = \cdots = f_r(X_1,\ldots,X_n) = 0,$$

we want to find connections between the qualitative properties (finiteness, density, etc.) of the rational solutions $V(\mathbf{Q})$ and the properties of the analytic or algebraic variety of the complex solutions $V(\mathbf{C})$. The existence of such a dictionary between algebraic, analytic geometry and arithmetic is very largely conjectural, but we nevertheless have some (deep) theorems and precise questions. You can consult [38] to acquire a deeper insight into this subject.

We will start by describing the predicted geometric properties.

To simplify things, we will assume in this section that the varieties are smooth and projective. By imitating differential geometry, we can give an algebraic definition of *regular differential forms* on an algebraic variety V as linear forms on the tangent space which are defined everywhere locally (i.e., for every point P on an open set containing the point P) by the form

$$\omega = \sum_i f_i dg_i,$$

where the f_i, g_i are algebraic functions on V without poles at P. We likewise define the space of *regular differential k-forms* as those which can be expressed locally as

$$\omega = \sum_{(i_1,\ldots,i_k)} f_{i_1,\ldots,i_k} dg_{i_1} \wedge \cdots \wedge dg_{i_k}.$$

2.1. Example. We will give a nontrivial example where the necessity of describing the differential form in various charts intervenes. Let E be an elliptic curve in $\mathbf{P}^2$ given by the equation $ZY^2 = X^3 + aXZ^2 + bZ^3$. Let $x = X/Z$ and $y = Y/Z$ so that we can define

$$\omega = \frac{dx}{y} = \frac{2dy}{3x^2 + a}.$$

Since y and $3x^2 + a$ do not simultaneously vanish (the curve is assumed to be smooth), the form ω does not have a pole outside of the point at infinity. We can also see that it is regular around the point at infinity. By setting $u := 1/x$ and $v := y/x^2$, the equation of the curve becomes $v^2 = u + au^3 + bu^4$, and we obtain

$$\frac{dx}{y} = -\frac{du}{v} = -\frac{2dv}{1 + 3au^2 + 4bu^3}$$

which is clearly regular at the point $(u, v) = (0, 0)$.

The set of regular differential k-forms forms a vector space denoted $\Omega^k[V]$. Whenever V is projective, this space has finite dimension denoted by $g_k(V)$. The following invariant is particularly important.

2.2. Definition. The *genus* of a smooth, projective algebraic variety of dimension r is the dimension of the space of regular differential r-forms:

$$g(V) := \dim \Omega^r[V].$$

Observe that any two differential r-forms, ω and ω', on V are "proportional" in the sense that there exists a function, f, such that $\omega' = f\omega$. By choosing a basis, $\omega_1, \ldots, \omega_{g(V)}$, for $\Omega^r[V]$, this allows us to define the *canonical* map $\Phi_V : V \cdots \to \mathbf{P}^{g(V)-1}$ given by $x \mapsto (\omega_1(x), \ldots, \omega_{g(V)}(x))$. This map is rational (cf. definition and notation in Appendix B, page 274): it is only defined on the open set $V \setminus Z$, where Z is the locus of the common zeros of the ω_i. More explicitly, this map can be described as follows (at least on an open set): there exist rational functions f_i such that $\omega_i = f_i\omega_1$, and therefore, the function Φ_V can be written as $\Phi_V(P) = (1, f_2(P), \ldots, f_{g(V)}(P))$ for P which are not poles of the f_i. By considering the tensor powers of differential forms, i.e., expressions of the form $\omega_1 \otimes \cdots \otimes \omega_m$, we obtain the space of differential k-forms of weight m, the plurigenera $g(m, V)$ which are the dimensions of the space of differential r-forms of weight m, and the pluricanonical maps $\Phi_{m,V} : V \cdots \to \mathbf{P}^{g(m,V)-1}$.

2.3. Definition. 1) The *Kodaira dimension* of a variety is $-\infty$ if $g(m, V) = 0$ for every m and if not is given by

$$\kappa(V) := \max_m \dim \Phi_{m,V}(V).$$

2) A projective variety is *pseudo-canonical* (or "*of general type*") if $\kappa(V) = \dim V$.

If V is a variety defined over a number field K, we can embed K into $\mathbf{C}$ and consider the complex points $V(\mathbf{C})$; we thus obtain an analytic variety, i.e., a variety defined by holomorphic functions.

2.4. Definition. A *Riemann surface* is a complex analytic variety of dimension 1. It is called algebraic if we can represent it as the set of complex points on an algebraic curve.[3]

It is fairly easy to show that if V is a projective curve, then $V(\mathbf{C})$ is compact. The converse is a deep theorem of Riemann: every compact Riemann surface is algebraic (and projective).

2.5. Examples. 1) The genus of a smooth projective curve can be any natural number. We have $g = 0$ if $V = \mathbf{P}^1$ (since $\Omega^1[\mathbf{P}^1] = 0$) and $g = 1$ if V is an elliptic curve (a basis for $\Omega^1[V]$ is given by $\omega := dx/y$); a smooth projective plane curve of degree d has genus $g = (d-1)(d-2)/2$. If V is defined over $\mathbf{C}$, then $V(\mathbf{C})$ is a compact Riemann surface, and $g(V)$ coincides with the number of handles or holes in the surface.

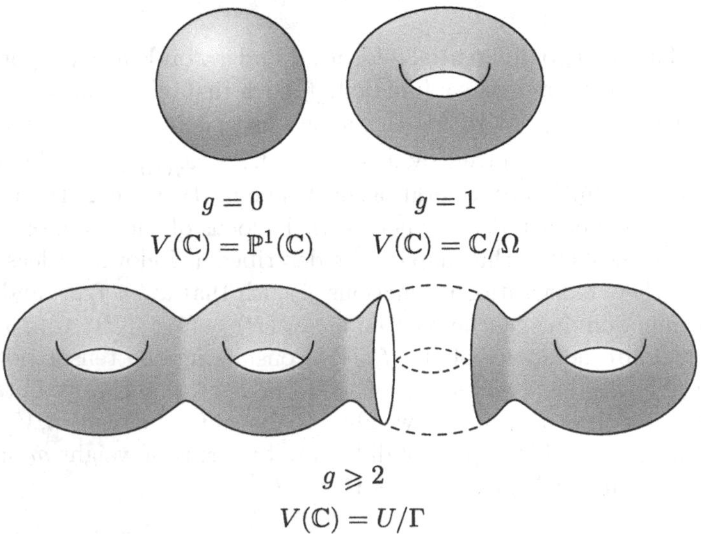

[3]We would like to point out the classical naming conflict of calling the same object a "curve" and a "surface".

For a smooth plane curve V of degree d, we find that V is isomorphic to $\mathbf{P}^1$ if $d = 1$ or 2; that $g = 1$ if $d = 3$, and therefore, $V(\mathbf{C}) \cong \mathbf{C}/\Lambda$ (i.e., V is an elliptic curve, see Chap. 5); and finally, that $g \geqslant 3$ if $d \geqslant 4$.

2) If now V is a smooth projective hypersurface in $\mathbf{P}^n$ of degree d, then the dimension is given by

$$\kappa(V) = \begin{cases} -\infty & \text{if } d \leqslant n, \\ 0 & \text{if } d = n+1, \\ \dim(V) & \text{if } d \geqslant n+2. \end{cases}$$

One of the deepest results linking the geometry of curves to their arithmetic properties was proven by Faltings in 1983 (see [16], [30], [38]). Faltings received the Fields Medal in 1986 for this work.

2.6. Theorem. (Mordell conjecture, Faltings's theorem) *Let C be a curve of genus $g \geqslant 2$ defined over a number field K. Then $C(K)$ is finite.*

This theorem completes the prior results of Siegel dating back to 1929 (see Theorem 5-4.1). In order to give a geometric statement of this theorem, it will be convenient to use the following notation.

2.7. Definition. Let C be a smooth projective curve of genus g and T a finite set of points. We denote by $U = C \setminus T$ the corresponding curve (which is affine if $T \neq \emptyset$). The *Euler-Poincaré characteristic* of U is defined by $\chi(U) := 2 - 2g - |T|$.

2.8. Theorem. (Siegel's theorem) *Let C be a smooth projective curve of genus g defined over a number field K and T a finite set of points. We denote by $U = C \setminus T$ the corresponding affine curve. If $\chi(U) < 0$, then the set of integral points $U(\mathcal{O}_K)$ is finite.*

Siegel's theorem was generalized to S-integral points by Mahler. If $g \geqslant 2$, Siegel's theorem is surpassed by Faltings's theorem, and if $g = 1$, we essentially obtain Theorem 5-4.1. If $g = 0$ and $|T| \geqslant 3$, the statement is equivalent to the unit theorem (Theorem 5-4.4), which affirms, for example, that the curve given by the equation $xy(y-1) = 1$ only has a finite number of S-integral points.

For the proof, see for example [38]. To illustrate the importance of Theorem 6-2.6, we will restate it as part *iii*) of the following theorem. If V is a projective subvariety of $\mathbf{P}^n$, we set:

$$N(V(\mathbf{Q}), H, B) := \operatorname{card}\{x \in V(\mathbf{Q}) \mid H(x) \leqslant B\}.$$

2.9. Theorem. *Let V be a projective variety defined over $\mathbf{Q}$. Then we have the following asymptotic estimates as B tends to infinity.*

i) If $V = \mathbf{P}^n$, then

$$N(\mathbf{P}^n(\mathbf{Q}), H, B) \sim \frac{2^n}{\zeta(n+1)} B^{n+1}.$$

ii) If $V = E$ is an elliptic curve of rank $r = \operatorname{rank} E(\mathbf{Q})$, then

$$N(E(\mathbf{Q}), H, B) \sim c_E (\log B)^{r/2}.$$

iii) If V is a curve of genus $\geqslant 2$, then $N(V(\mathbf{Q}), H, B)$ becomes constant for large enough B.

2.10. Remark. We underline the fact that from an arithmetical point of view, we have the following trichotomy of curves, V:

1) genus 0 curves with many rational points and $\kappa(V) = -\infty$;
2) genus 1 curves with few rational points and $\kappa(V) = 0$;
3) genus $\geqslant$ 2 curves with a finite number of rational points and $\kappa(V)$ maximal.

It is this trichotomy that the Lang conjectures are trying to generalize.

Proof. Point *iii)* is a reformulation of Theorem 6-2.6. To prove *i)*, we introduce the functions $F(B) := \operatorname{card}\{x \in \mathbf{Z}^{n+1} \mid 0 < \max|x_i| \leqslant B\}$ and $G(B) := \operatorname{card}\{x \in \mathbf{Z}^{n+1} \mid 0 < \max|x_i| \leqslant B,\ \gcd(x_0, \ldots, x_n) = 1\}$. We can see that $F(B) = (2\lfloor B\rfloor + 1)^{n+1} - 1 = (2B)^{n+1} + O(B^n)$. By regrouping the elements of $\mathbf{Z}^{n+1}$ according to the gcd of their coordinates, we see that

$$F(B) = \sum_{d \leqslant B} G(B/d).$$

By using the Möbius formula (see Exercise 4-6.5), we obtain

$$\begin{aligned} G(B) &= \sum_{d \leqslant B} \mu(d) F(B/d) \\ &= (2B)^{n+1} \sum_{d \leqslant B} \frac{\mu(d)}{d^{n+1}} + O\left(B^n \sum_{d \leqslant B} \frac{1}{d^n}\right) \\ &= (2B)^{n+1} \sum_{d=1}^{\infty} \frac{\mu(d)}{d^{n+1}} + O(B^n \log B) \\ &= \frac{2^{n+1}}{\zeta(n+1)} B^{n+1} + O(B^n \log B), \end{aligned}$$

where the term $\log B$ can be omitted if $n \geqslant 2$.

To prove *ii*), we use the quadraticity (see Theorem 5-2.2.2) of the (logarithmic) height on an elliptic curve. If $||.||$ is a Euclidean norm on $\mathbf{R}^r$, we will prove that $\mathrm{card}\{x \in \mathbf{Z}^n \mid ||x|| \leqslant X\} \sim cX^r$ using the following argument. Let $\gamma = \max_{x\in[0,1]^r} ||x||$. Then we have the inclusions

$$\{x \in \mathbf{R}^r \ ; \ ||x|| \leqslant X-\gamma\} \subset \bigcup_{\substack{x \in \mathbf{Z}^r \\ ||x|| \leqslant X}} x+[0,1]^r \subset \{x \in \mathbf{R}^r \ ; \ ||x|| \leqslant X+\gamma\}.$$

By considering the volumes (where the volume of the ball $||\cdot|| \leqslant 1$ is denoted by v_r), this yields:

$$v_r(X-\gamma)^r \leqslant \mathrm{card}\{x \in \mathbf{Z}^n \ ; \ ||x|| \leqslant X\} \leqslant v_r(X+\gamma)^r,$$

and hence $\mathrm{card}\{x \in \mathbf{Z}^r \ ; \ ||x|| \leqslant X\} \sim v_r X^r$. By applying this to the Néron-Tate height $\hat{h} : E(\mathbf{Q}) \to \mathbf{R}$ and its associated real quadratic form $\hat{h}_{\mathbf{R}} : E(\mathbf{Q}) \otimes \mathbf{R} \to \mathbf{R}$ (see Chap. 5), we obtain

$$\begin{aligned} &\mathrm{card}\{P \in E(\mathbf{Q}) \mid \hat{h}(P) \leqslant X\} \\ &\quad = |E(\mathbf{Q})_{\mathrm{tor}}| \left|\{x \in E(\mathbf{Q}) \otimes \mathbf{R} \mid \hat{h}_{\mathbf{R}}(x) \leqslant X\}\right| \sim cX^{r/2}. \end{aligned}$$

By choosing $\hat{H} := \exp \hat{h}$, we thus have

$$N(E(\mathbf{Q}), \hat{H}, B) \sim c_E(\log B)^{r/2},$$

and we can easily see that the estimate still holds if we replace H by $\hat{H}$ since for every P, $C^{-1}H(P) \leqslant \hat{H}(P) \leqslant CH(P)$. □

We will now consider an algebraic curve over the field of complex numbers. As we have seen, its complex points form a *Riemann surface.* Simply connected Riemann surfaces were classified by Riemann. There are three of them: the sphere or projective line $\mathbf{P}^1(\mathbf{C})$, the affine plane or line $\mathbf{C} = \mathbf{P}^1(\mathbf{C}) \setminus \{\infty\}$ and the unit disk $U := \{z \in \mathbf{C} \mid |z| < 1\}$. The universal covering of a projective curve of genus 0 (resp. genus 1, resp. genus $\geqslant 2$) is the projective line (resp. plane, resp. disk). Therefore, the corresponding analytic variety is either $\mathbf{P}^1(\mathbf{C})$, $\mathbf{C}/\Omega$ (where Ω is a lattice in $\mathbf{C}$) or U/Γ (where Γ is a discrete subgroup of $\mathrm{Aut}(U)$). The parallel notion in complex geometry can be underlined using the following observation based on Picard's theorem, which states that *an entire, non-constant function $f : \mathbf{C} \to \mathbf{C}$ takes all complex values except for at most one* (think about the exponential function). We will also use the following topological property of the universal covering: if $\pi : \tilde{S} \to S$ is the universal covering of S, every holomorphic map $f : \mathbf{C} \to S$ can be factored through

π, in other words, there exists a holomorphic map $\tilde{f} : \mathbf{C} \to \tilde{S}$ such that $f = \pi \circ \tilde{f}$.

2.11. Proposition. *Let S be an algebraic Riemann surface, in other words, an algebraic curve of genus g minus s points, and let $\chi(S) := 2 - 2g - s$ be its Euler-Poincaré characteristic. There exists a non-constant, holomorphic map $f : \mathbf{C} \to S$ if and only if $\chi(S) \geqslant 0$.*

Proof. If $g = 0$, then $S = \mathbf{P}^1(\mathbf{C}) \setminus \{P_1, \ldots, P_s\}$, and by Picard's theorem, a non-constant, holomorphic map can only exist if $s = 0, 1$ or 2, i.e., if $S = \mathbf{P}^1(\mathbf{C})$ or $S = \mathbf{P}^1(\mathbf{C}) \setminus \{P_1\} = \mathbf{C}$ or $S = \mathbf{P}^1(\mathbf{C}) \setminus \{P_1, P_2\} = \mathbf{C} \setminus \{0\}$. If $g = 1$, then $S = E(\mathbf{C}) \setminus \{P_1, \ldots, P_s\}$ where $E(\mathbf{C}) = \mathbf{C}/\Omega$. If $s = 0$, we have a holomorphic map from $\mathbf{C} \to \mathbf{C}/\Omega$, but if $s > 0$, we would obtain, by lifting the map to the universal covering $\mathbf{C}$, an entire function which does not take infinitely many values and is therefore necessarily constant. Finally, if $g \geqslant 2$, every holomorphic map $f : \mathbf{C} \to S$ can be factored through the universal covering, which is the disk, and hence f is constant. □

This suggests the following definition, due to Brody.

2.12. Definition. A complex analytic variety X is *hyperbolic* (in Brody's meaning) if every holomorphic map $\mathbf{C} \to X(\mathbf{C})$ is constant.

With this definition, we can see that projective curves which are hyperbolic are exactly those of genus $\geqslant 2$, in other words, those for which the finiteness of the number of rational points was proven by Faltings. Affine curves which are hyperbolic are those of genus $\geqslant 1$ or of genus 0 with at least three points at infinity, in other words, those for which the finiteness of the number of integral points was proven by Siegel.

In order to take into account the case where the images of holomorphic maps are contained in a subvariety, Lang introduced the following definitions.

2.13. Definition. 1) Let X be an algebraic variety defined over $\mathbf{C}$. The *analytic special set* is the closure (for the usual topology) of the union of the images of non-constant holomorphic maps $f : \mathbf{C} \to X(\mathbf{C})$.

2) Let V be an algebraic variety. The *algebraic special set* is the closure (for the Zariski topology) of the union of the images of non-constant maps from an algebraic group to V.

Let us point out that since $\mathbf{P}^1$ is the image of, for example, an elliptic curve, the special set contains all of the rational curves.

The idea of the conjecture, due to Serge Lang, is to attempt to generalize the good dictionary between arithmetic, algebraic and geometric properties to varieties of higher dimension.

2.14. Conjecture. (Lang [45]) *Let V be a projective algebraic variety defined over a number field.*

1) The following three properties are equivalent.

i) The variety V has a finite number of rational points over every number field.

ii) Every subvariety of V (including itself) is pseudo-canonical.

iii) The analytic variety $V(\mathbf{C})$ is hyperbolic.

2) The complement of the algebraic special set of V has a finite number of rational points over every number field.

This conjecture is therefore a theorem if $\dim(V) = 1$, thanks essentially to the result of Faltings. Here is a very concrete open problem: is it true that the set of rational points over $\mathbf{Q}$ of the surface V, defined in $\mathbf{P}^3$ by

$$X_0^5 + X_1^5 + X_2^5 + X_3^5 = 0,$$

lie on a finite set of curves? For example, the lines $X_i + X_j = X_k + X_\ell = 0$ lie on V; are there other curves having infinitely many rational points?

We could state a variation, started by Lang and completed by Vojta [75], of this conjecture concerning the integral points on affine varieties. To do this, it will be convenient to use the following definitions and conventions. We will always be considering an affine variety $U \subset \mathbf{A}^n$ as a projective variety $V \subset \mathbf{P}^n$ minus a *hyperplane section* $D := V \cap H$ where $H := \mathbf{P}^n \setminus \mathbf{A}^n$ is the hyperplane "at infinity". We will assume that D has "*normal crossing*", which means that D is the union of r irreducible smooth components $D_1, D_2, \ldots, D_r$ and the D_i intersect transversely. In particular,

$$\dim D_{i_1} \cap \ldots D_{i_s} \leqslant \dim V - s,$$

and the tangent spaces intersect with the same dimensions. Hironaka's desingularization theorem tells us that an affine variety can always be represented as such.

If $r = \dim U = \dim V$, we will now look at the differential r-forms on V which are regular on U and which have at most a simple pole along D. We denote by $\Omega(V)[D]$ the vector space of these forms, by $g(m, V, D)$ its dimension and by $\Phi_{m,V,D} : U \cdots \to \mathbf{P}^{g(m,V,D)-1}$ the induced map. The *logarithmic Kodaira dimension* is therefore defined as

$$\kappa(V, D) := \max_m \dim \Phi_{m,V,D}(U).$$

In fact, with the given conventions, this integer only depends on U. The space U is said to be *log canonical* if $\kappa(V, D) = \dim U$.

2.15. Conjecture. (Lang-Vojta) *Let V be a smooth projective algebraic variety defined over a number field, D a hyperplane section with normal crossing and $U := V \setminus D$ the corresponding affine variety.*

1) The following three properties are equivalent.

- *i) The variety U possesses a finite number of S-integral points for every number field and finite set of places S.*
- *ii) Every subvariety of U (including itself) is log canonical.*
- *iii) The analytic variety $U(\mathbf{C})$ is hyperbolic.*

2) The complement of the special set of U possesses a finite number of S-integral points for every number field and finite set of places S.

This conjecture is equivalent to Siegel's theorem in the case of curves. Very few cases are known in dimension $\geqslant 2$. For example, according to the conjecture, the surface given by the equation

$$-1 - x^4 + y^4 + z^4 = 0$$

should have a finite number of integral points outside of a finite number of curves, such as the line $x - y = z - 1 = 0$ for example.

We will finish with an example due to Vojta which illustrates the necessity of the "normal crossing" hypothesis. Consider, in the projective plane with coordinates (x, y, z), the hyperplane section D composed of two lines D_1 and D_2 whose equations are given by $x = 0$ and $y = 0$ and the conic D_3 given by $z(x - y) - (x + y)^2 = 0$. We point out that the divisor $D = D_1 + D_2 + D_3$ does not have normal crossing, since $D_1 \cap D_2 \cap D_3 = \{(0, 0, 1)\}$. If it did have normal crossing, the Lang-Vojta conjecture would predict that the S-integral points are not Zariski dense. For $U := \mathbf{P}^2 \setminus D$, the algebra of coordinates of U is generated by the functions $f_1 = \frac{x}{y}$, $f_2 = \frac{z}{y}$, $f_3 = \frac{y}{x}$, $f_4 = \frac{z}{x}$ and $f_5 = \dfrac{4y^2}{z(x-y) - (x+y)^2}$. The S-integral points are therefore the points where the functions take S-integral values. If $k \in \mathbf{Z}$ and $\epsilon \in \mathscr{O}_{K,S}^*$, we define the point

$$P_{k,\epsilon} := \left(\epsilon, 1, \epsilon + 3 - \frac{4(\epsilon^k - 1)}{\epsilon - 1} \right).$$

We can check that $f_5(P_{k,\epsilon}) = -\epsilon^{-k}$, and the points $P_{k,\epsilon}$ are thus all S-integral. If $\mathscr{O}_{K,S}^*$ is infinite, it is clear that the points form a dense set in the plane for the Zariski topology.

3. p-adic Numbers

The technique of going from discrete to continuous—by embedding **Z** *or* **Q** *into* **R***—is classical. There exist completions other than* **R***. In fact, there exists exactly one, up to isomorphism, for each prime number p. These completions, far from being exotic, are actually more rich in arithmetic and topological content. We will briefly describe them here, and we recommend the following texts for further study: [2] and [8].*

3.1. Definition. A *p-adic integer* is an equivalence class[4] of sequences $x := \{x_0, x_1, \dots, x_n, \dots\}$ of integers such that $\forall n$, $x_n \equiv x_{n+1} \bmod p^n$. Two sequences are equivalent if $x_n \equiv x'_n \bmod p^n$.

We can also write the sequence of integers in the form $\{a_0, a_0 + a_1 p, a_0 + a_1 p + a_2 p^2, \dots\}$ and, if we want to, bound it by taking the integers a_i to be in $[0, p-1]$. This suggests the following notation for a p-adic integer, "$x = \sum_{i=0}^{\infty} a_i p^i$", to which we will soon give a more precise meaning.

We can naturally define the operations of sum and product, which endow the p-adic integers with a ring structure, denoted $\mathbf{Z}_p$. Divisibility is particularly simple.

3.2. Lemma. *The ring* $\mathbf{Z}_p$ *is integral. Furthermore, it satisfies the following properties.*

i) $\mathbf{Z}_p^* = \{x = \{x_n\}_{n \geqslant 0} \mid x_0 \not\equiv 0 \bmod p\}$.

ii) Every non-zero element can be written uniquely $x = p^m u$, *where* $m \in \mathbf{N}$ *and* $u \in \mathbf{Z}_p^*$.

Proof. Let $x = \{x_n\}_n \in \mathbf{Z}_p^*$. Since $x_n \equiv x_0 \not\equiv 0 \bmod p$, the integers x_n are relatively prime to p, and we can choose integers x'_n such that x'_n is the inverse of x_n modulo p^n. Therefore, $x'_n \equiv x'_{n+1} \bmod p^n$, and $x' := \{x'_n\}_n$ thus defines a p-adic integer such that $xx' = 1$. If $x = \{x_n\}_n \in \mathbf{Z}_p$, we set $m := \max\{n \mid x_n \equiv 0 \bmod p^n\}$. Then $x_{m+k} = p^m u_k$, where p does not divide u_k. The factorization given in the statement follows from this. □

We denote by $m = \operatorname{ord}_p(x)$ the maximal power of p which divides x.

3.3. Definition. We denote by $\mathbf{Q}_p$ the field of fractions of $\mathbf{Z}_p$; it is called the field of p-adic numbers.

[4] In more scholarly terms, we could define the p-adic integers as $\mathbf{Z}_p = \varprojlim_n \mathbf{Z}/p^n\mathbf{Z}$.

We point out that every p-adic integer is congruent mod p^n to a natural number, i.e., $\mathbf{Z}_p/p^n\mathbf{Z}_p \cong \mathbf{Z}/p^n\mathbf{Z}$. We will now introduce a topology which will make it clear that $\mathbf{Z}_p$ is a completion of $\mathbf{Z}$.

3.4. Definition. We define the *p-adic absolute value* by $|x|_p = p^{-\operatorname{ord}_p(x)}$ (and $|0|_p = 0$). We say that the sequence (u_n) tends to ℓ if $\lim_n |u_n - \ell|_p = 0$.

3.5. Lemma. *For the p-adic topology, the following properties hold.*

i) The closure of $\mathbf{Z}$ *is* $\mathbf{Z}_p$*, which is compact. The field* $\mathbf{Q}_p$ *is locally compact.*

ii) A sequence $(u_n) \in \mathbf{Z}_p$ *converges if and only if* $\lim_n(u_{n+1} - u_n) = 0$*. Likewise, a series* $\sum_n u_n$ *converges if and only if* $\lim_n u_n = 0$.

Proof. Let $x = \{x_n\}_n \in \mathbf{Z}_p$. Then $|x - x_n| \leqslant p^{-n}$, and the sequence of integers therefore converges to x. Next, $\mathbf{Q}_p = \cup_{m\geqslant 0} p^{-m}\mathbf{Z}_p$. The map $x \mapsto (x \bmod p^n)_{n\geqslant 1}$ from $\mathbf{Z}_p$ to $\prod_{n\geqslant 1} \mathbf{Z}/p^n\mathbf{Z}$ is injective and continuous, and its image is closed. The compactness of $\mathbf{Z}_p$ follows from the compactness of the product $\prod_{n\geqslant 1} \mathbf{Z}/p^n\mathbf{Z}$. The second assertion comes from the ultrametric inequality $|u_M + \cdots + u_N|_p \leqslant \max_{M\leqslant n\leqslant N}\{|u_n|_p\}$. □

3.6. Examples. If $a_n \in \mathbf{Z}$, then the series $\sum_n a_n t^n$ converges for $|t|_p < 1$, in other words, for $t \in p\mathbf{Z}_p$. Thus, the series $\sum_n a_n p^n$ indeed defines a p-adic number (it is the analogue of the decimal expansion of a real number). The "logarithm" series, $\sum_{n\geqslant 1} \frac{t^n}{n}$, also converges for $|t|_p < 1$, because $\operatorname{ord}_p(t^n/n) = n\operatorname{ord}_p(t) - \operatorname{ord}_p(n) \geqslant n\operatorname{ord}_p(t) - \log n/\log p$. The "exponential" series, $\sum_n \frac{t^n}{n!}$, converges if $\operatorname{ord}_p(t) > 1/(p-1)$ or $|t|_p < p^{-\frac{1}{p-1}}$. In fact,

$$\operatorname{ord}_p(n!) = \left\lfloor \frac{n}{p} \right\rfloor + \cdots + \left\lfloor \frac{n}{p^m} \right\rfloor + \ldots \leqslant n\left(\frac{1}{p} + \cdots + \frac{1}{p^m} + \ldots\right) = \frac{n}{p-1}.$$

3.7. Theorem. *Let* $F \in \mathbf{Z}[X_1, \ldots, X_n]$*. The following statements are equivalent.*

i) $\forall m, \exists x \in \mathbf{Z}^n$ *such that* $F(x) \equiv 0 \bmod p^m$.

ii) $\exists x \in (\mathbf{Z}_p)^n$ *such that* $F(x) = 0$ *(in* $\mathbf{Z}_p$*).*

Proof. If $x \in (\mathbf{Z}_p)^n$ satisfies $F(x) = 0$, then x is congruent modulo p^m to an n-tuple of integers. Conversely, if we had $x^{(m)} \in \mathbf{Z}^n$ such that $F(x^{(m)}) \equiv 0 \bmod p^m$, then we could extract a sequence such that $x^{(m+1)} \equiv x^{(m)} \bmod p^m$. We could therefore define, in the p-adics, $x = \lim_m x^{(m)}$

and would then have $F(x) \equiv F(x^{(m)}) \equiv 0 \bmod p^m$ for every m, and hence $F(x) = 0$. □

We denote by $\nabla F(x) := \left(\dfrac{\partial F}{\partial X_1}(x), \ldots, \dfrac{\partial F}{\partial X_n}(x)\right)$ the gradient of F. We will now introduce the p-adic analogue of Newton's method for finding the zeros of functions or polynomials.

3.8. Theorem. (Hensel's lemma) *Let $F \in \mathbf{Z}_p[X_1, \ldots, X_n]$, $\delta \geqslant 0$ and $x_0 \in (\mathbf{Z}_p)^n$ such that:*

i) $F(x_0) \equiv 0 \bmod p^{2\delta+1}$,

ii) $\nabla F(x_0) \equiv 0 \bmod p^{\delta}$, *but* $\nabla F(x_0) \not\equiv 0 \bmod p^{\delta+1}$.

Then there exists $x \in (\mathbf{Z}_p)^n$ such that $x \equiv x_0 \bmod p^{\delta+1}$ and $F(x) = 0$. In particular, a smooth point on the hypersurface $F = 0$ modulo p lifts to $\mathbf{Z}_p$.

Proof. With the notation given in the statement, we can write $F(x_0) = p^{2\delta+1}a$ where $a \in \mathbf{Z}_p$, $\nabla F(x_0) = p^{\delta}b$, $b \in (\mathbf{Z}_p)^n$ and $b \not\equiv 0 \bmod p$. Then we have

$$F(x_0 + p^{\delta+1}u) \equiv F(x_0) + p^{\delta+1}\nabla F(x_0) \cdot u \equiv p^{2\delta+1}(a + b \cdot u) \bmod p^{2\delta+2}.$$

This yields a solution $x_1 = x_0 + p^{\delta+1}u$ such that $F(x_1) \equiv 0 \bmod p^{2\delta+2}$ as soon as $a + b \cdot u \equiv 0 \bmod p$, which is possible because $b \not\equiv 0 \bmod p$. By iterating this procedure, we obtain a sequence (x_m) where $x_{m+1} \equiv x_m \bmod p^{\delta+m+1}$ and $F(x_m) \equiv 0 \bmod p^{2\delta+m+1}$. The sequence therefore converges in $\mathbf{Z}_p$ to x, and since $F(x) \equiv 0 \bmod p^m$ for all m, we have indeed found x such that $F(x) = 0$. □

3.9. Example. The simplest application of this lemma is to a polynomial $P \in \mathbf{Z}[X]$ such that $P(a_0) \equiv 0 \bmod p$ but $P'(a_0) \not\equiv 0 \bmod p$. Hensel's lemma gives us a way to construct a root $a \in \mathbf{Z}_p$ of the polynomial P such that $a \equiv a_0 \bmod p$.

3.10. Remark. This theorem, together with the result of Lang-Weil (Theorem 6-1.6), allows us to find an algorithm for deciding if an equation is solvable $\bmod N$ for every integer N or, in the same fashion, if it is solvable in $\mathbf{Z}_p$ or $\mathbf{Q}_p$ for all p. To see this, take the case of a polynomial $F \in \mathbf{Z}[X_1, \ldots, X_n]$, which we will assume to be irreducible. The Lang-Weil estimates show that the equation modulo p possesses roughly p^{n-1} solutions, while the number of singular solutions is $O(p^{n-2})$. For large enough p, there will be a nonsingular solution modulo p and hence, by Hensel's lemma, a lifting to $\mathbf{Z}_p$ of this solution. For a given p, the previous theorem essentially provides an algorithm which tells us that either there exists a solution in $\mathbf{Z}_p$ or there exists δ such that the equation is not solvable modulo p^δ.

3.11. Remark. Hensel's lemma allows us to specify the structure of the group of p-adic units $U := \mathbf{Z}_p^*$. For this, we will introduce the subgroups $U_m := \{x \in U \mid x \equiv 1 \bmod p^m\}$.

3.12. Lemma. *Let p be odd. There exists a unique subgroup $\mu_{p-1} \subset \mathbf{Z}_p^*$ isomorphic to $U/U_1 \cong \mathbf{F}_p^*$. If $m \geqslant 1$, there is an isomorphism $U_m/U_{m+1} \cong \mathbf{Z}/p\mathbf{Z}$. In particular, as topological groups,*

$$\mathbf{Z}_p^* \cong \mathbf{Z}/(p-1)\mathbf{Z} \times \mathbf{Z}_p.$$

If $p = 2$, then $\mathbf{Z}_2^* \cong \{\pm 1\} \times U_2 \cong \mathbf{Z}/2\mathbf{Z} \times \mathbf{Z}_2$.

Proof. The proof immediately follows from Hensel's lemma: the solutions of $x^{p-1} \equiv 1 \bmod p$ can be lifted to $\mathbf{Z}_p$, and the map $x \to 1 + px$ induces a bijection from $\mathbf{Z}_p$ to U_1, then an isomorphism from U_1/U_2 to $\mathbf{Z}/p\mathbf{Z}$, which proves the second part of the statement when $p \neq 2$. The case $p = 2$ can be treated similarly. We could also notice that the "logarithm" map $U_1 \to p\mathbf{Z}_p$, given by $1 + px \mapsto \sum_n (-1)^{n+1} p^n x^n / n$, and the "exponential" map $p\mathbf{Z}_p \to U_1$ provide the desired isomorphism. □

Using Hensel's lemma, we can also completely study the p-adic squares.

3.13. Lemma. *The squares in $\mathbf{Q}_p^*$ can be described as follows.*

i) If p is odd, any unit $u \in \mathbf{Z}_p^*$ where $u \equiv 1 \bmod p$ is a square. Furthermore, $(\mathbf{Q}_p^* : \mathbf{Q}_p^{*2}) = 4$, and representatives of the classes are given by $1, \epsilon, p, \epsilon p$ where ϵ is not a square modulo p, i.e., $\left(\frac{\epsilon}{p}\right) = -1$.

ii) If $p = 2$, a unit $u \in \mathbf{Z}_2^*$ with $u \equiv 1 \bmod 8$ is a square. Also, $(\mathbf{Q}_2^* : \mathbf{Q}_2^{*2}) = 8$, and the representatives are $\{\pm 1, \pm 2, \pm 3, \pm 6\}$.

Proof. Consider the equation $F(x) = x^2 - u = 0$. For p odd, if $u \equiv 1 \bmod p$, then $F(1) \equiv 0 \bmod p$ and $F'(1) = 2 \not\equiv 0 \bmod p$. More generally, if $u \equiv v^2 \bmod p$, then $F(v) \equiv 0 \bmod p$ and $F'(v) = 2v \not\equiv 0 \bmod p$. Hensel's lemma therefore gives an $x \in \mathbf{Z}_p$ such that $x^2 = u$. We can thus see that a p-adic number $y = p^m u$ (where $m \in \mathbf{Z}$ and $u \in \mathbf{Z}_p^*$) is a square if and only if m is even and u is a square modulo p. If now $p = 2$, as soon as we have $x_0 \not\equiv 0 \bmod 2$, where $F(x_0) \equiv 0 \bmod 2^3$, we can apply Hensel's lemma (with "δ" equal to 1) and deduce that u is a square. The only remaining point to check is the congruence $x^2 \equiv u \bmod 8$ for odd u. □

In order to study quadratic forms over $\mathbf{Q}_p$, we can, as with a field of characteristic $\neq 2$, reduce to the case of diagonal forms, $a_1x_1^2 + \cdots + a_nx_n^2$ with $a_i \in \mathbf{Z}_p$, and then, by factoring $p^2x^2 = (px)^2$, reduce to the case where $\mathrm{ord}_p(a_i) = 0$ or 1. To summarize, it suffices to study forms of the type

$$\begin{aligned} Q(x_1, \ldots, x_n) &= Q_1(x_1, \ldots, x_s) + pQ_2(x_{s+1}, \ldots, x_n) \\ &= a_1x_1^2 + \cdots + a_sx_s^2 + p(a_{s+1}x_{s+1}^2 + \cdots + a_nx_n^2) \end{aligned}$$

where $a_i \in \mathbf{Z}_p^*$. We can easily see that there exists $x \neq 0$ such that $Q(x) = 0$ if and only if there exists a nontrivial zero of Q_1 or Q_2. Hensel's lemma yields the following result.

3.14. Lemma. *Let p be an odd prime. The equation $a_1x_1^2 + a_2x^2 + a_3x_3^2 = 0$, where $a_i \in \mathbf{Z}_p^*$, has a nontrivial zero in $\mathbf{Q}_p$.*

Proof. If $p \neq 2$, we can find a nonsingular point $\bmod p$ on the conic, which lifts to a p-adic point. □

3.15. Corollary. *Any quadratic form in $n \geqslant 5$ variables has a nontrivial zero in $\mathbf{Q}_p$.*

Proof. If p is odd, we can write $Q(x) = a_1x_1 + \cdots + a_sx_s^2 + p(a_{s+1}x_{s+1}^2 + \cdots + a_nx_n^2)$ where $a_i \not\equiv 0 \bmod p$. Since either $s \geqslant 3$ or $n - s \geqslant 3$, the result follows from the previous lemma. If $p = 2$, the proof can be deduced from the lemmas and remarks below. □

3.16. Remark. Quadratic forms are thus a little bit more complicated over $\mathbf{Q}_2$ as shown in the following example—essentially treated in the study of sums of squares in Chap. 3. The quadratic form $Q(x, y, z, t) := x^2 + y^2 + z^2 - 7t^2$ does not have any nontrivial zero in $(\mathbf{Q}_2)^4$. We can nevertheless prove, for example, the following lemma, whose proof is also based on Hensel's lemma and left to the reader.

3.17. Lemma. *a) Let $Q(x) = a_1x_1^2 + \cdots + a_sx_s^2$ with $a_i \in \mathbf{Z}_2^*$. A primitive solution to $Q(x) \equiv 0 \bmod 8$ can be lifted to $\mathbf{Z}_2^s$.*
b) Let $Q(x) = a_1x_1^2 + \cdots + a_sx_s^2 + 2a_{s+1}x_{s+1}^2 + \cdots + 2a_nx_n^2$ with $a_i \in \mathbf{Z}_2^*$. A primitive solution to $Q(x) \equiv 0 \bmod 16$ can be lifted to $\mathbf{Z}_2^n$.

The following theorem is an archetypal local-global arithmetic theorem.

3.18. Theorem. (Hasse-Minkowski) *If $Q(x) = \sum_{i,j} a_{i,j}x_ix_j$ is a quadratic form with rational coefficients, then it has a nontrivial rational zero if and only if it has a nontrivial zero in $\mathbf{R}$ and in every $\mathbf{Q}_p$.*

For the proof, see [2] or [8]. In reference to this theorem, a class of varieties defined over $\mathbf{Q}$ is said to "*satisfy the Hasse principle*" if the existence of a real point and a p-adic point for every p implies the existence of a rational point. We can also reformulate the Hasse-Minkowski theorem as saying that the quadrics or hypersurfaces of degree 2 satisfy the Hasse principle.

3.19. Corollary. *The quadratic form* $Q(x, y, z, t) = x^2 + y^2 + z^2 - mt^2$ *has a nontrivial rational zero if and only if* m *is not of the form* $4^a(8b+7)$.

Proof. We apply the preceding theorem by observing that the quadratic form $Q(x, y, z, t)$ always has, by Hensel's lemma, a nontrivial zero in $\mathbf{Q}_p$ for odd p. It also has a nontrivial real zero by assumption. Finally, it has a nontrivial zero in $\mathbf{Q}_2$ if and only if the condition from the corollary below is satisfied. □

3.20. Corollary. *A quadratic form in five variables* $Q(x, y, z, t, u)$ *has a nontrivial rational zero if and only if it is neither positive-definite nor negative-definite.*

The Hasse-Minkowski theorem cannot be generalized to hypersurfaces or varieties of higher degree. Exercises 1-6.23 and 3-6.24 give examples of equations having (nontrivial) solutions in $\mathbf{R}$ and in each $\mathbf{Q}_p$ but none in $\mathbf{Q}$. If we examine the case of cubic hypersurfaces, for which there is always a real zero, we know however how to prove the following result.

3.21. Proposition. (Lewis [52]) *A cubic hypersurface* $F(x_1, \ldots, x_n) = 0$ *has a nontrivial zero in* $\mathbf{Q}_p$ *whenever* $n \geqslant 10$. *There exist cubic hypersurfaces with no nontrivial zeros in* $\mathbf{Q}_p$ *when* $n = 9$.

3.22. Theorem. *Let* $F(x_1, \ldots, x_n) = 0$ *be a smooth cubic form with coefficients in* $\mathbf{Z}$.

1) (Heath-Brown [36]) *If* $n \geqslant 10$, *the form* F *has a nontrivial rational zero.*
2) (Hooley [39]) *If* $n = 9$ *and if the form* F *has a nontrivial* p*-adic zero for every* p, *then it has a nontrivial rational zero.*

It is actually fairly easy to construct cubic forms in 9 variables without any nontrivial zeros in $\mathbf{Q}_p$. We will start off with a cubic form in three variables over $\mathbf{F}_p$ having only the trivial zero in $\mathbf{F}_p$. It suffices to take, for example, $G_0(x, y, z) = \mathrm{N}(x + y\omega + z\omega^2)$ where $1, \omega, \omega^2$ is a basis for $\mathbf{F}_{p^3}$ over $\mathbf{F}_p$ and $\mathrm{N} = \mathrm{N}_{\mathbf{F}_p}^{\mathbf{F}_{p^3}}$. We then take a form $G(x, y, z)$ with coefficients in $\mathbf{Z}_p$ such that

$G \bmod p$ coincides with G_0, and we set

$$F(x_1, \dots, x_9) := G(x_1, x_2, x_3) + pG(x_4, x_5, x_6) + p^2G(x_7, x_8, x_9).$$

If F had a nontrivial zero $x \in (\mathbf{Z}_p)^9$, there would exist one such that $x \not\equiv 0 \bmod p$. But by reducing modulo p, we would have $G(x_1, x_2, x_3) \equiv 0 \bmod p$, hence p would divide x_1, x_2 and x_3. By reducing modulo p^2, we can infer that $pG(x_4, x_5, x_6) \equiv 0 \bmod p^2$, hence p would divide x_4, x_5 and x_6. Finally, by reducing modulo p^3, we would have $p^2G(x_7, x_8, x_9) \equiv 0 \bmod p^3$, hence p would divide x_7, x_8 and x_9, which yields a contradiction.

The following statement is part of the folklore.

3.23. Conjecture.

1) A cubic form in 10 or more variables represents zero over $\mathbf{Q}$.

2) A cubic form in 9 or more variables represents zero over $\mathbf{Q}$ *if and only if it represents zeros over every* $\mathbf{Q}_p$.

We know, by Davenport, that a cubic form in 16 or more variables represents zero over $\mathbf{Q}$. An optimistic version of the conjecture would be to replace 9 by 5 in *2)*. However, there do exist forms in 4 variables which contradict the Hasse principle. One of the first counterexamples is due to Cassels and Guy:

$$5x^3 + 12y^3 + 9z^3 + 10t^3 = 0.$$

Here is an even simpler one, due to Birch and Swinnerton-Dyer (see [13]):

$$-5x^3 + 22y^3 + 2z^3 + 4w^3 - 6zw(x + 4y) = 0.$$

The proof of the nonexistence of (nontrivial) rational solutions is given in Exercise 3-6.24, where this equation is written (with $\alpha := \sqrt[3]{2}$) in the form:

$$\mathrm{N}_{\mathbf{Q}}^{K}\left(x + 4y + z\alpha + w\alpha^2\right) - 6(x + y)(x^2 + xy + 7y^2) = 0.$$

3.24. Remark. No difficulties arise when generalizing what we have presented to p-adic completions of $\mathbf{Q}$ in the case of a number field K equipped with a non-zero prime ideal $\mathfrak{p}$. We define the $\mathfrak{p}$-adic absolute value by $|x|_{\mathfrak{p}} := \mathrm{N}\,\mathfrak{p}^{-\operatorname{ord}_{\mathfrak{p}}(x)}$ and the ring of $\mathfrak{p}$-adic integers (resp. the field of $\mathfrak{p}$-adic numbers) by:

$$\mathscr{O}_{\mathfrak{p}} := \varprojlim_{n} \left(\mathscr{O}_K/\mathfrak{p}^n\right), \ (\text{resp. } K_{\mathfrak{p}} = \mathrm{Frac}(\mathscr{O}_{\mathfrak{p}})).$$

3.25. Remarks. Adeles and ideles. We can regroup the p-adic completions into a global object which proves to be very interesting. We denote by M_K, as in Chap. 5, the set of places of K, i.e., the union of the set M_K^∞ of Archimedean places (real embeddings and pairs of complex embedding) and the set of (non-zero) prime ideals of $\mathcal{O}_K$. For every finite set S of places of K which contains the Archimedean places, we set

$$\mathrm{A}_K^S := \prod_{v\in S} K_v \times \prod_{v\notin S} \mathcal{O}_{K_v}.$$

These sets are endowed with the product topology, and their union therefore also inherits a topology.

3.26. Definition. The ring of *adeles* of K is the ring

$$\mathrm{A}_K := \bigcup_S \mathrm{A}_K^S.$$

We can also define the adeles as the set of sequences $x = (x_v)_{v\in M_K}$ such that $x_v \in K_v$ and for every $v \in M_K$, except for a finite number of them (dependent on x), we have $x_v \in \mathcal{O}_v$. The field K is embedded diagonally in the adeles, and one important property is that A/K is compact. Every local field K_v is also embedded in the adeles by $x_v \mapsto (0, \dots, 0, x_v, 0, \dots)$.

3.27. Definition. The group of *ideles* of K is the group

$$\mathrm{J}_K := \bigcup_S \mathrm{J}_K^S = \bigcup_S \left(\mathrm{A}_K^S\right)^* = \mathrm{A}_K^*.$$

Ideles are naturally endowed with the topology inherited from the product topology on $\mathrm{J}_K^S = \prod_{v\in S} K_v^* \times \prod_{v\notin S} \mathcal{O}_{K_v}^*$. We should however point out that this topology is different from the topology induced by the inclusion $\mathrm{J}_K \subset \mathrm{A}_K$. We can also define the ideles as the set of sequences $x = (x_v)_{v\in M_K}$ such that $x_v \in K_v^*$ and for every $v \in M_K$, except for a finite number of them (dependent on x), we have $x_v \in \mathcal{O}_v^*$. Every finite extension, L/K, has a natural "norm" map:

$$\mathrm{N}_K^L : \mathrm{J}_L \to \mathrm{J}_K.$$

The multiplicative group of the field K^* is embedded diagonally in the ideles. Every multiplicative group K_v^* is also embedded in the ideles by $x_v \mapsto (1, \dots, 1, x_v, 1, \dots)$. We have a numerical norm on the ideles defined for $x = (x_v)_{v\in M_K}$ by

$$\|x\| := \prod_{v\in M_K} |x_v|_v.$$

The kernel of the norm contains K^* (cf. Theorem 5-2.1.5) and is denoted J_K^0. An important property of J_K^0/K^* is that it is compact.

By considering the map from J_K to the fractional ideals of K which associates to $(x_v)_{v\in M_K}$ the ideal $\prod_{\mathfrak{p}} \mathfrak{p}^{\mathrm{ord}_{\mathfrak{p}} x_{\mathfrak{p}}}$, it can be shown that we have an isomorphism:

$$\mathrm{J}_K/\mathrm{J}_K^{S_\infty}K^* \cong C\ell_K. \tag{6.15}$$

In fact, the compactness of J_K^0/K^* is equivalent to the combination of the finiteness of the class group and Dirichlet's unit theorem.

4. Transcendental Numbers and Diophantine Approximation

A proof that a number, such as π or e, is transcendental resembles, at least formally, a proof in Diophantine approximation. This classic theme is expanded on in Baker's book [1] and in [12]. We offer you here a first taste of this theory.

It can be said that the theory of Diophantine approximation and transcendental numbers starts with Liouville's result, which says that if α is an algebraic number of degree $d := [\mathbf{Q}(\alpha) : \mathbf{Q}] > 1$, then there exists a constant $C = C(\alpha)$ such that for every rational number, we have $\left|\alpha - \frac{p}{q}\right| \geqslant \frac{C}{q^d}$.

We have seen in the proof of Siegel's theorem (Theorem 5-4.1) that it was essential to improve on the exponent d. The first result in this direction is due to Thue (1909).

4.1. Theorem. (Thue) *Let α be an algebraic number of degree $d = [\mathbf{Q}(\alpha) : \mathbf{Q}] > 1$ and $\epsilon > 0$. There exists a constant $C = C(\alpha, \epsilon)$ such that for every rational number p/q, we have*

$$\left|\alpha - \frac{p}{q}\right| \geqslant \frac{C}{q^{\frac{d}{2}+1+\epsilon}}. \tag{6.16}$$

The proof is sketched further down. Successive improvements are due to Siegel (1921), who showed that the exponent $\frac{d}{2} + 1 + \epsilon$ can be replaced by $2\sqrt{d} + \epsilon$ (which is an improvement on Thue's theorem when $d \geqslant 12$); Gel'fond and Dyson (1947), who proved that the exponent can be replaced by $\sqrt{2d} + \epsilon$; and finally Roth (1954), who proved a result which is essentially optimal considering Dirichlet's theorem (Corollary 3-3.8). The Fields Medal was awarded to Roth in 1958 for this proof.

4.2. Theorem. (Roth) *Let $\alpha \notin \mathbf{Q}$ be an algebraic number and $\epsilon > 0$. There exists a constant $C = C(\alpha, \epsilon) > 0$ such that for every rational number*

p/q we have

$$\left|\alpha - \frac{p}{q}\right| \geqslant \frac{C}{q^{2+\epsilon}}. \qquad (6.17)$$

One of the most important theorems which is purely about transcendence is due to Baker (1966). It concerns linear forms of logarithms. The Fields Medal was awarded to Baker in 1970 for this result and its numerous applications. We denote by $\log \alpha$ a complex number β such that $\exp(\beta) = \alpha$; for example, we can write $2i\pi = \log 1$.

4.3. Theorem. (Baker) *Let $\alpha_1, \ldots, \alpha_n$ be non-zero algebraic numbers. If $\log \alpha_1, \ldots, \log \alpha_n$ are $\mathbf{Q}$-linearly independent, then the numbers $1, \log \alpha_1, \ldots, \log \alpha_n$ are $\bar{\mathbf{Q}}$-linearly independent.*

4.4. Remark. We can recover as a corollary to this theorem a certain number of classical results on transcendence.

i) The number e is transcendental (Hermite, 1873); this is true because if it were algebraic then the number $1 = \log e$ would be transcendental.
ii) The number π is transcendental (Lindemann, 1882). This is true because the number $2\pi i = \log 1$ is transcendental.
iii) If $\alpha \in \bar{\mathbf{Q}} \setminus \{0, 1\}$ and $\beta \in \bar{\mathbf{Q}} \setminus \mathbf{Q}$, the number $\gamma := \alpha^{\beta}$ is transcendental (Gel'fond and Schneider, 1934). This is because if it were algebraic, since $\log \gamma - \beta \log \alpha = 0$, we could deduce that $\beta \in \mathbf{Q}$.

Further down, we will give (Theorem 6-4.15) a quantitative version of Baker's assertion, which has become a fundamental tool in the study of Diophantine equations.

Schematically, the proofs of all of these theorems follow the same pattern as the proof of Liouville's result (see [12] for a more complete picture).

- 1st step. We start by constructing a polynomial with integer coefficients which vanishes at designated points α or at least takes a very small value at them. In Liouville's proof, a minimal polynomial of the algebraic number α is chosen, and in the general case, an elementary lemma formalized by Siegel is used (see Lemmas 6-4.5 and 6-4.6).

- 2nd step. We use the fact that we can control the size of the coefficients of the polynomial F to conclude that F is still very small at an algebraic point β close to α. To do this, we make use of a lemma by Schwarz (Lemma 6-4.10) or also Taylor's formula (Lemma 6-4.7).

- 3rd step. We prove that the polynomial F must vanish at β or that the height of β must be large. To do this, we use Liouville's inequality, which says, in its most rudimentary form, that a non-zero integer has absolute value $\geqslant 1$ (see Corollary 6-4.12).

- 4th step. We prove a "zeros estimate" adapted to the situation, which allows us to control the location of the zeros of F. If $F \in \mathbf{Z}[X]$, we can in general settle for counting the zeros of F, but if $F \in \mathbf{Z}[X_1, \ldots, X_n]$, where $n \geqslant 2$, this step could prove to be very difficult.

We will now present some lemmas which are useful for putting this method into action.

4.5. Lemma. (Siegel's Lemma I) *Let $N > M$, and let the following be a system of linear equations with integer coefficients a_{ij} not all equal to zero:*

$$\begin{cases} a_{11}x_1 + \cdots + a_{1N}x_N & = 0 \\ \vdots \quad \vdots & \vdots \\ a_{M1}x_1 + \cdots + a_{MN}x_N & = 0. \end{cases}$$

Then there exists a nontrivial solution $(x_1, \ldots, x_N) \in \mathbf{Z}^N$ which satisfies

$$\max |x_i| \leqslant \left(N \max_{i,j} |a_{ij}| \right)^{\frac{M}{N-M}}.$$

Proof. The proof is another application of the pigeonhole principle. Set $a_{ij}^+ = \max(0, a_{ij})$, $a_{ij}^- = \max(0 - a_{ij})$ and $L_i := \sum_j |a_{ij}|$, and observe that we can assume that $L_i \geqslant 1$ (if not, the corresponding equation is trivial and can be omitted). Choose $X := \left\lfloor (L_1 \cdots L_M)^{\frac{1}{N-M}} \right\rfloor$, and consider the map from $[0, X]^N$ to $\mathbf{Z}^M$ given by

$$L(x_1, \ldots, x_N) = (a_{11}x_1 + \cdots + a_{1N}x_N, \ldots, a_{M1}x_1 + \cdots + a_{MN}x_N).$$

We clearly have:

$$-X \sum_j a_{ij}^- \leqslant a_{i1}x_1 + \cdots + a_{iN}x_N \leqslant X \sum_j a_{ij}^+.$$

The number of values taken by $L(x)$ is thus at most

$$\prod_{i=1}^{M} (X \sum_j a_{ij}^- + X \sum_j a_{ij}^+ + 1) = \prod_{i=1}^{M} (XL_i + 1).$$

By the initial choice of X, we have the inequality $(X+1)^{N-M} > L_1 \cdots L_M$, and hence

$$(X+1)^N > \prod_{i=1}^{M} (XL_i + L_i) \geqslant \prod_{i=1}^{M} (XL_i + 1).$$

Consequently, there exist two distinct elements $x', x'' \in [0, X] \cap \mathbf{Z}^N$ such that $L(x') = L(x'')$. The element $x := x' - x''$ is thus in $\mathbf{Z}^N \setminus \{0\}$ and satisfies $L(x) = 0$ and

$$0 < \max_i |x_i| \leqslant (L_1 \cdots L_M)^{\frac{1}{N-M}} \leqslant \left(N \max_{i,j} |a_{ij}|\right)^{\frac{M}{N-M}}. \qquad \square$$

By observing that $\log \max |x_i|$ is the (logarithmic) height of a solution, we can remember the upper bound (neglecting a term in $\log N$) as:

$$\text{height of a solution} \leqslant (\text{height of the equations}) \times \frac{\text{number of equations}}{\text{dimension of the solutions}}.$$

We will state, without proof (see for example [38]), the version where the coefficients a_{ij} are in a number field. The idea is the same: an equation with coefficients in K provides $d := [K : \mathbf{Q}]$ equations with coefficients in $\mathbf{Q}$.

4.6. Lemma. (Siegel's Lemma II) *Let K be a number field of degree $d := [K : \mathbf{Q}]$ and $a_{ij} \in K$ not all zero. Suppose that $dM < N$, and let $A := H(\ldots, a_{ij}, \ldots)$. Then the linear system*

$$\begin{cases} a_{11}x_1 + \cdots + a_{1N}x_N &= 0 \\ \vdots \quad \vdots & \vdots \\ a_{M1}x_1 + \cdots + a_{MN}x_N &= 0 \end{cases}$$

has a nontrivial solution $(x_1, \ldots, x_N) \in \mathbf{Z}^N$ such that

$$\max |x_i| \leqslant (NA)^{\frac{dM}{N-dM}}.$$

The following lemma is a version of the following principle: if a polynomial vanishes with a large order at a point, then the polynomial takes small values in a neighborhood of this point if the coefficients of the polynomial are not too large.

4.7. Lemma. (Application of Taylor's formula) *Let $P \in \mathbf{C}[X_1, \ldots, X_m]$ be a polynomial of degree $\leqslant D$ such that the absolute value of the coefficients is $\leqslant ||P||$. Suppose that P vanishes with order T at $\alpha = (\alpha_1, \ldots, \alpha_m)$. If $\beta = (\beta_1, \ldots, \beta_m)$ satisfies $|\alpha_i - \beta_i| \leqslant \epsilon$, then for $|i| = i_1 + \cdots + i_m < T$,*

$$\left| \frac{1}{i_1! \cdots i_m!} \frac{\partial^{|i|} P}{\partial X_1^{i_1} \cdots \partial X_m^{i_m}} (\beta) \right| \leqslant (3^m A)^D \epsilon^{T-|i|} ||P||$$

where $A := \max\{1, |\alpha_1|, \dots, |\alpha_m|\}$.

Proof. We write Taylor's formula at α for $Q := \dfrac{\partial^{|i|}P}{\partial X_1^{i_1}\cdots\partial X_m^{i_m}}$ as:

$$Q(\beta) = \sum_{|j|\geqslant T-|i|} \frac{1}{j_1!\cdots j_m!}\frac{\partial^{|j|}Q}{\partial X_1^{j_1}\cdots\partial X_m^{j_m}}(\alpha)(\beta_1-\alpha_1)^{j_1}\cdots(\beta_m-\alpha_m)^{j_m}.$$

The product of the $(\alpha_i - \beta_i)$ has absolute value $\leqslant \epsilon^{T-|i|}$. By expanding the sum as factors, we have an upper bound given by a sum of the type

$$\sum_j\sum_h \frac{h!}{i!j!(h-i-j)!}||P||A^{h-i-j} \leqslant 3^{mD}A^D||P||,$$

as in the statement of the lemma. □

4.8. Definition. Let $\alpha \in D(0,1)$. The *Blaschke factor* associated to α is the function

$$B_\alpha(z) := \frac{z-\alpha}{1-\bar{\alpha}z}\cdot \tag{6.18}$$

Some very simple properties of this factor are summarized in the following lemma, whose proof is left to the reader.

4.9. Lemma. *The Blaschke factor has the following properties.*

i) The function $B_\alpha(z)$ is holomorphic on the closed disk $\bar{D}(0,1)$ and has a unique simple zero at $z = \alpha$.

ii) If $|z| = 1$, then $|B_\alpha(z)| = 1$. In particular, $||B_\alpha||_1 : \sup_{|z|\leqslant r}|B_\alpha(z)| = 1$.

iii) For $z \in D(0,1)$, we have the upper bound

$$|B_\alpha(z)| \leqslant \frac{|z|+|\alpha|}{1-|\alpha||z|}\cdot$$

A classical lemma of Schwarz (see for example [74]) says that a holomorphic function $g(z)$ on $D(0,1)$ such that $g(0) = 0$ and $|g(z)| \leqslant 1$ satisfies $|g(z)| \leqslant |z|$ and $||g||_r \leqslant r$. The lemma stated below is a refinement of this.

4.10. Lemma. (Schwarz lemma) *Let $f(z)$ be a function which is holomorphic on the closed disk $D(0,R)$. Suppose that f vanishes with order T at $z = 0$, and we denote by $\|f\|_r = \sup_{|z|=r}|f(z)| = \sup_{|z|\leqslant r}|f(z)|$. Then*

$$\|f\|_r \leqslant \left(\frac{r}{R}\right)^T \|f\|_R.$$

Slightly more generally, if f has zeros of order T_i at $z_i \in D(0,r_0)$, with

$\sum_i T_i = T$, *then we have*

$$||f||_r \leqslant \prod_i \left(\frac{R(r+|z_i|)}{R^2 - r|z_i|} \right)^{T_i} ||f||_R \leqslant \left(\frac{R(r+r_0)}{R^2 - rr_0} \right)^T ||f||_R. \tag{6.19}$$

Proof. We will prove the second inequality, which is more general. We introduce the function $g(z) := f(Rz)$, which is holomorphic on the closed disk $\bar{D}(0,1)$ and has zeros of order T_i at $\alpha_i := z_i/R$. Next, we set

$$g^*(z) := \frac{g(z)}{\prod_i B_{\alpha_i}(z)^{T_i}}.$$

The function $g^*(z)$ is holomorphic on the closed disk, and on the circle $|z| = 1$, we have $|g^*(z)| = |g(z)|$ (by property *ii*) of Lemma 6-4.9). In particular, $||g^*||_1 = ||g||_1 = ||f||_R$. This implies the inequalities:

$$|g(z)| = \left| g^*(z) \prod_i B_{\alpha_i}(z)^{T_i} \right| \leqslant \prod_i \left(\frac{|z| + |\alpha_i|}{1 - |z||\alpha_i|} \right)^{T_i} ||g^*||_1,$$

from which we can deduce that

$$||f||_r = ||g||_{\frac{r}{R}} \leqslant \prod_i \left(\frac{\frac{r}{R} + \frac{|z_i|}{R}}{1 - \frac{r|z_i|}{R^2}} \right)^{T_i} ||f||_R \leqslant \left(\frac{R(r+r_0)}{R^2 - rr_0} \right)^T ||f||_R.$$

□

4.11. Lemma. (Liouville's Inequality) *Let K be a number field of degree d and v a normalized absolute value. If $\alpha \in K^*$, then*

$$|\alpha|_v \geqslant H(\alpha)^{-d} \geqslant H_K(\alpha)^{-1}.$$

Proof. This follows easily from the construction of the Weil height and from the observation that $H(\alpha) = H(\alpha^{-1})$. In fact,

$$H_K(\alpha) = H_K(\alpha^{-1}) = \prod_{v \in M_K} \max\left\{1, \left|\alpha^{-1}\right|_v\right\} \geqslant |\alpha|_v^{-1},$$

and the desired inequality follows from this. □

This lemma is most often used to prove that an algebraic number with a controlled height is zero whenever its absolute value is sufficiently small, or that "an arithmetic quantity cannot be too small without being zero". We will state an explicit corollary of this type.

4.12. Corollary. *Let α be an algebraic number in a number field K and let v be a place of K. If $|\alpha|_v < H_K(\alpha)^{-1}$, then $\alpha = 0$.*

As a zeros estimate, we will start with an (easy) example in many variables and prove an elementary lemma which will help us in the proof of Thue's theorem.

4.13. Lemma. *Let $P \in \mathbf{C}[X_1, \ldots, X_n]$ be a non-zero polynomial and S a finite set of complex numbers. If $d := \deg P < |S|$, then there exists $x \in S^n$ such that $P(x) \neq 0$.*

Proof. If $n = 1$, then the number of roots is $\leqslant d$. We will reason by induction on n and write $P = \sum_{j=0}^{d} P_j(X_1, \ldots, X_{n-1}) X_n^j$. Let T be the set of elements $(x_1, \ldots, x_{n-1}) \in S^{n-1}$ such that there exists j where $P_j(x_1, \ldots, x_{n-1}) \neq 0$. By induction, we know that $|T| \geqslant 1$. The number of zeros in S^n is therefore $\leqslant (|S^{n-1}| - |T|)|S| + d|T| = |S|^n - |T|(|S| - d) < |S|^n$. □

4.14. Lemma. *Let $P \in \mathbf{Z}[X]$ be a non-zero polynomial and $\beta := p/q \in \mathbf{Q}$. Then the order of vanishing, $\operatorname{ord}_\beta(P)$, satisfies*

$$\operatorname{ord}_\beta P \leqslant \frac{m(P)}{\log \max(|p|, |q|)} \leqslant \frac{h(P) + \log \deg P}{h(\beta)}$$

where $m(P) := \int_0^1 \log |P(\exp(2\pi i t)| \, dt \leqslant h(P) + \log \deg P$.

Proof. We denote by $r := \operatorname{ord}_\beta(P)$. By the hypotheses, we know that there exists $Q \in \mathbf{Z}[X]$ such that $P = (qX - p)^r Q$. Since $m(P_1 P_2) = m(P_1) + m(P_2)$, we can deduce that

$$r \log \max(|p|, |q|) \leqslant r m(pX - q) + m(Q) = m(P).$$

The elementary inequality linking $m(P)$ and $h(P)$ is proven in Appendix A in the form $m(P) \leqslant \log \|P\|_2 \leqslant h(P) + \frac{1}{2} \log(\deg P + 1)$. □

Proof. (of Thue's theorem (6-4.1)) We are considering an algebraic number α, which we can assume is an algebraic integer, and we let $d := [\mathbf{Q}(\alpha) : \mathbf{Q}]$. We want to know whether there exist rational approximations which satisfy

$$\left| \alpha - \frac{p}{q} \right| \leqslant q^{-\delta}. \tag{6.20}$$

If the set of solutions of this inequality is infinite, we can assume that there exists a first solution, $\beta_1 := p_1/q_1$, where q_1 is very large, and a second solution $\beta_2 := p_2/q_2$, where $\log q_2 / \log q_1$ is very large. We also fix an ϵ such that $0 < \epsilon < 1/4$. The first step is to construct an auxiliary polynomial by choosing the following parameters (the proof will justify this choice):

$$T := \left\lfloor \frac{\log q_2}{\log q_1} \right\rfloor, \quad D := \left\lfloor \frac{dT(1+\epsilon)}{2} \right\rfloor.$$

Throughout the different steps in the proof, we will denote by C, C_1, C_2, etc., constants which only depend on d and α.

1st step. (Construction of an auxiliary polynomial) *There exists a polynomial $F(X,Y) = P(X) - YQ(X) \in \mathbf{Z}[X,Y]$ such that* $\deg P, \deg Q \leqslant D$ *and*

$$\frac{1}{h!}\frac{\partial^h F}{\partial X^h}(\alpha,\alpha) = 0 \text{ for } 0 \leqslant h \leqslant T-1,$$

and whose coefficients have absolute value $\leqslant C_1^{\frac{T}{\epsilon}}$.

The proof of this is a direct application of Siegel's lemma (Lemmas 6-4.5 and 6-4.6), where the number of free coefficients is $2(D+1)$, the number of equations (over $\mathbf{Z}$) is dT and the height of the equations involves binomial coefficients.

2nd step (Application of Taylor's formula) *Let $j \leqslant T/2$. Then we have the inequality:*

$$\left| \frac{1}{j!}\frac{\partial^j F}{\partial X^j}(\beta_1,\beta_2) \right| \leqslant \max\left\{ q_1^{-\delta(T-j)}, q_2^{-\delta} \right\} C_2^{D+\frac{T}{\epsilon}}.$$

Using Taylor's formula in one variable (cf. Lemma 6-4.7) and the hypotheses, we have

$$\begin{aligned}
\frac{\partial F}{\partial X^j}(\beta) &= P^{(j)}(\beta_1) - \beta_2 Q^{(j)}(\beta_1) \\
&= \sum_{h\geqslant 0} \frac{P^{(j+h)}(\alpha)}{h!}(\beta_1-\alpha)^h - \beta_2 \sum_{h\geqslant 0} \frac{Q^{(j+h)}(\alpha)}{h!}(\beta_1-\alpha)^h \\
&= \sum_{h\geqslant T-j} \frac{1}{h!}\frac{\partial^{j+h} F}{\partial X^{j+h}}(\alpha,\alpha)(\beta_1-\alpha)^h \\
&\quad - (\beta_2-\alpha)\sum_{h\geqslant 0} \frac{Q^{(j+h)}(\alpha)}{h!}(\beta_1-\alpha)^h.
\end{aligned}$$

The first sum, divided by $j!$, is bounded above by $C^D q_1^{-\delta(T-j)}||F||$, while the second is bounded above by $C^D q_2^{-\delta}||Q||$. By using the estimate obtained in the 1st step: $||F|| = \max(||P||, ||Q||) \leqslant C_1^{\frac{T}{\epsilon}}$, we obtain the desired upper bound.

3rd step. (Extrapolation by Liouville's inequality) *We have* $\frac{\partial^j F}{\partial X^j}(\beta_1, \beta_2) = 0$ *for* $0 \leqslant j \leqslant T \times \dfrac{\delta - \frac{d}{2} - 1 - \frac{d\epsilon}{2} - \frac{C_3}{\epsilon \log q_1}}{(\delta+1)}$.

Let j be such that $\frac{1}{j!}\frac{\partial^j F}{\partial X^j}(\beta_1, \beta_2) \neq 0$. Then since it is a rational number with denominator $q_1^{D-j} q_2$, its absolute value is larger than $q_1^{-D+j} q_2^{-1}$ (cf. Lemma 6-4.11). By combining this with the previous step, we obtain

$$q_1^{-D+j} q_2^{-1} \leqslant \max\left\{q_1^{-\delta(T-j)}, q_2^{-\delta}\right\} C_2^{D+\frac{T}{\epsilon}}.$$

By our choice of T, we have $q_1^{T+1} \geqslant q_2 \geqslant q_1^T$. We can then deduce from this that $-D + j - T - 1 \leqslant -\delta(T-j) + (D + T/\epsilon)\frac{\log C_2}{\log q_1}\cdot$ By our choice of D, we have $D \leqslant \dfrac{dT(1+\epsilon)}{2} \leqslant D+1$, hence the inequality

$$T\left(\delta - \frac{d}{2} - 1 - \frac{d\epsilon}{2} - \frac{C_3}{\epsilon \log q_1}\right) \leqslant j(\delta+1).$$

4th step. (Zeros estimate) *There exists* $j \leqslant C_4 T/\epsilon \log q_1$ *such that*

$$\gamma_j := \frac{1}{j!}\frac{\partial^j F}{\partial X^j} \neq 0.$$

We introduce the Wronskian $W(X) := P'(X)Q(X) - P(X)Q'(X)$. Observe that W is not identically zero, for if so, P and Q would be proportional and hence divisible by $(X-\alpha)^T$ and likewise by $P_\alpha(X)^T$ (where P_α is the minimal polynomial of α). We know $dT > D$, and therefore, $P = 0$. Lemma 6-4.14 then allows us to prove that

$$\mathrm{ord}_{\beta_1} W \leqslant C_4 \frac{T}{\epsilon \log q_1}\cdot$$

The conclusion is now clear: we obtain a contradiction if q_1 is too large and $\delta > \frac{d}{2} + 1 + 2\epsilon$, which finishes the proof of Thue's theorem. □

While technically more elaborate, the proof of Roth's theorem relies essentially on the same ingredients: we construct a polynomial, $P \in \mathbf{Z}[X_1, \ldots, X_m]$ which vanishes with large order at $(\alpha, \ldots, \alpha)$, and the zeros estimate (the most difficult part) also relies on the use of Wronskians. The result obtained is by its nature not computationally effective in the two cases since the bounds obtained depend on the size of solutions of inequalities whose existence is not known and whose nonexistence is practically denied by the conclusion of the theorem.

The following statement is the promised computationally effective version of Baker's theorem on linear forms of logarithms (Theorem 6-4.3).

4.15. Theorem. (Baker [1]) *Let $\alpha_1, \dots, \alpha_n$ be non-zero algebraic numbers. There exists $C > 0$, which can be computed and only depends on d, n and the α_i, such that for every $\beta_0, \dots, \beta_n$ which are algebraic and of degree at most d and height at most B, we have*

$$|\beta_0 + \beta_1 \log \alpha_1 + \cdots + \beta_n \log \alpha_n| \geqslant B^{-C}, \tag{6.21}$$

whenever the quantity is non-zero.

The most widely used version of this theorem is the following (see also [1] for the proof).

4.16. Corollary. *Let $\alpha_1, \dots, \alpha_n$ be non-zero algebraic numbers. Then there exists $C > 0$, which can be computed and only depends on n and the α_i, such that for all $b_1, \dots, b_n$, integers with absolute value at most B, we have*

$$\left|\alpha_1^{b_1} \cdots \alpha_n^{b_n} - 1\right| \geqslant B^{-C}, \tag{6.22}$$

whenever the quantity is non-zero.

The corollary can be deduced, of course, from an inequality of the type $|\exp(z) - 1| \geqslant C|z|$. The assertion is given with the Archimedean absolute value, but it remains valid for a p-adic absolute value and is moreover proved in a similar manner (by conveniently defining p-adic logarithms). By assuming that the constant $C = C(n, d, \alpha_1, \dots, \alpha_n)$ exists and can be computed, we can see how this theorem allows us to explicitly find a solution to the S-unit equation.

Proof. (that Corollary 6-4.16 implies Theorem 5-4.4) For convenience sake, we let S be the set of Archimedean places of the field K, plus a finite set of finite places and $r := |S|-1$. Consider the embedding $L : \mathscr{O}_{K,S}^*/\mu_K \hookrightarrow \mathbf{R}^{|S|}$ given by $L(\alpha) := (\log|\alpha|_v)_{v \in S}$. Let $\epsilon_1, \dots, \epsilon_r$ be a basis for the S-units modulo roots of unity, in other words, every element $u \in \mathscr{O}_{K,S}^*$ can be written uniquely as

$$u = \zeta \epsilon_1^{m_1} \cdots \epsilon_r^{m_r} \quad \text{where } \zeta \in \mu_K \text{ and } m_i \in \mathbf{Z}.$$

We can define two norms on the lattice $L(\mathscr{O}_{K,S}^*)$: the norm induced by the sup-norm on $\mathbf{R}^{|S|}$ and the norm $M(u) = \max_i |m_i|$. Since these two norms are comparable, we obtain a constant c_1 such that for every $u \in \mathcal{O}^*_{K,S}$, we have

$$c_1^{-1} M(u) \leqslant \max_{v \in S} |\log |u|_v| \leqslant c_1 M(u).$$

Since we know that $\sum_{v \in S} \log |u|_v = 0$ (for $u \in \mathcal{O}^*_{K,S}$), we can see that $\max_{v \in S} \log |u|_v \leqslant \max_{v \in S} |\log |u|_v| \leqslant (|S|-1) \max_{v \in S} \log |u|_v$, and we can therefore conclude that there exists c_2 such that

$$c_2^{-1} M(u) \leqslant \max_{v \in S} \log |u|_v \leqslant c_2 M(u). \tag{6.23}$$

We will now arrive at a solution $u, v \in \mathcal{O}^*_{K,S}$ to the equation $u + v = 1$, where $v = \zeta' \epsilon_1^{m'_1} \cdots \epsilon_r^{m'_r}$. On the one hand, by the previous considerations, we obtain:

$$\min_{v \in S} \left| -\frac{v}{u} - 1 \right|_v = \min_{v \in S} \left| \frac{1}{u} \right|_v = \exp\left(-\log \max_v |u|_v \right) \leqslant \exp\left(-\frac{M(u)}{c_2} \right). \tag{6.24}$$

On the other hand, a direct application of the corollary to Baker's theorem (6-4.16) provides the existence of a constant c dependent on the field K, on the set S and on the fundamental units $\epsilon_1, \ldots, \epsilon_r$ such that we have the inequality

$$\max\{M(u), M(v)\}^{-c} \leqslant \min_{v \in S} \left| \zeta' \zeta^{-1} \epsilon_1^{m'_1 - m_1} \cdots \epsilon_r^{m'_r - m_r} - 1 \right|_v . \tag{6.25}$$

Up to reversing the roles of u and v, we can suppose that $M(v) \leqslant M(u)$ and deduce from (6.24) and (6.25) that $M(u)^{-c} \leqslant \exp\left(-\frac{M(u)}{c_2} \right)$, which clearly bounds $M(u)$ and therefore leaves only a finite number of possibilities for u and hence also for v. □

4.17. Remark. We point out that to prove the finiteness of the set of solutions of the S-unit equation, it suffices to have a weaker version of Baker's theorem of the form:

$$|m_1 \log \alpha_1 + \cdots + m_n \log \alpha_n| \geqslant \exp(-\psi(M)) \text{ with } \psi(M) = o(M),$$

where $M := \max |m_i|$. In particular, Liouville's inequality would only give $|m_1 \log \alpha_1 + \cdots + m_n \log \alpha_n| \geqslant \exp(-cM)$ and is therefore simply insufficient! Finally, it is clear that Baker's argument is computationally effective, under the condition that the function $\psi(M)$ is given explicitly, even if it often leads to very large bounds.

We will finish this chapter, a good part of it dedicated to problems of effectiveness (in a computational sense), by pointing out that it is unrealistic to expect to find computationally effective solutions to all of the Diophantine problems. The famous proof concerning Hilbert's 10 problem by Matijasevic (see [24]) proves that we will never find a universal algorithm which decides whether a Diophantine equation has an integer solution. Nevertheless, we can hope that the problem of determining the finite set of integer points on a curve (Siegel's theorem) or even the finite set of rational points on a curve (Faltings's theorem) has a computationally effective solution. For integral (or S-integral) points, the problem is solved for curves of genus 0 and 1 by Baker. Baker's method can also be applied to curves of genus 2 which can written as $y^2 = P(x)$ (where $\deg P = 5$ or 6), but not in general to curves of genus 2 of the following type:

$$y^4 + f_3(x,y) + f_2(x,y) = 0 \tag{6.26}$$

where the f_i are homogeneous of degree i.

5. The a, b, c Conjecture

This section is a little peculiar, since it discusses the consequences of a conjecture, which has not yet been proven and whose formulation was presented in the 1980's by Masser and Oesterlé. Moreover, all of the assertions in this section are conditional upon it. Nevertheless, the elementary character of the a, b, c conjecture and the depth of its implications make it a very active subject of investigation and experimentation. A surprise is provided by the "dictionary" between such elementary statements and the theory of elliptic curves. To deepen your understanding of the subject and its connections, we recommend the presentation of Oesterlé in séminaire Bourbaki [56].

We will begin by proving the following easy theorem.

5.1. Theorem. *Let A, B, C be non-constant polynomials which are relatively prime to each other and such that $A + B + C = 0$. Then*

$$\max\{\deg(A), \deg(B), \deg(C)\} \leqslant r_0(ABC) - 1, \tag{6.27}$$

where $r_0(P)$ denotes the number of distinct zeros of P.

Proof. We first write the factorizations of A, B, C:

$$A = a\prod_{i=1}^{r}(T-\alpha_i)^{\ell_i}, \; B = b\prod_{i=1}^{s}(T-\beta_i)^{m_i}, \; C = c\prod_{i=1}^{t}(T-\gamma_i)^{n_i}.$$

We then introduce the determinant given by $\Delta = \det\begin{pmatrix} A & B \\ A' & B' \end{pmatrix}$. We can easily see that $\Delta = -\det\begin{pmatrix} A & C \\ A' & C' \end{pmatrix} = \det\begin{pmatrix} B & C \\ B' & C' \end{pmatrix}$ and thus that

$\prod_{i=1}^{r}(T-\alpha_i)^{\ell_i-1}$ divides Δ and likewise $\prod_{i=1}^{s}(T-\beta_i)^{m_i-1}$ and $\prod_{i=1}^{t}(T-\gamma_i)^{n_i-1}$. Suppose, for example, that $\deg(C)$ is the largest of the degrees. Then we have:

$$(\deg(A)-r)+(\deg(B)-s)+(\deg(C)-t) \leqslant \deg(\Delta) \leqslant \deg(A)+\deg(B)-1,$$

hence $\deg(C) \leqslant r+s+t-1$, which is what we wanted to prove. □

The following conjecture is suggested by analogy.

5.2. Conjecture. (Masser-Oesterlé) *Let $\epsilon > 0$. There exists a constant C_ϵ such that if a, b, c are relatively prime integers which satisfy the equation $a+b+c=0$, then*

$$\max\{|a|,|b|,|c|\} \leqslant C_\epsilon \left(\prod_{p\,|\,abc} p\right)^{1+\epsilon}. \tag{6.28}$$

If we introduce the notation $\mathrm{Rad}(n) := \prod_{p\,|\,n} p$ (resp. $\mathrm{rad}(n) = \log \mathrm{Rad}(n)$), we can rewrite the previous inequality in the form

$$h(a,b,c) \leqslant (1+\epsilon)\,\mathrm{rad}(abc) + C_\epsilon.$$

We are going to see that this apparently innocent assertion—christened "the a, b, c conjecture"—has surprisingly deep consequences. A stronger form—christened "the effective a, b, c conjecture"—requires that the constant C_ϵ be computable (in terms of ϵ).

5.3. Remark. Let S be a finite set of prime numbers, and let $u, v \in \mathbf{Z}_S$ be S-unit solutions to the equation $u+v=1$. If we reduce the expressions to relatively prime integers, $u=a/c$, $v=b/c$, and apply the a, b, c conjecture to the equation $a+b=c$, we obtain

$$\max\{h(u),h(v)\} \leqslant (1+\epsilon)\sum_{p\in S} \log p + C_\epsilon.$$

By reversing the argument, we see that we can reformulate the a, b, c conjecture as a uniform version of the bound on the heights of solutions of the S-unit equation.

5.4. Proposition. *Assume that Conjecture 6-5.2 is true. Let $\ell, m, n \geqslant 2$ be integers such that $\ell^{-1}+m^{-1}+n^{-1} < 1$. Furthermore, let S be a finite set of prime numbers, and let u, v and w be S-unit integers. Then the number* *of solutions to*

$$uX^\ell + vY^m + wZ^n = 0, \tag{6.29}$$

where X, Y, Z *are relatively prime, is finite and bounded uniquely in terms of* S.

Proof. By applying the a, b, c conjecture to (6.29), we obtain

$$\max\left(|X|^\ell, |Y|^m, |Z|^n\right) \leqslant C_\epsilon \operatorname{Rad}(uvwX^\ell Y^m Z^n)^{1+\epsilon} \leqslant C_{\epsilon,S}\, |XYZ|^{1+\epsilon},$$

from which we can easily deduce

$$|XYZ| \leqslant C'_{\epsilon,S}\, |XYZ|^{(\ell^{-1}+m^{-1}+n^{-1})(1+\epsilon)}.$$

The last inequality is clearly a bound on the integer $|XYZ|$ whenever $\ell^{-1} + m^{-1} + n^{-1} < 1$. □

5.5. Remarks. In the remaining cases, i.e., up to permutation $(\ell, m, n) = (2,2,m)$, $(2,3,3)$, $(2,3,4)$, $(2,3,5)$, $(2,3,6)$, $(2,4,4)$ or $(3,3,3)$, it can be shown that there are, at least for certain u, v and w, infinitely many integer solutions. This statement gives a proof—modulo the a, b, c conjecture—of Faltings's theorem (Theorem 6-2.6) for Fermat curves given by the homogeneous equation (for $m \geqslant 4$):

$$uX^m + vY^m + wZ^m = 0.$$

We can considerably strengthen these statements, still modulo the a, b, c conjecture. First observe that we can reformulate the a, b, c conjecture by expressing it in terms of relatively prime a and b, thus forgetting c, in the form

$$\max\{|a|, |b|\}^{1-\epsilon} \leqslant C_\epsilon \operatorname{Rad}(ab(a+b)).$$

The generalization that we have in mind is the following.

5.6. Proposition. *Suppose that the* a, b, c *conjecture is true.*

i) Let $F \in \mathbf{Z}[X, Y]$ *be homogeneous of degree* d *with no multiple factors and* $\epsilon > 0$. *Then there exists a constant* $C_{F,\epsilon} > 0$ *such that for all relatively prime integers* a *and* b, *we have:*

$$\max\{|a|, |b|\}^{d-2-\epsilon} \leqslant C_{F,\epsilon} \operatorname{Rad}(F(a,b)).$$

ii) Let $f \in \mathbf{Z}[X]$ *be of degree* d *with no multiple factors and* $\epsilon > 0$. *Then there exists a constant* $C_{f,\epsilon} > 0$ *such that for every integer* a, *we have:*

$$|a|^{d-1-\epsilon} \leqslant C_{f,\epsilon} \operatorname{Rad}(f(a)).$$

To prove the proposition, we can rely on (a particular case of) a theorem of Belyi and the Riemann-Hurwitz formula, which are stated below. A morphism ϕ of degree d from $\mathbf{P}^1$ to $\mathbf{P}^1$ is given by two homogeneous polynomials, A and B, of degree d and which are relatively prime: $(x_0, x_1) \to (A(x_0, x_1), B(x_0, x_1))$. For almost all points $x = (x_0, x_1) \in \mathbf{P}^1$, the cardinality of $\phi^{-1}\{x\}$ is constant, equal to d. The morphism ϕ is *ramified* above x precisely when $|\phi^{-1}\{x\}| < d$. For the proof of Formula 6-5.8, we refer you to [35] or [38].

5.7. Proposition. (Belyi) *Let S be a finite subset of $\mathbf{P}^1(\bar{\mathbf{Q}})$. There exists a finite morphism $\phi : \mathbf{P}^1 \to \mathbf{P}^1$ such that*

i) ϕ *is unramified over* $\mathbf{P}^1 \setminus \{0, 1, \infty\}$,
ii) $\phi(S) \subset \{0, 1, \infty\}$.

5.8. Proposition. (Riemann-Hurwitz formula for $\mathbf{P}^1$) *Let $\phi : \mathbf{P}^1 \to \mathbf{P}^1$ be of degree d. Then $|\phi^{-1}\{x\}| = d$ for almost all points x of $\mathbf{P}^1$, and*

$$2d - 2 = \sum_{x \in \mathbf{P}^1} \left(d - |\phi^{-1}\{x\}|\right).$$

Proof. (of Proposition 6-5.7) Let d be the degree of the field generated by an irrational point of S. By applying the minimal polynomial which vanishes at this point, we send it to $0 \in \mathbf{P}^1$, and the new ramification points are now defined over a field of degree $< d$. By iterating this procedure, we can reduce to assuming that $S \subset \mathbf{P}^1(\mathbf{Q})$. We finish the proof by repeatedly using morphisms of the type:

$$\phi(x) := \frac{b^b}{a^a (b-a)^{b-a}} x^a (1-x)^{b-a}.$$

Any such morphism is unramified over $\mathbf{P}^1 \setminus \{0, 1, \infty\}$ and sends $\{0, 1, \infty, \frac{a}{b}\}$ to $\{0, 1, \infty\}$. □

Proof. (of Proposition 6-5.6) First observe that statement *ii)* can be deduced from statement *i)* applied to the polynomial $F(X,Y) = Y^{d+1} f(X/Y)$. To prove statement *i)*, we take $S \subset \mathbf{P}^1(\bar{\mathbf{Q}})$, the set of zeros of $F(X,Y)$, and we find a Belyi map (by Proposition 6-5.7), $\phi : \mathbf{P}^1 \to \mathbf{P}^1$, given by two polynomials, $A(X,Y)$ and $B(X,Y)$, of degree δ. We set $C(X,Y) = A(X,Y) - B(X,Y)$. The Riemann-Hurwitz formula can therefore be written

$$\delta + 2 = \left|\phi^{-1}\{0, 1, \infty\}\right|.$$

We point out that, for $x \in \mathbf{P}^1$, we have
1) $A(x) = 0$ if and only if $\phi(x) = \infty$,

2) $B(x) = 0$ if and only if $\phi(x) = 0$,
3) $C(x) = (A - B)(x) = 0$ if and only if $\phi(x) = 1$.

Thus, if we set $D(X,Y) = ABC(X,Y)$ and $D_0(X,Y) = \operatorname{Rad} D(X,Y)$, we have the inclusion $S \subset \{\text{zeros of } D\} = \{\text{zeros of } D_0\}$, and thus, since F is square-free, $D_0 = FG$. Likewise, we have $\deg(D_0) = |\phi^{-1}\{0,1,\infty\}| = \delta+2$, hence $\deg(G) = \delta + 2 - d$. Moreover, $A(X,Y)$ and $B(X,Y)$ are relatively prime. Then there exists an integer R (essentially a resultant) and polynomials with integer coefficients such that $A(X,Y)U(X,Y) + B(X,Y)V(X,Y) = RX^m$ and $A(X,Y)U'(X,Y) + B(X,Y)V'(X,Y) = RY^{m'}$. If now a and b are relatively prime integers, we can deduce that $e := \gcd(A(a,b), B(a,b))$ divides R and hence is bounded independently of a and b. We therefore apply the statement of the a, b, c conjecture to the triple $\dfrac{A(a,b)}{e} - \dfrac{B(a,b)}{e} = \dfrac{C(a,b)}{e}$. We then obtain

$$\max\left\{\left|\frac{A(a,b)}{e}\right|, \left|\frac{B(a,b)}{e}\right|, \left|\frac{C(a,b)}{e}\right|\right\} \leqslant C_\epsilon \operatorname{Rad}(D(a,b))^{1+\epsilon}.$$

We can easily see that $\max(|A(a,b)|, |B(a,b)|) \geqslant C \max(|a|,|b|)^\delta$ and, on the other hand, that $\operatorname{Rad}(D(a,b)) = \operatorname{Rad}(D_0(a,b))$ and

$$\operatorname{Rad}(D_0(a,b)) \leqslant \operatorname{Rad}(F(a,b))|G(a,b)| \leqslant C \operatorname{Rad}(F(a,b)) \max(|a|,|b|)^{\delta+2-d}.$$

By combining the obtained inequalities and by simplifying by $\max(|a|,|b|)^\delta$, we get exactly the desired assertion. □

5.9. Remark. Proposition 6-5.6 allows us to show that if we assume the a, b, c conjecture, then the set of rational points on the projective curve given by the homogeneous equation

$$F(X,Y) = mZ^d$$

has a finite number of rational points whenever $d := \deg(F) \geqslant 4$, which corresponds to $g = \dfrac{(d-1)(d-2)}{2} \geqslant 2$. Elkies [29] extended this argument by using Belyi's results and proved that the a, b, c conjecture allows us to recover Faltings's theorem for every curve. Thus, in particular, a computationally effective solution to the a, b, c conjecture would allow us to effectively compute the rational points on curves of genus $\geqslant 2$.

We will now prove that the a, b, c conjecture can be formulated in terms of elliptic curves (see Chap. 5 for notations and notions).

5.10. Conjecture. (Szpiro) *Let $\epsilon > 0$. There exists a constant C_ϵ such that for every elliptic curve $E/\mathbf{Q}$ with minimal discriminant Δ_E and* *conductor N_E, we have*

$$|\Delta_E| \leqslant C_\epsilon N_E^{6+\epsilon}. \tag{6.30}$$

We can formulate a slightly stronger variation.

5.11. Conjecture. (Frey-Szpiro) *Let $\epsilon > 0$. There exists a constant C_ϵ such that for every elliptic curve $E/\mathbf{Q}$ with minimal discriminant Δ_E and conductor N_E, we have*

$$\max\{H(j_E), |\Delta_E|\} \leqslant C_\epsilon N_E^{6+\epsilon}, \tag{6.31}$$

where $H(j_E)$ designates the height of the invariant j_E.

Since $j_E = c_4^3/\Delta_E$ and $1728\Delta_E = c_4^3 - c_6^2$, we can replace the left-hand side of the inequality by $\max(|c_4|^3, |\Delta_E|)$ or $\max(|c_4|^3, |c_6|^2, |\Delta_E|)$, up to modifying the constant C_ϵ.

5.12. Proposition. *The a, b, c conjecture is equivalent to the following assertion: for every positive $\epsilon > 0$, there exists a constant $C_\epsilon > 0$ such that for every elliptic curve E defined over $\mathbf{Q}$, we have*

$$\max\left(|\Delta_E|, |c_4^3|, |c_6^2|\right) \leqslant C_\epsilon \left(N_E\right)^{6+\epsilon}. \tag{6.32}$$

Thus, the a, b, c conjecture and the Frey-Szpiro conjecture are equivalent.

Proof. We will first show that the a, b, c conjecture implies the Frey-Szpiro conjecture. We assume the inequality $c_4^3 - c_6^2 = 1728\Delta$, and we set $d = \gcd(c_4^3, c_6^2)$ and $R = \mathrm{Rad}(c_4^3c_6^2 1728\Delta/d^3)$. We also denote by $\lceil x\rceil$ the smallest integer which is an upper bound for the real number x. If we factor $d = \prod_{i=1}^s p_i^{r_i}$, then $p_i^{3\lceil r_i/3\rceil}$ divides c_4^3 (resp. $p_i^{2\lceil r_i/2\rceil}$ divides c_6^2), and we can write:

$$\begin{aligned}
R &= \mathrm{Rad}\left\{\left(\frac{c_4}{\prod p_i^{\lceil r_i/3\rceil}}\right)^3\left(\frac{c_6}{\prod p_i^{\lceil r_i/2\rceil}}\right)^2\left(\frac{1728\Delta}{\prod p_i^{r_i}}\right)\prod p_i^{3\lceil r_i/3\rceil+2\lceil r_i\rceil-2r_i}\right\}\\
&\leqslant \mathrm{Rad}\left\{6\left(\frac{c_4}{\prod p_i^{\lceil r_i/3\rceil}}\right)\left(\frac{c_6}{\prod p_i^{\lceil r_i/2\rceil}}\right)\frac{N}{\prod p_i}\right\}\\
&\leqslant \frac{6|c_4c_6N|}{\prod p_i^{\lceil r_i/3\rceil+\lceil r_i/2\rceil+1}}.
\end{aligned}$$

The second to last inequality is true since, on the one hand, $\mathrm{Rad}(\Delta) = \mathrm{Rad}(N)$ and, on the other hand, if $\ell \neq 2, 3$ divides d, then there is additive reduction and ℓ^2 divides N, thus ℓ indeed appears in $N/\prod_i p_i$. Therefore, we use the a, b, c conjecture with $a = c_4^3/d$, $b = c_6^2/d$ and $c = 1728\Delta/d$, and we obtain

$$\max\left(|c_4^3|, |c_6^2|, |\Delta|\right) \leqslant C_\epsilon \left\{|c_4 c_6 N| \prod p_i^{\alpha(r_i)}\right\}^{1+\epsilon},$$

where we let $\alpha(r) := r - \lceil r/3 \rceil - \lceil r/2 \rceil - 1$. Observe that $\alpha(r) \leqslant 0$ for $r \leqslant 10$ (whereas $\alpha(12) = 1$), and we use the following elementary computation: if ℓ divides d, then the order of d at ℓ is at most 10. In fact, if ℓ^4 divided c_4 and ℓ^6 divided c_6, the model that we started with would not be minimal, contrary to the hypotheses. Thus either $\mathrm{ord}_\ell(c_4) \leqslant 3$ or $\mathrm{ord}_\ell(c_6) \leqslant 5$. We can therefore conclude that

$$\max\left(|c_4^3|, |c_6^2|, |\Delta|\right) \leqslant C_\epsilon |c_4 c_6 N|^{1+\epsilon}.$$

We will allow ourselves from now on to denote by ϵ and C_ϵ the successive constants (a priori different). We first obtain the inequalities $|c_4^2| \leqslant C_\epsilon |c_6 N|^{1+\epsilon}$ and $|c_6| \leqslant C_\epsilon |c_4 N|^{1+\epsilon}$, which imply the inequalities $|c_4| \leqslant C_\epsilon N^{2+\epsilon}$ and $|c_6| \leqslant C_\epsilon N^{3+\epsilon}$, which yields the first implication of the proposition.

For the converse, let a, b, c satisfy $a + b + c = 0$ and $\gcd(a, b, c) = 1$. We consider the associated Frey-Hellegouarch curve: $y^2 = x(x - a)(x + b)$, and we can easily compute that

$$j = 2^8 \frac{(a^2 + ab + b^2)^3}{(abc)^2} \quad \text{and} \quad \Delta = 2^4 (abc)^2.$$

The Frey-Szpiro conjecture applied to this curve can therefore be written as

$$\log\max\left(|a^2 + ab + b^2|^3, |abc|^2\right) \leqslant (6 + \epsilon)\log(\mathrm{Rad}(abc)) + C_\epsilon,$$

which of course implies that $\log\max(|a|, |b|, |c|) \leqslant (1 + \epsilon)\log(\mathrm{Rad}(abc)) + C_\epsilon$. □

5.13. Remark. We can also prove that Szpiro's conjecture implies the a, b, c conjecture with exponent $6/5$. To see this, let $a + b = c$ where a, b and $c \geqslant 0$ and $a \geqslant b$. Consider the elliptic curve $E : y^2 = x^3 - 2(a - b)x^2 + (a + b)^2 x$. It is a curve isogenous to the Frey-Hellegouarch curve $y^2 = x(x + a)(x - b)$ (in other words, there exists an *isogeny*, which is a surjective, algebraic homomorphism with a finite kernel between the two curves). The isogeny with kernel equal to the group of order 2 generated by $P = (0, 0)$ is given by the formulas: $(x, y) \mapsto (y^2/x^2, -y(ab + x^2)/x^2)$. We can check that the model of E is minimal and even semi-stable, except perhaps in 2 (the curve is semi-stable in 2 if and only if 2^4 divides abc) and that the discriminant of the model equals $D = -2^8 abc^4$. Szpiro's conjecture therefore implies that

$$a^5 \leqslant abc^4 \leqslant C_\epsilon R(abc)^{6+\epsilon},$$

which gives the desired inequality.

5.14. Remark. Diverse approaches have been proposed to prove the a, b, c conjecture. Philippon proposed an approach based on very strong lower bounds of linear forms of logarithms. Today, because of Wiles's theorem [80] (see Theorem 6-6.14, for the statement and this same section for the definition of $X_0(N)$) which guarantees the existence of a modular parametrization

$$\phi_E : X_0(N_E) \to E,$$

the most exciting approach is to try to bound the *degree* of the parametrization ϕ_E. In particular, the following conjecture implies the a, b, c conjecture (see [38] and [54]).

5.15. Conjecture. (Degree conjecture) *For every $\epsilon > 0$, there exists a constant C_ϵ such that for every elliptic curve E defined over $\mathbf{Q}$, there exists a modular parametrization $\phi_E : X_0(N_E) \to E$ which satisfies*

$$\deg(\phi_E) \leqslant C_\epsilon N_E^{2+\epsilon}.$$

6. Some Remarkable Dirichlet Series

We have already encountered many Dirichlet series. In this section, we will introduce various generalizations of them. We will succinctly describe the connections—proven and conjectural—between some of these "ζ" or "L" functions: the series associated to modular forms and their generalizations, the series associated to (families of) Galois representations and finally the Hasse-Weil series associated to algebraic varieties. This leads us to the border of the "automorphic world". To go further, we recommend consulting [27], [32], [68], as well as [20], [26], [66] and [65].

We will start by stating a generalization from the Riemann zeta function to the Dedekind zeta function of the analytic continuation with a functional equation (Theorem 4-5.6).

To do this, it will be convenient to introduce some small modifications to the Gamma function.

Notation. We denote the modifications to the Gamma function as follows:

$$\Gamma_{\mathbf{R}}(s) := \pi^{-s/2}\Gamma\left(\frac{s}{2}\right) \qquad \text{and} \qquad \Gamma_{\mathbf{C}}(s) := (2\pi)^{-s}\Gamma(s). \qquad (6.33)$$

6.1. Definition. Let K be a number field, r_1 (resp. r_2) the number of real (resp. complex) embeddings. Let $r = r_1 + r_2 - 1$, and choose $\epsilon_1, \ldots, \epsilon_r$ to be a basis for $\mathscr{O}_K^*/\mu_K$. The *regulator* of units of a number field is defined as the absolute value of any $r \times r$ determinant taken from the $r \times (r+1)$ matrix of coefficients $\log|\epsilon_i|_v$ for $1 \leqslant i \leqslant r$, where v is an Archimedean place.

6.2. Theorem. (Hecke) *Let K be a number field containing w_K roots of unity and with r_1 real embeddings, r_2 pairs of complex embeddings, discriminant Δ_K, class number h_K and regulator of units R_K. The function $\zeta_K(s)$, initially defined for $\mathrm{Re}(s) > 1$, can be analytically continued to the whole complex plane, except for a simple pole at $s = 1$ with residue:*

$$\lim_{s\to 1}(s-1)\zeta_K(s) = \frac{2^{r_1}(2\pi)^{r_2} R_K h_K}{w_K\sqrt{|\Delta_K|}}. \tag{6.34}$$

Furthermore, if we let

$$\xi_K(s) := |\Delta_K|^{s/2}\Gamma_{\mathbf{R}}(s)^{r_1}\Gamma_{\mathbf{C}}(s)^{r_2}\zeta_K(s),$$

then we can write the functional equation of $\zeta_K(s)$ in the form

$$\xi_K(s) = \xi_K(1-s). \tag{6.35}$$

Finally, $\xi_K(s)$ is bounded in every vertical strip (outside of a neighborhood of 0 and 1).

We should also point out that $\zeta_K(s) \neq 0$ for $\mathrm{Re}(s) > 1$ (look at the Euler product), and hence $\xi_K(s) \neq 0$ as well. Because of the functional equation, we also have that $\xi_K(s) \neq 0$ for $\mathrm{Re}(s) < 0$. By observing that $\Gamma(s)$ has a simple pole at the negative integers, we can deduce that, in the half-plane $\mathrm{Re}(s) < 0$, the function $\zeta_K(s)$ only vanishes at negative integers, with order r_2 at odd negative integers and order $r_1 + r_2$ at even negative integers, the order at zero being $r_1 + r_2 - 1$.

Let χ be a *primitive* Dirichlet character modulo $N \geqslant 2$ (see Exercise 1-6.12). We set $\epsilon = 0$ (resp. $\epsilon = +1$) if $\chi(-1) = 1$ (resp. if $\chi(-1) = -1$), and

$$\Lambda(\chi, s) := N^{s/2}\Gamma_{\mathbf{R}}(s+\epsilon)L(\chi, s).$$

The function $L(\chi, s)$ can therefore be continued to an entire function and satisfies the functional equation (where w_χ is a complex number of absolute value 1):

$$\Lambda(\chi, s) = w_\chi \Lambda(\bar{\chi}, 1-s). \tag{6.36}$$

Furthermore, $\Lambda(\chi, s)$ is bounded in every vertical strip.

We have seen in Chap. 4 that if $K = \mathbf{Q}(\sqrt{d})$, then there exists a character χ_d modulo Δ_K such that $\zeta_K(s) = \zeta(s)L(\chi_d, s)$. From this, we can deduce the following formulas.

i) If K is real quadratic and $\epsilon > 1$ is its fundamental unit, then

$$L(\chi_d, 1) = \frac{2h_K \log \epsilon}{\sqrt{|\Delta_K|}}.$$

ii) If K is imaginary quadratic (other than $\mathbf{Q}(i)$ and $\mathbf{Q}(j)$, for which $w_K =$ 4 and 6, respectively), then

$$L(\chi_d, 1) = \frac{\pi h_K}{\sqrt{|\Delta_K|}}.$$

By adding these formulas to those proven for $L(\chi, 1)$ in Exercise 4-6.6, we can infer some interesting properties concerning h_K and $\log \epsilon$ from them. In the case where we take $K = \mathbf{Q}(\exp(2\pi i/\ell)$ (for an odd prime ℓ), we have essentially proven, during the proof of Lemma 4-4.18, the formula

$$\zeta_K(s) = \zeta(s) \prod_{\chi} L(\chi, s),$$

where the product is taken over the nontrivial Dirichlet characters modulo ℓ. We can, of course, also deduce the formula:

$$\frac{(2\pi)^{(\ell-1)/2} R_K h_K}{2\ell^{(\ell+1)/2}} = \prod_{j=1}^{\ell-1} L(\chi_j, 1).$$

The Artin L-functions associated to a representation ρ, which are defined in Appendix C, provide another example. They satisfy a functional equation of the same type, but we do not in general know whether the meromorphic continuation is in fact holomorphic.

Modular forms. *We now define some other Dirichlet series coming from a world apparently far away from the previous ones, namely the automorphic world. It gives us an opportunity to briefly introduce modular functions and curves to which we have already alluded.*

6.3. Definition. The *Poincaré half-plane* is $\mathscr{H} := \{z \in \mathbf{C} \mid \operatorname{Im}(z) > 0\}$, and the *extended Poincaré half-plane* is $\mathscr{H}^* := \mathscr{H} \cup \mathbf{P}^1(\mathbf{Q})$.

The group $\mathrm{GL}_2^+(\mathbf{R})$ of 2×2 matrices with positive determinant, as well as the group $\mathrm{SL}_2(\mathbf{R})$ of matrices with determinant 1, acts on $\mathscr{H}$ by the action

$\left(\begin{pmatrix} a & b \\ c & d \end{pmatrix}, z\right) \mapsto (az+b)/(cz+d)$. The groups $\mathrm{GL}_2^+(\mathbf{Q})$ and $\mathrm{SL}(2,\mathbf{Z})$ act on $\mathscr{H}^*$. The action of the latter group is discrete and we can therefore form the quotients $Y := \mathrm{SL}(2,\mathbf{Z})\backslash\mathscr{H}$ and $X := \mathrm{SL}(2,\mathbf{Z})\backslash\mathscr{H}^*$ and endow them with the structure of a Riemann surface (see [68]). In fact, $Y \cong \mathbf{A}^1(\mathbf{C})$ and $X \cong \mathbf{P}^1(\mathbf{C})$.

The group $\mathrm{SL}(2,\mathbf{Z})$ and its finite index subgroups play an important role in arithmetical questions. In what follows, we will introduce a whole family of such subgroups.

6.4. Definition. A subgroup $\Gamma \subset \mathrm{SL}(2,\mathbf{Z})$ is a *congruence subgroup* if it contains $\Gamma(N)$ for a certain N where

$$\Gamma(N) := \left\{ A = \begin{pmatrix} a & b \\ c & d \end{pmatrix} \in \mathrm{SL}(2,\mathbf{Z}) \mid A \equiv I \bmod N \right\}.$$

We denote by $Y(N) := \Gamma(N)\backslash\mathscr{H}$ and $X(N) := \Gamma(N)\backslash\mathscr{H}^*$.

Besides $\Gamma(N)$ itself, two other congruence subgroups deserve to be mentioned.

i) The congruence group

$$\Gamma_1(N) := \left\{ A = \begin{pmatrix} a & b \\ c & d \end{pmatrix} \in \mathrm{SL}(2,\mathbf{Z}) \mid A \equiv \begin{pmatrix} 1 & * \\ 0 & 1 \end{pmatrix} \bmod N \right\}$$

where $Y_1(N) := \Gamma_1(N)\backslash\mathscr{H}$ and $X_1(N) := \Gamma_1(N)\backslash\mathscr{H}^*$;

ii) The congruence group

$$\Gamma_0(N) := \left\{ A = \begin{pmatrix} a & b \\ c & d \end{pmatrix} \in \mathrm{SL}(2,\mathbf{Z}) \mid c \equiv 0 \bmod N \right\}$$

where $Y_0(N) := \Gamma_0(N)\backslash\mathscr{H}$ and $X_0(N) := \Gamma_0(N)\backslash\mathscr{H}^*$.

We can easily see that $\Gamma_1(N)$ is normal in $\Gamma_0(N)$ and that $\Gamma_0(N)/\Gamma_1(N) \cong (\mathbf{Z}/N\mathbf{Z})^*$ by the map $\begin{pmatrix} a & b \\ c & d \end{pmatrix} \mapsto d \bmod N$.

In can be shown (see [27] or [68]) that $Y_0(N)$ (resp. $Y_1(N)$, $Y(N)$) are affine algebraic curves, whereas $X_0(N)$ (resp. $X_1(N)$, $X(N)$) are projective algebraic curves. Furthermore, $X_0(N)$ and $X_1(N)$ are defined over $\mathbf{Q}$, whereas $X(N)$ is defined over $\mathbf{Q}(\exp(2\pi i/N))$.

6.5. Definition. Let Γ be a congruence subgroup. A *modular form* of weight k with respect to Γ is a holomorphic function $f : \mathscr{H} \to \mathbf{C}$ such that the following properties hold.

i) For $\begin{pmatrix} a & b \\ c & d \end{pmatrix} \in \Gamma$ and $z \in \mathscr{H}$, we have

$$f\left(\frac{az+b}{cz+d}\right) = (cz+d)^k f(z). \tag{6.37}$$

ii) The function f is holomorphic on $\mathscr{H}^*$, in other words, for every $\begin{pmatrix} a & b \\ c & d \end{pmatrix} \in \mathrm{SL}(2, \mathbf{Z})$, the limit of $f\left(\frac{az+b}{cz+d}\right)(cz+d)^{-k}$ as $\operatorname{Im} z$ tends to infinity exists. If this limit is always zero, f is said to be a *cusp form* or *parabolic.*

We denote by $M_k(\Gamma)$ the vector space of these modular forms and $S_k(\Gamma)$ the subspace of cusp forms. We are talking essentially about modular forms for $\Gamma_0(N)$, but we can introduce a variation of them, with the help of a Dirichlet character, χ, modulo N. A function f is called a modular form for $\Gamma_0(N)$ twisted by χ if it is modular when the (6.37) is replaced by

$$\forall \begin{pmatrix} a & b \\ c & d \end{pmatrix} \in \Gamma_0(N), \quad f\left(\frac{az+b}{cz+d}\right) = \chi(d)(cz+d)^k f(z). \tag{6.38}$$

We denote by $M_k(N, \chi)$ (resp. $S_k(N, \chi)$) the space of these forms (resp. cusp forms). It can be shown that $S_k(\Gamma_1(N)) = \oplus_\chi S_k(N, \chi)$.

6.6. Remarks. 1) Observe that, since $-Id \in \Gamma_0(N)$, an element f of $M_k(N, \chi)$ must satisfy $f(z) = \chi(-1)(-1)^k f(z)$. Thus $M_k(N, \chi) = \{0\}$, except maybe if $\chi(-1) = (-1)^k$.

2) Every congruence subgroup Γ contains an element $T_h := \begin{pmatrix} 1 & h \\ 0 & 1 \end{pmatrix}$ with h non-zero and minimal: for example, $T_1 \in \Gamma_1(N)$. Therefore, every $f \in M_k(\Gamma)$ satisfies $f(z+h) = f(z)$, which allows us to write its Fourier series expansion as:

$$f(z) = \sum_{n \in \mathbf{Z}} a_n q_h^n, \qquad \text{where} \qquad q_h := \exp\left(\frac{2\pi i z}{h}\right). \tag{6.39}$$

Moreover, the condition of being holomorphic on $\mathscr{H}^*$ imposes that $a_n = 0$ for $n < 0$, whereas its vanishing at ∞ is written $a_n = 0$ for $n \leqslant 0$ (n.b. this is a necessary condition, but for f to be a form, holomorphy must be tested at all points in $\mathbf{P}^1(\mathbf{Q}) = \mathscr{H}^* \setminus \mathscr{H}$).

3) If $\gamma = \begin{pmatrix} a & b \\ c & d \end{pmatrix} \in \mathrm{GL}_2(\mathbf{R})^+$, therefore $\delta := ad - bc > 0$, and if we let $\gamma' = \delta^{-1/2}\gamma$, then $\gamma' \in \mathrm{SL}_2(\mathbf{R})$ and $\gamma' \cdot z = \gamma \cdot z$. Therefore, if f is modular

of weight k for γ', then

$$f\left(\frac{az+b}{cz+d}\right) = \delta^{k/2}(cz+d)^k f(z). \tag{6.40}$$

6.7. Definition. Let $f \in S_k(\Gamma_0(N))$, and let $f(z) = \sum_{n=1}^{\infty} a_n \exp(2\pi i n z) = \sum_{n=1}^{\infty} a_n q^n$ be its Fourier expansion. The *Dirichlet series* associated to f is defined by:

$$L(s,f) := \sum_{n=1}^{\infty} a_n n^{-s}. \tag{6.41}$$

We point out that we have the relation (called the "Mellin transform"):

$$\Gamma_{\mathbf{C}}(s)L(f,s) = (2\pi)^{-s}\Gamma(s)L(f,s) = \int_0^{\infty} f(it)t^{s-1}dt.$$

6.8. Definition. Let $f = \sum_n a_n(f)q^n \in M_k(\Gamma_0(N))$. *Hecke operators* are defined as follows.

1) If p does not divide N, we define the operator $f \mapsto T_p f$ by:

$$a_n(T_p f) := a_{np}(f) + p^{k-1}a_{n/p}(f)$$

where, by convention, $a_{n/p} = 0$ if p does not divide n.

2) If p divides N, we define the operator $f \mapsto U_p f$ by:

$$a_n(U_p f) := a_{np}(f).$$

A small generalization which is often useful consists of defining the T_p on all of $M_k(\Gamma_1(N))$. This can be done by setting, for $f \in M_k(N,\chi)$:

$$a_n(T_p f) := a_{np}(f) + \chi(p)p^{k-1}a_{n/p}(f).$$

Note that, since $\chi(p) = 0$ when p divides N, we can consider the previous formula to also define U_p when p divides N.

6.9. Theorem. (Hecke, see [68]) *Hecke operators commute with each other. If $f = \sum_n a_n(f)q^n \in S_k(\Gamma_0(N))$ is an eigenvalue for each of the Hecke operators, i.e., $T_p f = \lambda_p f$ and $U_p f = \lambda_p f$, then $a_p(f) = \lambda_p a_1(f)$, and if we normalize f by the condition $a_1(f) = 1$, the function $L(s,f)$ can be factored as an Euler product in the form:*

$$\begin{aligned} L(s,f) &= \sum_{n=1}^{\infty} a_n(f)n^{-s} \\ &= \prod_{p \mid N}(1 - a_p(f)p^{-s})^{-1} \prod_{p \nmid N}(1 - a_p(f)p^{-s} + p^{k-1-2s})^{-1}. \end{aligned} \tag{6.42}$$

Therefore, we see the appearance of the Euler product, which, for $k = 2$, much resembles the L-function associated to an elliptic curve. To underline this resemblance, we will see how, under certain conditions, $L(s, f)$ satisfies a functional equation.

Observe that the matrix $W_N := \begin{pmatrix} 0 & 1 \\ -N & 0 \end{pmatrix}$, which is not in $\mathrm{SL}(2, \mathbf{Z})$ but in $\mathrm{GL}_2^+(\mathbf{Q})$, nevertheless normalizes the subgroup $\Gamma_0(N)$ because

$$W_N \begin{pmatrix} a & b \\ c & d \end{pmatrix} W_N^{-1} = \begin{pmatrix} d & -c/N \\ -bN & a \end{pmatrix}.$$

We can deduce from this that W_N acts on $M_k(\Gamma_0(N))$ (resp. $S_k(\Gamma_0(N))$), and since $W_N^2 = -NId$, we see that the spaces $M_2(\Gamma_0(N))$ (resp. $S_2(\Gamma_0(N))$) can be decomposed into the sum of two eigenspaces in which:

$$f\left(-\frac{1}{Nz}\right) = f(W_N \cdot z) = \pm N^{k/2} z^k f(z) \tag{6.43}$$

This remark can be used as a motivation for the following assertion.

6.10. Theorem. (Hecke) *Let $\epsilon = \pm 1$, and let $f(\tau) = \sum_{n \geqslant 1} a_n \exp(2\pi i n \tau)$ be a modular cusp form for $\Gamma_0(N)$ of weight k such that*

$$f\left(-\frac{1}{N\tau}\right) = \epsilon N^{k/2} \tau^k f(\tau). \tag{6.44}$$

Let $\Lambda(s, f) := N^{s/2}(2\pi)^{-s} L(s, f)$, where $L(s, f) := \sum_{n=1}^{\infty} a_n n^{-s}$. Then the function $\Lambda(s, f)$ can be analytically continued to the complex plane and satisfies the functional equation

$$\Lambda(s, f) = i^k \epsilon \Lambda(k - s, f). \tag{6.45}$$

Furthermore, $\Lambda(s, f)$ is bounded in every vertical strip.

Proof. We first note that for $\tau = it$ (where $t \in \mathbf{R}_+$), (6.44) is written

$$f\left(\frac{i}{Nt}\right) = (i)^k \epsilon N^{k/2} t^k f(it).$$

We can therefore perform the following computation by using the variable change $t \mapsto 1/Nt$:

$$\begin{aligned} \Lambda(s, f) &= N^{s/2} \int_0^{\infty} f(it) t^{s-1} dt \\ &= N^{s/2} \int_0^{\frac{1}{\sqrt{N}}} f(it) t^{s-1} dt + N^{s/2} \int_{\frac{1}{\sqrt{N}}}^{\infty} f(it) t^{s-1} dt \\ &= i^k \epsilon N^{\frac{1}{2}(k-s)} \int_{\frac{1}{\sqrt{N}}}^{\infty} f(it) t^{k-1-s} dt + N^{s/2} \int_{\frac{1}{\sqrt{N}}}^{\infty} f(it) t^{s-1} dt. \end{aligned}$$

We can easily see that the latter expression defines an entire function, because by showing first that $|a_n| = O(n^c)$, we can see that $|f(it)| = O(\exp(-2\pi t))$ as t tends to infinity. Furthermore, the expression is clearly $(i^k\epsilon)$-symmetric when we change s to $k-s$. The last assertion of the theorem is clear by the expression for $\Lambda(s,f)$ as an integral. □

We will end this introduction to modular forms by indicating several connections between modular forms and Galois representations.

First of all, we know how to associate, thanks to Deligne and Deligne-Serre (see [26]), Galois representations to modular forms.

6.11. Theorem. (Deligne) *Let ℓ be a prime number and $f = \sum_n a_n q^n \in M_k(N,\chi)$ a modular Hecke eigenform which is normalized (i.e., $a_1 = 1$). We know that the field K generated by the values of χ and the a_n is a number field. Then there exists an ℓ-adic representation (with coefficients in a completion K_v, an extension of $\mathbf{Q}_\ell$),*

$$\rho : G_{\mathbf{Q}} \longrightarrow \mathrm{GL}_2(K_v),$$

which satisfies the following properties.

i) ρ is unramified outside of $N\ell$.
ii) For p not dividing $N\ell$, we have the formulas

$$\mathrm{Tr}\,\rho\,(\mathrm{Frob}_p) = a_p \qquad \text{and} \qquad \det\rho\,(\mathrm{Frob}_p) = \chi(p)p^{k-1}. \tag{6.46}$$

Moreover, if $f \in S_k(N,\chi)$, then the representation is irreducible.

We can, of course, deduce some representations $\bar\rho$ modulo ℓ from this by composing with $\mathscr{O}_v \to \mathbf{F}_{\ell^s}$. In the fairly special case of forms of weight 1 (i.e. $k=1$), Deligne and Serre proved that we can lift these representations to characteristic zero and thus obtain Artin representations. Recall also that if $f \neq 0$ is in $M_1(N,\chi)$, then $\chi(-1) = -1$.

6.12. Theorem. (Deligne-Serre) *Let $f = \sum_n a_n q^n \in S_1(N,\chi)$ be a modular Hecke eigenform which is normalized (i.e., $a_1 = 1$). Then there exists an Artin representation,*

$$\rho_f : G_{\mathbf{Q}} \longrightarrow \mathrm{GL}_2(\mathbf{C}),$$

which satisfies the following properties.

i) ρ_f is unramified outside of $N\ell$.
ii) For p not dividing $N\ell$, we have the formulas

$$\mathrm{Tr}\,\rho_f\,(\mathrm{Frob}_p) = a_p \qquad \text{and} \qquad \det\rho_f\,(\mathrm{Frob}_p) = \chi(p). \tag{6.47}$$

Furthermore, the representation ρ_f is irreducible and odd (i.e., if c denotes complex conjugation, then $\det\rho_f(c) = -1$*).*

Let us point out that the representation ρ_f is continuous; this would not, of course, be the case if we had simply embedded K_v into $\mathbf{C}$ and thus obtained a representation $G_{\mathbf{Q}} \to \mathrm{GL}_2(K_v) \to \mathrm{GL}_2(\mathbf{C})$.

6.13. Conjecture. *Every Artin representation of dimension 2 which is irreducible and odd is associated to a modular form of weight 1.*

L-functions associated to algebraic varieties.

We will now return to the Hasse-Weil zeta function associated to an algebraic variety V of dimension r, which we will assume, for the sake of simplicity, to be smooth, projective and defined over $\mathbf{Q}$. We know (see Proposition B-1.22) that for p outside of a finite set S, the reduction modulo p of V remains smooth; we will denote it by $\tilde{V}_p$, and it is a projective variety defined over $\mathbf{F}_p$. We then have a natural definition (see the first section of this chapter) for the zeta function of $V/\mathbf{Q}$ by omitting the Euler factors for $p \in S$[5]:

$$\zeta_S(V/\mathbf{Q}, s) := \prod_{p\notin S} Z(\tilde{V}_p/\mathbf{F}_p, p^{-s}) = \prod_{j=0}^{2r}\prod_{p\notin S} P_j(\tilde{V}_p/\mathbf{F}_p, p^{-s})^{(-1)^{i+1}}. \quad (6.48)$$

This suggests that we let

$$L_{j,S}(V/\mathbf{Q}, s) := \prod_{p\notin S} P_j(\tilde{V}_p/\mathbf{F}_p, p^{-s})^{-1} = \prod_{p\notin S}\prod_{i=1}^{B_j}\left(1 - \alpha_{p,i}p^{-s}\right)^{-1}. \quad (6.49)$$

Then we have $\zeta(V/\mathbf{Q}, s) = \prod_{i=0}^{2r} L_j(V/\mathbf{Q}, s)^{(-1)^i}$. Since $|\alpha_{p,i}| = p^{j/2}$, the Euler product is convergent for $\mathrm{Re}(s) > 1 + j/2$.

It is always true that $L_0(V, s) = \zeta(s)$ and $L_{2r}(V, s) = \zeta(s - r)$, since $P_0(\tilde{V}_p, T) = 1 - T$ and $P_{2r}(\tilde{V}_p, T) = 1 - p^rT$. By using relation (6.14), we see that

$$L_{2r-i}(V/\mathbf{Q}, s) = L_i(V/\mathbf{Q}, s - r + i).$$

The zeta function of a curve $C/\mathbf{Q}$ is written

$$\zeta(C/\mathbf{Q}, s) = \frac{\zeta(s)\zeta(s-1)}{L_1(C/\mathbf{Q}, s)}. \quad (6.50)$$

[5]There exists a more sophisticated procedure than the theory introduced here for defining the local factors for all p—see for example [65].

Therefore, we see that what we called the "L-function" of an elliptic curve $E/\mathbf{Q}$ in Chap. 5 is denoted here by $L_1(E/\mathbf{Q}, s)$.

We will reformulate Wiles's theorem (Theorem 5-7.6, the Shimura-Taniyama-Weil conjecture) in two (nontrivially) equivalent forms.

6.14. Theorem. *Let $E/\mathbf{Q}$ be an elliptic curve with conductor $N = N_E$.*

1) There exists a modular cusp form $f \in S_2(\Gamma_0(N))$ such that

$$L(E, s) = L(f, s). \tag{6.51}$$

2) There exists a non-constant morphism $\phi_E : X_0(N) \to E$.

Commentary. Wiles actually proved this result with some supplementary hypotheses, which were subsequently shown to be unnecessary. Some extraordinary features of this result deserve to be pointed out. The function $L(E, s)$ is constructed starting with local information—actually it only suffices to know card $E(\mathbf{F}_p)$—and the theorem indicates that the obtained L-function comes from a global object—a modular form—which determines its characteristics. The link between these two objects, the elliptic curve defined over $\mathbf{Q}$ and the modular form for $\Gamma_0(N)$, is achieved through the Galois representations associated to each of these objects (see Appendix C). The existence of such a link is actually suggested by the L-series associated to their functional equations (proved or conjectured). This program has been vastly generalized and is today called *the Langlands program.* Without being able to explain the details (see for example [20] and [32] for an introduction and references), we will only say that Langlands theory associates to each irreducible *automorphic* representation a function $L(\pi, s)$ defined by an Euler product and which has a functional equation (relating $L(\pi, s)$ and $L(\check{\pi}, 1-s)$). These representations are obtained as factors of the space $L^2(Z_{\mathrm{A}_\mathbf{Q}} \mathrm{GL}_n(\mathbf{Q}) \backslash \mathrm{GL}_n(\mathrm{A}_\mathbf{Q}))$ (here, $\mathrm{A}_\mathbf{Q}$ denotes the ring of adeles, Z the center of GL_n and $Z_{\mathrm{A}_\mathbf{Q}}$ the points of the center with values in the adeles) and have infinite dimension (see for example [20] or [32]). Langlands conjectures, for example, that every Artin L-function associated to a representation of dimension n coincides with the $L(\pi, s)$ function associated to a representation of $\mathrm{GL}_n(\mathrm{A}_\mathbf{Q})$. In this context, modular forms are associated to representations of dimension 2 (the group GL_2).

This suggests that we thus describe what we should expect from a "nice" zeta function.

Expected properties of zeta or L functions.

i) They are defined by a Dirichlet series in a half-plane $\mathrm{Re}(s) > a$:

$$L(M, s) = \sum_{n=1}^{\infty} a_n n^{-s}.$$

ii) They are written, in a half-plane $\mathrm{Re}(s) > b$, as an Euler product:

$$L(M,s) = \prod_p L_p(M,s) = \prod_p \left(1 + a_{p,1}p^{-s} + \cdots + a_{p,d}p^{-ds}\right)^{-1}.$$

We call d the degree of the Euler product. Moreover, we require in general that $1 + a_{p,1}T + \cdots + a_{p,d}T^d = \prod_{j=1}^d (1 - \alpha_{p,j}T)$ where, for almost every p, we have the equality $|\alpha_{p,j}| = p^{w/2}$ and, in particular, $a_{p,d} \neq 0$. The integer w is called the *weight* of M.

iii) The function $L(M,s)$ can be analytically continued to the complex plane, except for at a finite number of poles. It satisfies a functional equation of the type $\Lambda(M,s) = w_M \Lambda(\check{M}, 1-s)$, where w_M is a complex number with absolute value 1,

$$\Lambda(M,s) = A^{s/2} \prod_{j=1}^t \Gamma_{\mathbf{R}}(s+t_j)^{h_j} \Gamma_{\mathbf{C}}(s+t'_j)^{h'_j} L(M,s),$$

$L(\check{M},s)$ is a function of the same type and A, t_j, t'_j, h_j, h'_j are some constants.

iv) Outside of a neighborhood of its possible poles, the function $\Lambda(M,s)$ is bounded in every vertical strip $\sigma_1 \leqslant \mathrm{Re}(s) \leqslant \sigma_2$.

6.15. Remark. With some optimism, we could add as a property the analogue of the Riemann hypothesis (abbreviated GRH):

"The zeros of $\Lambda(M,s)$ are situated on the line $\mathrm{Re}(s) = (w+1)/2$." (GRH)

We should point out that the given hypotheses imply that the Euler product which defines $L(M,s)$ is absolutely convergent for $\mathrm{Re}(s) > 1 + w/2$ and non-zero in this half-plane. Therefore, $\Lambda(M,s)$ does not vanish for $\mathrm{Re}(s) > 1 + w/2$ and, because of the functional equation, for $\mathrm{Re}(s) < w/2$. Just like the function $\zeta(s)$, the function $L(M,s)$ has "trivial" zeros in the half-plane $\mathrm{Re}(s) < w/2$, these being governed by the Gamma factors in the functional equation. Thus, the generalized Riemann hypothesis describes the location of the zeros in the "critical strip" $w/2 \leqslant \mathrm{Re}(s) \leqslant 1 + w/2$.

6.16. Conjecture. (Hasse-Weil, see [65]) *Let $V/\mathbf{Q}$ be a smooth projective variety. To every $0 \leqslant j \leqslant 2\dim V$ we can associate an integer A_j, local factors $L_{p,j}(V,s)$ at places $p \in S$ of bad reduction and a Gamma factor $L_{\infty,j}(V,s)$ such that the product*

$$\Lambda_j(V,s) := A^{s/2} L_{\infty,j}(V,s) \left(\prod_{p \in S} L_{p,j}(V,s) \right) L_{S,j}(V/\mathbf{Q}, s)$$

satisfies the previous properties and, in particular, the functional equation

$$\Lambda_j(V/\mathbf{Q}, s) = \pm\Lambda_j(V/\mathbf{Q}, j+1-s). \tag{6.52}$$

This conjecture has only been proven in a few cases.

Appendix A

Factorization

"'Four thousand two hundred and seven, that's the exact number,' the King said, referring to his book."
LEWIS CARROLL (THROUGH THE LOOKING GLASS)

In this chapter we take another look at the factorization problem that we started in Chap. 2 by explicitly describing a method for factoring polynomials and by sketching two of the most powerful algorithms developed during the last two decades for factoring integers: an algorithm due to Lenstra which uses elliptic curves [49] and the number field sieve *algorithm originally due to Pollard (see [19]). For those who are put off by probabilistic or heuristic estimation methods, keep in mind that once a factorization is found, it is very quick and easy to check it. It would be appropriate to complete this introduction by citing [22],* the *reference for algorithmic number theory. Furthermore, most of these algorithms are already implemented and available, for example with the* PARI/GP *package.*

1. Polynomial Factorization

We begin with the observation that it is fairly easy to find polynomial roots whose multiplicity is greater than 1: we just need to compute $D(X) := \gcd(P(X), P'(X))$. We can therefore essentially concentrate on factoring polynomials without any multiple roots.

There are many factorization algorithms in $\mathbf{F}_p[X]$ (or even $\mathbf{F}_q[X]$). We present one of them, due to Berlekamp, which is very efficient as long as p is not too large. It is based on the following two lemmas.

1.1. Lemma. *Let $f(X) \in \mathbf{F}_q[X]$ be a square-free polynomial of degree n.*

M. Hindry, *Arithmetics*, Universitext,
DOI 10.1007/978-1-4471-2131-2,

We define the $n \times n$ *matrix,* $B = ((b_{i,j}))_{0 \leqslant i,j \leqslant n-1}$, *by the polynomials:*

$$b_{0,j} + b_{1,j}X + \cdots + b_{n-1,j}X^{n-1} \equiv X^{jq} \mod f(X).$$

Then for any polynomial $h(X) = h_0 + h_1X + \cdots + h_{n-1}X^{n-1}$, *the following are equivalent:*

$$h(X)^q - h(X) \equiv 0 \mod f(X) \quad \Leftrightarrow \quad (B - I)\begin{pmatrix} h_0 \\ \vdots \\ h_{n-1} \end{pmatrix} = 0.$$

Furthermore, the dimension of $\mathrm{Ker}(B - I)$ *is the number of irreducible factors of* f.

Proof. The map from $\mathbf{F}_q[X]/(f(X))$ onto itself given by $P \mapsto P^q$ is linear and B is by definition the matrix of this linear transformation in the basis $1, \ldots, X^{n-1}$. The first statement follows directly from this observation. For the second, write f as the product $f = f_1 \cdots f_r$. Then

$$\mathbf{F}_q[X]/(f(X)) \cong \mathbf{F}_q[X]/(f_1(X)) \times \cdots \times \mathbf{F}_q[X]/(f_r(X)).$$

The equation $h(X)^q - h(X) \equiv 0 \mod f(X)$ translates to $h(X)^q - h(X) \equiv 0 \mod f_i(X)$, for $i = 1, \ldots, r$. Since $\mathbf{F}_q[X]/(f_i(X))$ is a finite extension of $\mathbf{F}_q$, this is therefore equivalent to $h(X) \equiv \lambda_i \mod f_i(X)$ with $\lambda_i \in \mathbf{F}_q$. We therefore have a total of q^r solutions, which proves that the vector space of solutions has dimension r. □

1.2. Lemma. *Let* $f(X) \in \mathbf{F}_q[X]$ *be a polynomial of degree* n *and let* $h(X)$ *be a polynomial of degree* $\leqslant n - 1$ *such that* $h(X)^q - h(X) \equiv 0 \mod f(X)$. *Then we have the following factorization:*

$$f(X) = \prod_{c \in \mathbf{F}_q} \gcd(f(X), h(X) - c).$$

Proof. The product on the right hand side clearly divides $f(X)$. Conversely, since $\prod_{c \in \mathbf{F}_q}(X - c) = X^q - X$, we see that $f(X)$ divides $h(X)^q - h(X) = \prod_{c \in \mathbf{F}_q}(h(X) - c)$. □

Let us now summarize the steps of a factorization algorithm for polynomials in $\mathbf{F}_q[X]$. By calculating $\gcd(f(X), f'(X))$ and factoring it out, we are back to the case where f is square-free. By applying the division algorithm to X^{jq} divided by $f(X)$, we construct the matrix B from Lemma A-1.1, and we calculate a solution to the linear system which gives a polynomial $h(x)$. Finally, we successively calculate $\gcd(f(X), h(X) - c)$ for $c \in \mathbf{F}_q$ until we find a nontrivial factor.

Even though we now see that there exists an algorithm for factoring polynomials in $\mathbf{F}_p[X]$, it is less easy to see in $\mathbf{Z}[X]$. It is nevertheless a priori possible to bound the size of a factor of a given polynomial. One such bound is given by the Gel'fond inequality or by the following lemma.

1.3. Lemma. *Let $P(X) = p_0 + p_1X + \cdots + p_dX^d$ and $Q(X) = q_0 + q_1X + \cdots + q_eX^e$ be two polynomials with integer coefficients. If Q divides P, then*

$$\sum_{j=0}^{e} |q_j| \leqslant 2^d \left(\sum_{i=0}^{d} |p_i|^2 \right)^{1/2}. \tag{A.1}$$

Proof. We define the *Mahler measure* of a polynomial $P = a_0(X-\alpha_1)\cdots(X-\alpha_d)$ to be

$$M(P) := \exp \int_0^1 \log\left|P(e^{2\pi it})\right| dt = |a_0| \prod_{i=1}^{d} \max(1, |\alpha_i|),$$

where the second equality is equivalent to the formula $\int_0^1 \log\left|e^{2\pi it} - \alpha\right| dt = \log\max(1, |\alpha|)$, which is well-known and is a particular case of Jensen's formula. We clearly have $M(PQ) = M(P)M(Q)$, and for all polynomials with integer coefficients, $M(P) \geqslant 1$. By using the relation between coefficients and roots, we can show that

$$|p_k| \leqslant |a_0| \left| \sum_{j_1 < \cdots < j_k} \alpha_{j_1} \cdots \alpha_{j_k} \right| \leqslant \binom{d}{k} M(P).$$

Jensen's convexity inequality gives

$$M(P) := \exp \frac{1}{2} \int_0^1 \log\left|P(e^{2\pi it})\right|^2 dt \tag{A.2}$$

$$\leqslant \left(\int_0^1 \left|P(e^{2\pi it})\right|^2 dt \right)^{1/2} = \left(\sum_{i=0}^{d} |p_i|^2 \right)^{1/2}. \tag{A.3}$$

We can therefore suppose that we have a factorization $P = QR$. Then we have

$$M(Q) \leqslant M(Q)M(R) = M(P)$$

and can therefore conclude that

$$\sum_{j=0}^{e} |q_j| \leqslant 2^e M(Q) \leqslant 2^d M(P) \leqslant 2^d \left(\sum_{i=0}^{d} |p_i|^2 \right)^{1/2}.$$

□

To make things simpler, we will assume that $f(X) \in \mathbf{Z}[X]$ is *monic*. From a theoretical point of view, we may make this assumption without loss of generality because if $f(X) = aX^d + \cdots \in \mathbf{Z}[X]$, then the polynomial $\tilde{f}(X) := a^{d-1}f(X/a) = X^d + \ldots$ will be monic with integer coefficients, and a factorization of $\tilde{f}$ will give a factorization of f. In order to find a factorization $f(X) = f_1(X)f_2(X)$, let us start with the trivial remark that one such factorization stays the same modulo N. We can therefore start by factoring $\bar{f}(X)$ in $\mathbf{F}_p[X]$ for some primes p and compare the degrees of the factors. If by any chance we find that some degrees are incompatible, we can use this to prove that $f(X)$ is irreducible. In general, we proceed by making use of the following variation of Hensel's lemma.

1.4. Lemma. *Let $f(X) \in \mathbf{Z}[X]$ be monic and p prime. Assume that $f(X) \equiv f_1(X)g_1(X) \bmod p$, with $\gcd(\bar{f}_1, \bar{g}_1) = 1$ in $\mathbf{F}_p[X]$. Then for all $m \geqslant 1$, there exist two monic polynomials $f_m, g_m \in \mathbf{Z}[X]$ such that $f_m \equiv f_1 \bmod p$, $g_m \equiv g_1 \bmod p$ and*

$$f(X) \equiv f_m(X)g_m(X) \bmod p^m.$$

Proof. The hypothesis that the reductions modulo p of f_1 and g_1 are relatively prime is essential and is used in the assertion that there exist $U, V \in \mathbf{Z}[X]$ such that $Uf_1 + Vg_1 \equiv 1 \bmod p$. Let us suppose that we have constructed the polynomials f_m and g_m so that $f = f_m g_m + p^m C$. In looking for polynomials in step $m+1$ of the form $f_{m+1} = f_m + p^m A$, $g_{m+1} = g_m + p^m B$, we find that $f - f_{m+1}g_{m+1} = p^m(C - Bf_m - g_m A) + p^{2m}AB \equiv p^m(C - Bf_1 - Ag_1) \bmod p^{m+1}$. It follows from the initial remark that there exist $A, B \in \mathbf{Z}[X]$ such that $Bf_1 + Ag_1 \equiv C \bmod p$. □

The factorization algorithm can be described as follows: a bound B for the size of the coefficients of a possible factor is computed according to Lemma A-1.3. Then f is reduced modulo a prime p, and the result is factored using the preceding algorithm (if a multiple root is found, then reduce modulo another prime). The factorization modulo p is lifted to a factorization modulo p^m where m is chosen so that $p^m > B$. After that, we check to see if that factorization comes from a factorization over $\mathbf{Z}$.

You can refer to Cohen's book [22] for a more detailed discussion of this algorithm and its variations.

2. Factorization and Elliptic Curves

Let us start by presenting a relatively inefficient algorithm (in certain cases), but which lends itself well to being generalized: Pollard's "$p-1$" method.

2.1. Definition. Let Y be an integer. A number is said to be *Y-smooth* if all of its prime divisors are smaller than Y. A number is said to be *Y-powersmooth* if every prime power which divides it is smaller than Y.

Let N be a number which we would like to factor and p a prime divisor of N. Suppose that $p-1$ is Y-powersmooth for some Y which is not too large. Then $p-1$ divides $m(Y) := \operatorname{lcm}(2,3,\ldots,Y)$. If a is an integer which is relatively prime to N, we would have $a^{m(Y)}-1 \equiv 0 \bmod p$, and therefore

$$\gcd\left(a^{m(Y)}-1, N\right) \neq 1,$$

which would very likely produce a non-trivial factor of N.

The method would therefore be efficient if N had a prime factor p such that $p-1$ is Y-smooth (or Y-powersmooth) for some Y which is not too large. The problem with this method is that large prime numbers p where $p-1$ is Y-smooth are fairly rare (see Proposition A-2.3 for an estimate of how rare). Likewise, one could hope to find an a which has a period significantly smaller than $p-1$. It is not difficult to see that this case is likewise fairly rare. The key idea of Lenstra's algorithm is to observe that we are actually working in $\mathbf{F}_p^*$, which is cyclic of cardinality $p-1$. If we can then apply the same type of reasoning to other groups of varying cardinality, we would have a better chance of factoring N. This is precisely what elliptic curve theory gives us.

If E is a curve over $\mathbf{F}_p$ and $P \in E(\mathbf{F}_p)$, its order n_P divides $\operatorname{card} E(\mathbf{F}_p) \in [p+1-2\sqrt{p}, p+1+2\sqrt{p}]$. If $\operatorname{card} E(\mathbf{F}_p)$ is Y-powersmooth, we analogously have

$$[m(Y)](P) = O_E.$$

Now we have the advantage of being able to try multiple elliptic curves. We only need to find sufficiently many curves such that the orders of their groups of rational points are Y-smooth. Let us see how this procedure, once properly formulated, gives a factorization algorithm.

We first need to discuss points in the projective plane or on an elliptic curve over $\mathbf{Z}/N\mathbf{Z}$. Let us point out that we do not lose anything by assuming that the integer N that we would like to factor is not divisible by 2 or by 3. It is not difficult to generalize the notions considered over a field. We do it in an ad hoc manner.

2.2. Definition. The *projective plane* over $\mathbf{Z}/N\mathbf{Z}$ is defined as the quotient $\mathscr{A} := \{(x_0,x_1,x_2) \in (\mathbf{Z}/N\mathbf{Z})^3 \mid \gcd(x_0,x_1,x_2,N) = 1\}$ by the relation $(x_0,x_1,x_2) \sim (ux_0,ux_1,ux_2)$ for $u \in (\mathbf{Z}/N\mathbf{Z})^*$.

If we assume that $\gcd(N, 6) = 1$, an elliptic curve over $\mathbf{Z}/N\mathbf{Z}$ is given by the Weierstrass equation,

$$zy^2 = x^3 + axz^2 + bz^3 \text{ with } a, b \in \mathbf{Z}/N\mathbf{Z} \text{ and } 4a^3 + 27b^2 \in (\mathbf{Z}/N\mathbf{Z})^*.$$

We will now carry out the calculations on the points of this elliptic curve, using the addition formulas 5-1.7. It seems like this might be problematic, because (for example) in order to add $P_1 = (x_1, y_1)$ and $P_2 = (x_2, y_2)$ one should invert $(x_1 - x_2)$, but if we keep in mind that we are trying to factor N, it is enough to calculate $\gcd(x_1 - x_2, N)$ and to observe that if this is nontrivial, we have found a factorization. Another possibility would be to use the projective coordinate formulas, and in this case, we would verify that $E(\mathbf{Z}/N\mathbf{Z})$ is a group. It turns out that from an algorithmic point of view, it is more economical to do the calculations in affine coordinates.

It is important to notice that the computation of $[m](P)$ does not require m addition steps (which would be restrictive), but by quick exponentiation only needs $O(\log m)$ addition steps (or duplication). Here, $\log m(Y) = \psi(Y) \sim Y$.

In order to effectively construct an elliptic curve and a point modulo N "at random", we could use many methods. One of the simplest ones consists of randomly choosing three integers modulo N, say x_0, y_0 and a, and setting $b := y_0^2 - x_0^3 - ax_0$. One needs to check that $\Delta := 4a^3 + 27b^2$ is invertible modulo N (if it is not, we have almost surely found a factorization of N), and we have a point $P = (x_0, y_0)$ on the elliptic curve $y^2 = x^3 + ax + b$. We therefore try to calculate $[m(Y)]P$; if the calculation does not produce a factor of N, we choose a new elliptic curve and start over. Notice that we can carry out the calculations on many elliptic curves simultaneously.

To analyze the performance of these algorithms, we need to estimate the number of Y-smooth integers. The following useful proposition is due to Canfield, Erdös and Pomerance [21].

2.3. Proposition. *If $2 \leqslant y \leqslant x$, we define:*

$$\psi(x, y) := \operatorname{card}\{n \leqslant x \mid n \text{ is } y\text{-smooth}\},$$

and let $u := \log x / \log y$. Then we have the formula

$$\psi(x, y) = xu^{-u(1+o(1))},$$

where the term $o(1)$ tends uniformly to 0 as x tends to infinity, and for a given $\epsilon > 0$, y satisfies: $(\log x)^\epsilon < \log y < (\log x)^{1-\epsilon}$.

In particular, if $y := \exp\left[C(\log x)^b(\log\log x)^c\right]$, we have:

$$\psi(x, y) = x \exp\left[-\frac{1-b}{C}(\log x)^{1-b}(\log\log x)^{1-c}(1 + o(1))\right].$$

The statement given here corresponds to the domain $u \in [(\log x)^\epsilon, (\log x)^{1-\epsilon}]$. We find a precise description of the asymptotic behavior of the function $\psi(x,y)$ in Tenenbaum's book [72].

Actually, we would need to know how many Y-smooth numbers an interval of type $[p-2\sqrt{p}, p+2\sqrt{p}]$ contains; in other words, we need a lower estimate of the value $\psi(p+2\sqrt{p}, Y) - \psi(p-2\sqrt{p}, Y)$. We do not know how to prove the estimate given below, in the same way that we do not know how to prove the existence of prime numbers in very small intervals, but it has been confirmed experimentally.

Set $L(x) := \exp\sqrt{\log x \log\log x}$. Proposition A-2.3 says that the probability that a random number $\leqslant x$ is $L(x)^a$-smooth is $L(x)^{-\frac{1}{2a}+o(1)}$. It is therefore natural to conjecture that this statement is still true on a sufficiently large interval.

2.4. Conjecture. *The ratio of $L(x)^a$-smooth numbers in the interval $[x-\sqrt{x}, x+\sqrt{x}]$ is $\geqslant L(x)^{-\frac{1}{2a}+o(1)}$.*

If the conjecture is true, in order to find an elliptic curve where the number of points over $\mathbf{F}_p$ is $L(p)^a$-smooth, we have to try $L(p)^{\frac{1}{2a}+o(1)}$ of them. We should perform $L(p)^a$ operations on each curve, hence a complexity on the order of $L(p)^{a+1/2a}$. By choosing $a = 1/\sqrt{2}$ (in other words Y on the order of $L(p)^{1/\sqrt{2}}$), we therefore obtain a complexity on the order of

$$L(p)^{\sqrt{2}} = \exp\sqrt{2\log p \log\log p}.$$

The complexity of the algorithm depends on the size of the smallest factor of N. This property is not very useful for factoring RSA type numbers but is a major advantage for most other integers.

Another property of the algorithm is that it does not require too much memory: in fact, we only have to save data which are polynomial in $\log N$.

3. Factorization and Number Fields

We will sketch the *number field sieve* algorithm originally suggested by Pollard and developed by Buhler, Lenstra and Pomerance (cf. [19]).

We are looking for an irreducible, monic polynomial $f(X) \in \mathbf{Z}[X]$ and an integer m such that $f(m) \equiv 0 \bmod N$. One handy and efficient method is to choose an integer d (usually $2 \leqslant d \leqslant 5$), then to look at $m := \lfloor N^{1/d} \rfloor$, to write N in base m (in other words calculate $a_i \in [0, m-1]$ such that

$N = a_0 + a_1 m + \cdots + a_{d-1} m^{d-1} + m^d$) and to choose

$$f(X) := a_0 + a_1 X + \cdots + a_{d-1} X^{d-1} + X^d.$$

Let us point out that the fact that the m expansion starts with m^d is equivalent to $m^d \leqslant N \leqslant 2m^d - 1$. The first inequality is true by construction and the second will be true if, for example, $N > 2^{d(d+1)}$. Finally, there is no saying a priori that $f(X)$ will be irreducible, but if that is not the case a factorization $f(X) = f_1(X) f_2(X)$ will give $N = f_1(m) f_2(m)$, which would be of course exactly what we want.

We will now construct a ring $A := \mathbf{Z}[X]/(f(X)) = \mathbf{Z}[\alpha]$ (where α is a root of f) and its field of fractions $K = \mathbf{Q}(\alpha)$. The idea of the algorithm is to look for a set S of pairs of integers (a, b) such that

i) $\prod_{(a,b)\in S}(a + bm)$ is a square (in $\mathbf{Z}$),
ii) $\prod_{(a,b)\in S}(a + b\alpha)$ is a square (in $\mathbf{Z}[\alpha]$).

If we have succeeded in doing that, we can consider the ring homomorphism $\phi : A \to \mathbf{Z}/N\mathbf{Z}$ given by $\phi(\alpha) = m$. We then find $\beta \in \mathbf{Z}[\alpha]$ such that $\beta^2 = \prod_{(a,b)\in S}(a + b\alpha)$, then $\phi(\beta)$ and $u \in \mathbf{Z}$ such that $u^2 = \prod_{(a,b)\in S}(a + bm)$. By construction, $\phi(\beta)^2 = u^2$ in $\mathbf{Z}/N\mathbf{Z}$. We then compute $\gcd(\phi(\beta) + u, N)$ and $\gcd(\phi(\beta) - u, N)$, which will very likely give us a factorization.

The main difficulties are, on the one hand, to construct a set of pairs (a, b) which satisfy the conditions i) and ii) and, on the other hand, to compute the square root of $\gamma := \prod_{(a,b)\in S}(a + b\alpha)$ in $\mathbf{Z}[\alpha]$.

To find a "simultaneous root", the idea is to choose a parameter Y, then to choose (by way of a number field sieve) integer pairs (a, b) such that $a + bm$ and $a + \alpha b$ are Y-smooth. We define an algebraic integer $\gamma \in \mathbf{Z}[\alpha]$ to be Y-smooth if $\mathrm{N}_{\mathbf{Q}}^{K}(\gamma)$ is itself Y-smooth. Having constructed a large enough set, say T, of pairs (a, b) (we need $\mathrm{card}(T)$ to be greater than $\pi(Y)$), we perform Gaussian elimination over $\mathbf{F}_2$ in order to find an adequate subset S. Initially, we will get an S such that

$$\prod_{(a,b)\in S}(a + bm) \quad \text{and} \quad \mathrm{N}_{\mathbf{Q}}^{K}\left(\prod_{(a,b)\in S}(a + b\alpha)\right) \quad \text{are squares (in } \mathbf{Z}).$$

This of course is not enough to guarantee that $\gamma := \prod_{(a,b)\in S}(a + b\alpha)$ is a square. Now we see how to refine the number field sieve so that at least the ideal generated by γ is a square. In general, the fact that the norm of an ideal is a square does not at all imply that the ideal is a square. We can however take advantage of the particular form of the algebraic numbers that we have produced.

3.1. Lemma. *Let p be a prime number which does not divide $(\mathcal{O}_K : \mathbf{Z}[\alpha])$. An ideal $\mathfrak{p}$ above p and which divides $a + b\alpha$ is an ideal which has norm p (i.e., of degree 1) and corresponds to the root $r \bmod p$ of $f(X) \bmod p$ such that $a + br \equiv 0 \bmod p$.*

Proof. In fact, according to the hypothesis, if the factorization in $\mathbf{F}_p[X]$ is written as $f(X) = f_1(X)^{e_1} \cdots f_g(X)^{e_g} \in \mathbf{F}_p[X]$, this corresponds to a decomposition $p\mathcal{O}_K = \mathfrak{p}_1^{e_1} \cdots \mathfrak{p}_g^{e_g}$ with $\mathrm{N}(\mathfrak{p}_i) = p^{\deg(f_i)}$ (see Exercise 3-6.20). □

We therefore refine the decomposition of $a + \alpha b$ by introducing

$$R(p) := \{r \in \mathbf{Z}/p\mathbf{Z} \mid f(r) \equiv 0 \bmod p\}$$

and the exponent corresponding to each $r \in R(p)$:

$$e_{p,r}(a + b\alpha) = \begin{cases} \operatorname{ord}_p \mathrm{N}(a + b\alpha) & \text{if } a + br = 0 \bmod p, \\ 0 & \text{if not.} \end{cases}$$

Therefore, we have $(a+b\alpha)\mathcal{O}_K = \prod_p \prod_{r\in R(p)} \mathfrak{p}_r^{e_{p,r}(a+b\alpha)}$ (ignoring the factors where p divides $(\mathcal{O}_K : \mathbf{Z}[\alpha])$). Hence $\mathrm{N}(a + b\alpha) = \pm \prod_p \prod_{r\in R(p)} p^{e_{p,r}(a+b\alpha)}$. Most importantly, the ideal generated by $\gamma := \prod_{(a,b)\in S}(a + b\alpha)$ will be a square if and only if

$$\sum_{(a,b)\in S} e_{p,r}(a + b\alpha) = 0 \bmod 2 \quad \text{(for every } p \text{ and } r \in R(p)).$$

The fact that $\gamma\mathcal{O}_K$ is a square (of ideals) does not always imply that γ is a square, but we have gotten closer. In order to measure how close, we introduce the group

$$\mathscr{C} := \{\gamma \in K^* \mid \text{there exists a fractional ideal } \mathscr{A} \text{ such that } (\gamma) = \mathscr{A}^2\}.$$

If we denote by $C\ell_K$ the ideal class group and $C\ell_K[2]$ the subgroup of elements killed by 2, we have the following exact sequence:

$$0 \longrightarrow \mathcal{O}_K^*/\mathcal{O}_K^{*2} \longrightarrow \mathscr{C}/K^{*2} \longrightarrow C\ell_K[2] \longrightarrow 0$$

($\gamma \in \mathscr{C}$ maps to the class $\mathscr{A}$ such that $\mathscr{A}^2 = \gamma\mathcal{O}_K$). In particular according to the unit theorem (Theorem 3-5.6), we have

$$\operatorname{rank} \mathscr{C}/K^{*2} = r_1 + r_2 + \operatorname{rank} C\ell_K[2].$$

The computation $C\ell_K[2]$ is simply too large to carry out. However, in order to increase the chances that γ is a square, we could calculate a small number of generalized Legendre symbols: we choose some prime ideals $\mathfrak{p}_1, \ldots, \mathfrak{p}_s$

and compute

$$\left(\frac{\gamma}{\mathfrak{p}}\right) := \begin{cases} +1 & \text{if } \gamma \text{ is a non-zero square modulo } \mathfrak{p}, \\ -1 & \text{if } \gamma \text{ is not a square modulo } \mathfrak{p}. \end{cases}$$

In this way, we can refine the sieve that produces candidates for squares by insisting that $\prod_{(a,b)\in S}(a+bm)$ is a square in $\mathbf{Z}$ and that $\prod_{(a,b)\in S}(a+b\alpha)$ generates a square ideal in $\mathscr{O}_K$ and $\left(\frac{\gamma}{\mathfrak{p}_i}\right) = +1$ for $i = 1, \ldots, s$.

We need to observe that the "sieve" part will use a lot of memory: we should in fact calculate and save the prime numbers smaller than Y, then arrange the pairs of numbers $a + bm$ for a and b in a chosen interval, test their divisibility by primes smaller than Y keeping only those which reduce to ± 1, and then start over again to sieve the $N_{\mathbf{Q}}^K(a + b\alpha)$.

In order to compute the square root of an algebraic integer $\gamma \in \mathbf{Z}[\alpha]$, let us suppose (to make things simpler) that we know its minimal polynomial $F(X) \in \mathbf{Z}[X]$. Observe that if $G(X)$ is the minimal polynomial of $\sqrt{\gamma}$, then $G(X)$ and $G(-X)$ divide $F(X^2)$. This suggests that we should use the following procedure.

- We factor the polynomial $F(X^2) = G(X)G(-X)$ (in $\mathbf{Z}[X]$);
- We perform division algorithm $G(X) = (X^2 - \gamma)Q(X) + R(X)$ in $\mathbf{Z}[\alpha][X]$.
- If $R(X) = aX + b$, set $\beta = -b/a$.

Then $\beta^2 = \gamma$.

Let us point out that the polynomial $F(X^2)$ is necessarily factored according to its given form and that if the remainder $R(X)$ is constant, then the number γ is not a square! That is to say that we have made a false assumption (for example, we might have neglected the index $(\mathscr{O}_K : \mathbf{Z}[\alpha])$ or might have not compensated enough for $\mathscr{C}/K^{*2}$). We therefore need to start all over again with another set S.

Let us also point out that in general $\mathbf{Z}[\alpha]$ is not integrally closed and $\sqrt{\gamma}$ might not be an element of $\mathbf{Z}[\alpha]$. Nevertheless, it is easy to overcome this obstacle: in fact, if f is the minimal polynomial of α, then $f'(\alpha)^2\mathscr{O}_K \subset \mathbf{Z}[\alpha]$, and we can safely replace γ by $f'(\alpha)^2\gamma$.

We refer you to the original article [19] or to [22] for an analysis of the complexity, which, modulo a "reasonable" conjecture, is on the order of

$$O\left(\exp(C(\log N)^{1/3}(\log\log N)^{2/3})\right).$$

For very large numbers which do not have any medium-sized factors, this algorithm is therefore more powerful than the elliptic curve algorithm.

3.2. Remark. (Factorization and quantum computers) We have seen in Exercise 2-7.15 that a quick calculation of the period (or order) of $a \bmod N$ will provide a fast factorization algorithm. In 1997 Shor [69] proved that if one had access to a "quantum computer", one could calculate this period in polynomial time. It is not known whether such a computer (with the required properties) could be built, but this discovery stimulated a field of research which is still very active.

Appendix B

Elementary Projective Geometry

"La línea consta de un número infinito de puntos; el plano, de un número infinito de líneas; el volumen, de un número infinito de planos; el hipervolumen, de un número infinito de volúmenes...
No, decididamente no es éste, more geométrico, el mejor modo de iniciar mi relato."

JORGE LUIS BORGES (EL LIBRO DE ARENA)

We will give an introduction to projective algebraic geometry: lines, conics, quadrics, cubics and Bézout's theorem on the number of points of intersection of two plane curves. We will clarify the notion of smoothness from a purely algebraic point of view. The projective context allows us to introduce the notion of "reduction modulo p" of a rational point on an algebraic variety. We will finish with some allusions to intersection theory.

1. Projective Space

1.1. Definition. Let K be a field. The affine space over K of dimension n, denoted $\mathbf{A}^n$ or $\mathbf{A}^n(K)$, is the set K^n. The projective space over K of dimension n, denoted $\mathbf{P}^n$ or $\mathbf{P}^n(K)$, is the set of lines through the origin in the vector space $E = K^{n+1}$ or the quotient of $K^{n+1} \setminus \{0\}$ by the equivalence relations $(x_0, \ldots, x_n) \sim (y_0, \ldots, y_n)$ if there exists $u \in K^*$ such that $(x_0, \ldots, x_n) = (uy_0, \ldots, uy_n)$. If P is the equivalence class of $(x_0, \ldots, x_n)$, we say that $(x_0, \ldots, x_n)$ are projective coordinates of P, and we simply write $P = (x_0, \ldots, x_n)$.

M. Hindry, *Arithmetics*, Universitext,
DOI 10.1007/978-1-4471-2131-2,

1.2. Remark. The value of a polynomial at $P = (x_0, \ldots, x_n) \in \mathbf{P}^n$ does not have any meaning, but if the polynomial $F(x_0, \ldots, x_n)$ is homogeneous the fact that $F(x_0, \ldots, x_n) = 0$ or $F(x_0, \ldots, x_n) \neq 0$ is independent of the projective coordinates. This allows us to make the following definition.

1.3. Definition. An *algebraic subset* of $\mathbf{A}^n$ (resp. of $\mathbf{P}^n$) is the set of common zeros of a family of polynomials (resp. homogeneous polynomials). A *linear subvariety* or *linear subspace* is the set of zeros of a family of homogeneous linear polynomials. The *dimension* of a linear subvariety of $\mathbf{P}^n$ is the dimension of the corresponding vector space minus one. A *conic* (resp. *cubic*) in $\mathbf{P}^2$ is the set of zeros of a homogeneous polynomial in (x_0, x_1, x_2) of degree 2 (resp. of degree 3).

1.4. Remark. More generally, we can define the *dimension* of an algebraic subset $V \subset \mathbf{P}^n$ as follows: let s be the maximal dimension of a linear subvariety L such that $V \cap L = \emptyset$, then $\dim V := n - s - 1$. We will freely use the natural vocabulary of calling a *curve* an algebraic subset of dimension 1 and a *surface* an algebraic subset of dimension 2.

1.5. Proposition. *Let L_1 and L_2 be two linear subspaces of dimension n_1 and n_2 such that $n_1 + n_2 \geqslant n$. Then $L_1 \cap L_2$ is non-empty. Moreover, this intersection is a linear subspace of dimension $\geqslant n_1 + n_2 - n$. If the dimension is equal to $n_1 + n_2 - n$, we say that L_1 and L_2 intersect transversally.*

Proof. Consider the map $\pi : K^{n+1} \setminus \{0\} \to \mathbf{P}^n(K)$. The linear subspaces L_i are images of vector subspaces (minus the origin) E_i of dimension $n_i + 1$. From linear algebra, we know that $\dim(E_1 \cap E_2) \geqslant (n_1+1)+(n_2+1)-(n+1)$; thus $F := E_1 \cap E_2$ is a vector subspace of dimension $\geqslant n_1 + n_2 - n + 1$, and consequently the image $L_1 \cap L_2 = \pi(F \setminus \{0\})$ is non-empty and of dimension $\geqslant n_1 + n_2 - n$. □

Remark. This statement contains the classical fact that two lines in the projective plane always either meet at one point or are coincident.

The following procedure, called a *Segre embedding*, allows us to consider the product of projective spaces or projective varieties as a projective variety.

1.6. Proposition. (Segre embedding) *The map $S : \mathbf{P}^n \times \mathbf{P}^m \to \mathbf{P}^{mn+m+n}$ given by $((x_0, \ldots, x_n), (y_0, \ldots, y_m)) \mapsto (x_i y_j)_{0 \leqslant i \leqslant n, 0 \leqslant j \leqslant m}$ is a bijection between $\mathbf{P}^n \times \mathbf{P}^m$ and an algebraic subset of $\mathbf{P}^{nm+m+n}$.*

Proof. Let $z_{i,j}$ be coordinates of $\mathbf{P}^{mn+m+n}$. We can immediately see that the image of S is contained in the variety defined by $z_{i,j} z_{k,\ell} - z_{i,\ell} z_{k,j} = 0$.

Conversely, let a point R with coordinates $z_{i,j}$ satisfy these equations. If, for example, $z_{0,0} \neq 0$, then we can assume that $z_{0,0} = 1$, hence, $z_{k,\ell} = z_{k,0}z_{0,\ell}$, and therefore $R = S((1, z_{1,0}, \dots, z_{n,0})(1, z_{0,1}, \dots, z_{0,m}))$. □

The affine space $\mathbf{A}^n$ can be seen as a subspace of $\mathbf{P}^n$ by considering $(x_1, \dots, x_n) \mapsto (1, x_1, \dots, x_n)$. The image is the subset

$$U_0 := \{(x_0, x_1, \dots, x_n) \in \mathbf{P}^n \mid x_0 \neq 0\}.$$

The complement is the hyperplane $x_0 = 0$, which can be viewed as $\mathbf{P}^{n-1}$ and which is often called "the hyperplane at infinity". We can therefore write:

$$\mathbf{P}^n = \mathbf{A}^n \sqcup \mathbf{P}^{n-1} = \mathbf{A}^n \sqcup \mathbf{A}^{n-1} \sqcup \cdots \sqcup \mathbf{A}^1 \sqcup \mathbf{A}^0.$$

We can actually cover the projective space by open affine sets by setting $U_i := \{P \in \mathbf{P}^n \mid x_i(P) \neq 0\}$. We see that $\mathbf{P}^n = U_0 \cup \cdots \cup U_n$ and that, on the one hand, the map from $\mathbf{A}^n$ to U_i given by the formula $(x_1, \dots, x_n) \mapsto (x_1, \dots, x_i, 1, x_{i+1}, \dots, x_n)$ and, on the other hand, the map from U_i to $\mathbf{A}^n$ given by $(x_0, \dots, x_n) \mapsto (x_0/x_i, \dots, x_{i-1}/x_i, x_{i+1}/x_i, \dots, x_n/x_i)$ are reciprocal bijections.

1.7. Definition. The *Zariski topology* on $\mathbf{A}^n$ (resp. on $\mathbf{P}^n$) is the topology whose closed sets are algebraic subsets, i.e., common zeros of a family of polynomials (resp. homogeneous polynomials).

We can immediately verify that it is indeed a topology: if V_i is the set of zeros of a homogeneous ideal I_i (generated by homogeneous polynomials), then $V_1 \cup V_2$ is the set of zeros of the ideal I_1I_2, and $\cap_{i \in S} V_i$ is the set of zeros of the ideal $\sum_i I_i$. Notice that the sets $U_i \subset \mathbf{P}^n$ are open and dense.

1.8. Definition. An affine (or projective) algebraic subset V is *irreducible* if it is not possible to write it as the union of two closed proper subsets, i.e., if $V = V_1 \cup V_2$, with V_i closed, then either $V = V_1$ or $V = V_2$. An irreducible algebraic subset is called an *algebraic variety*.

We can easily show that every algebraic set can be written as a finite union of irreducible algebraic subsets. If we eliminate redundancies, this decomposition is unique, and the maximal irreducible subsets are called *irreducible components*. The following lemmas allows us to clarify this phrase which is often seen and used in geometry, "It suffices to verify this in the general case."

1.9. Lemma. *Let V be an affine (resp. projective) variety and Z a proper algebraic subset (i.e., $Z \neq V$). Let F be a polynomial (resp. homogeneous* *polynomial), and suppose that F vanishes on $V \setminus Z$. Then it vanishes on V.*

Proof. The ring of polynomial functions on V is integral since V is irreducible; this is true because if fg is zero on V and if we had $f \not\equiv 0$ and $g \not\equiv 0$ on V, then we could deduce that we had a nontrivial decomposition $V = (V \cap \{f = 0\}) \cup (V \cap \{g = 0\})$. So let $G = 0$ be a nontrivial equation of Z, hence $FG = 0$ on V. Since $G \not\equiv 0$ on V, we can indeed deduce that $F \equiv 0$ on V. □

We can define in an intuitive manner the *morphisms* or *algebraic maps* between two varieties as "maps defined by polynomials". A more precise definition is as follows.

1.10. Definition. Let $V \subset \mathbf{A}^m$ and $W \subset \mathbf{A}^n$ be two affine varieties defined over a field K. A *morphism* or *algebraic map* $f : V \to W$ defined over K is given by n polynomials $f_1, \ldots, f_n \in K[X_1, \ldots, X_m]$ such that for every $x \in V$, the point $f(x) = (f_1(x), \ldots, f_n(x))$ is in W. We say that f is an *isomorphism* if there exists another morphism $g : W \to V$ such that $f \circ g = id_W$ and $g \circ f = id_V$.

A *rational map* from V to W defined over K is given by n rational functions $f_1, \ldots, f_n \in K(X_1, \ldots, X_m)$ (where V is not contained in the subset of poles of the f_i) such that for every x in V, outside of the poles, the point $f(x) = (f_1(x), \ldots, f_n(x))$ is in W. Such a map is denoted

$$f : V \cdots \to W,$$

to indicate that it is not necessarily defined everywhere. The map f is called a *birational* map if there exists another rational map $g : W \cdots \to V$ such that $f \circ g = id_W$ and $g \circ f = id_V$ (wherever they are defined).

Let $V \subset \mathbf{P}^m$ and $W \subset \mathbf{P}^n$ be two projective varieties defined over a field K. A *morphism* or *algebraic map* $f : V \to W$ defined over K is a map $f : V \to W$ such that for every point $x \in V$, there exists an affine open set U in V which contains x and an affine open set U' in W such that $f_{|U} : U \to U'$ is a morphism of affine varieties.

Note that it is possible to globally define a morphism of affine varieties or even from $\mathbf{P}^n$ to $\mathbf{P}^n$ by polynomials, but in general, we need many charts to define a morphism of projective varieties. We will now give you some examples of isomorphisms.

After linear subspaces, the most elementary algebraic varieties are the *quadrics*, i.e., the hypersurfaces defined by a homogeneous polynomial of degree 2 which has the form

$$Q(x_0, \ldots, x_n) = \sum_{0 \leqslant, i,j \leqslant n} q_{i,j} x_i x_j.$$

If the characteristic of the base field is different from 2, we can assume that $q_{i,j} = q_{j,i}$. The quadric is therefore nondegenerate if and only if $\det(Q) = \det(q_{i,j}) \neq 0$. After a linear transformation, we can also assume that $Q(x_0, \ldots, x_n) = a_0 x_0^2 + \cdots + a_n x_n^2$. Note also that Q is either irreducible or a product of two linear forms. A reducible quadric is degenerate, but the converse is only true for conics in $\mathbf{P}^2$. We will now classify the quadrics (up to linear transformation) of $\mathbf{P}^2$ and $\mathbf{P}^3$ over an algebraically closed field.

1.11. Theorem. *Let K be an algebraically closed field (of characteristic $\neq 2$). All of the nondegenerate projective conics over K are equivalent and, in particular, isomorphic to the conic given by the equation $y_0 y_1 - y_2^2 = 0$. The latter is isomorphic to the projective line by the map $(x_0, x_1) \mapsto (x_0^2, x_1^2, x_0 x_1)$.*

If K is not algebraically closed and C is a conic, then $C(K) = \emptyset$ or C is isomorphic over K to $\mathbf{P}^1$.

Thus the usual classification of conics into ellipses, hyperbolas, and parabolas is valid in real affine geometry, whereas in the projective plane over an algebraically closed field, there is only one conic.

Proof. We will start by constructing a hyperbolic plane, i.e., a plane endowed with a basis in which the quadratic form is written $Q(x, y) = xy$. To do this, we choose an isotropic vector e_0, then another isotropic vector e_1 not orthogonal to e_0, and by adjusting by a scalar, we obtain an appropriate basis. We then choose a vector e_2 orthogonal to the hyperbolic plane and such that $Q(e_2) = -1$. After a linear transformation, we indeed have in the new basis $Q(x_0, x_1, x_2) = x_0 x_1 - x_2^2$. For the second statement, let $P \in C(K)$, and consider the set of lines passing through P. This set is parametrized by $\mathbf{P}^1$. We can easily verify (see the more general proof given below) that every line D passing through P intersects the conic in a second point P_D: the map $D \mapsto P_D$ provides the needed isomorphism. □

1.12. Theorem. *Let K be an algebraically closed field. All nondegenerate projective quadrics from $\mathbf{P}^3$ to K are equivalent and isomorphic to $\mathbf{P}^1 \times \mathbf{P}^1$ by the Segre map*

$$((x_0, x_1)(y_0, y_1)) \mapsto (x_0 y_0, x_1 y_0, x_0 y_1, x_1 y_1),$$

whose image is the quadric given by the equation $z_{0,0} z_{1,1} - z_{0,1} z_{1,0} = 0$. In particular, the surface is ruled in two ways.

Proof. After a linear transformation (over K which is algebraically closed), we can effectively assume that $Q(x_0, x_1, x_2, x_3) = x_0x_1 - x_2x_3$ (this amounts to writing the space as a direct sum of two hyperbolic planes). The isomorphism of the quadric to $\mathbf{P}^1 \times \mathbf{P}^1$ is thus a particular case of Proposition B-1.6. □

The following two lemmas are special cases of Bézout's theorem, proven further down (Theorem B-2.4).

1.13. Lemma. *Let C be a curve of degree d (i.e., defined by a homogeneous polynomial of degree d) in the projective plane and not containing the line D of $\mathbf{P}^2$. Then $C \cap D$ is composed of d points (counted with multiplicity).*

Proof. Let $F(x_0, x_1, x_2) = 0$ be the equation of degree d of C and $a_0x_0 + a_1x_1 + a_2x_2 = 0$ that of D. One of the a_i is non-zero, so we can take it to be a_0. The equation of points of intersection of C and D is therefore written $x_0 = -\frac{a_1}{a_0}x_1 - \frac{a_2}{a_0}x_2$ and

$$F\left(-\frac{a_1}{a_0}x_1 - \frac{a_2}{a_0}x_2, x_1, x_2\right) = 0,$$

which factors as $a\prod_i(\alpha_i x_1 - \beta_i x_2)^{m_i}$ with $\sum_i m_i = d$. □

1.14. Lemma. *If C is a curve of degree d in the projective plane with no components in common with the conic D of $\mathbf{P}^2$, then $C \cap D$ is composed of $2d$ points (counted with multiplicity).*

Proof. If the conic is composed of two lines, this lemma can be deduced from the previous lemma. We can thus assume that the conic is irreducible. Up to a linear change of coordinates, we can assume that the conic is written as $x_1x_0 - x_2^2 = 0$ and hence that it is parametrized by the map from $\mathbf{P}^1$ to $\mathbf{P}^2$ given by $(y_0, y_1) \mapsto (y_0^2, y_1^2, y_0y_1)$. Let $F(x_0, x_1, x_2) = 0$ be the equation of C. The equation of the points of intersection of C and D is thus written $P = (y_0^2, y_1^2, y_0y_1)$ and

$$F\left(y_0^2, y_1^2, y_0y_1\right) = 0,$$

which factors into $a\prod_i(\alpha_i y_1 - \beta_i y_0)^{m_i}$ with $\sum_i m_i = 2d$. □

Notation. We denote by $S_{n,d}$ the vector space of homogeneous polynomials of degree d in $x_0, \ldots, x_n$, and if $P_1, \ldots, P_r$ are points of $\mathbf{P}^n$, we denote by $S_{n,d}(P_1, \ldots, P_r)$ the subspace of $S_{n,d}$ formed of polynomials which vanish at each P_i.

1.15. Definition. A *linear system of hypersurfaces* S of degree d in $\mathbf{P}^n$ is a vector subspace S of $S_{n,d}$.

The set of hypersurfaces corresponding to the polynomials of S can be seen as a linear subvariety of dimension $\dim(S) - 1$ in the projective space corresponding to $S_{n,d}$.

1.16. Lemma. *We have the following formulas:*

$$\dim S_{n,d} = \binom{n+d}{d} \qquad \text{and} \qquad \dim S_{n,d}(P_1, \dots, P_r) \geqslant \dim S_{n,d} - r.$$

The lemma is obvious by noticing that vanishing at point P is a linear condition on the coefficients of a polynomial. The computation of the exact dimension of $S_{n,d}(P_1, \dots, P_r)$ can however be tricky.

1.17. Examples. We have

$$\dim S_{2,d} = \frac{(d+2)(d+1)}{2} \text{ and } \dim S_{2,d}(P_1, \dots, P_r) \geqslant \frac{(d+2)(d+1)}{2} - r$$

and, in particular, $\dim S_{2,2} = 6$ and $\dim S_{2,2}(P_1, \dots, P_r) \geqslant 6 - r$. Thus there always passes at least one conic through any five given points. We can specify under which conditions such a conic is unique.

1.18. Lemma. *Through any five points $P_1, \dots, P_5$ in the projective plane, there always passes a conic. Furthermore, if no four of the points are colinear, the conic is unique, i.e., $\dim S_{2,2}(P_1, \dots, P_5) = 1$.*

Proof. We will first treat the case where three of the points, P_1, P_2, P_3, are colinear. The conic must contain the line $L = 0$ defined by the three points. Hence, we have $S_{2,2}(P_1, \dots, P_5) = LS_{2,1}(P_4, P_5)$ since P_4 and P_5 are not on the line $L = 0$. There is only one line which passes through P_4 and P_5, hence $\dim S_{2,1}(P_4, P_5) = 1$ and $\dim S_{2,2}(P_1, \dots, P_5) = 1$. We will now treat the case where no three of the P_i are colinear. Suppose $\dim S_{2,2}(P_1, \dots, P_5) > 1$, and let P_6 be a point distinct from P_4 and P_5 on the line $L = 0$ defined by these two points. We would then have $\dim S_{2,2}(P_1, \dots, P_6) \geqslant 1$, and a corresponding conic containing P_4, P_5, P_6 must contain the whole line hence be composed of two lines, and then P_1, P_2, P_3 would be colinear. □

The dimension of $S_{2,3}$ is 10. Therefore, there is always a cubic passing through any nine points in the projective plane plane. If 4 of these points are colinear, the cubic must contain the corresponding line, and if 7 of these points are on the same conic, the cubic must contain the corresponding conic.

1.19. Definition. A point $P = (x_0, \dots, x_n)$ on a hypersurface $V = \{P \in$

$\mathbf{P}^n \mid F(P) = 0\}$ is *singular* if $\dfrac{\partial F}{\partial x_i}(P) = 0$ for $0 \leqslant i \leqslant n$. The hypersurface V is *singular* if such a point exists and *smooth* if not.

1.20. Remark. We can define the notion of smoothness for subvarieties of any dimension. To do this, let $V \subset \mathbf{P}^n$ be a subvariety of dimension m and codimension $r := n - m$. If I is the ideal of polynomials vanishing on V and $\mathscr{G}_I$ a finite generating set of I, a point P in V is *smooth* if the rank of the matrix

$$\left(\frac{\partial F}{\partial x_i}(P)\right)_{0 \leqslant i \leqslant n,\, F \in \mathscr{G}_I}$$

is equal to r (it is always $\leqslant r$). The projective tangent space (at a point P on a hypersurface with equation $F = 0$) is $\sum_{i=0}^n \dfrac{\partial F}{\partial x_i}(P) x_i = 0$. The map which associates to a nonsingular point its tangent hyperplane is classically known as *the Gauss map*.

1.21. Definition. The map "*reduction modulo p*" is defined from $\mathbf{P}^n(\mathbf{Q})$ to $\mathbf{P}^n(\mathbf{F}_p)$ as follows. If $P \in \mathbf{P}^n(\mathbf{Q})$, we choose coordinates $x_i \in \mathbf{Z}$ such that $\gcd(p, x_0, \ldots, x_n) = 1$, and we set $r_p(P) = (\tilde{x}_0, \ldots, \tilde{x}_n)$ where $\tilde{x}$ designates the class of x in $\mathbf{F}_p = \mathbf{Z}/p\mathbf{Z}$. If V is a projective subvariety of $\mathbf{P}^n$ defined over $\mathbf{Q}$, we define $\tilde{V}$ to be the "*reduction modulo p*" of V as follows. Let I_V be the ideal in $\mathbf{Q}[x_0, \ldots, x_n]$ of polynomials which vanish on V, $I_{V,\mathbf{Z}} := I_V \cap \mathbf{Z}[x_0, \ldots, x_n]$ and $\tilde{I}_V$ the image of $I_{V,\mathbf{Z}}$ in $\mathbf{F}_p[x_0, \ldots, x_n]$. Then $\tilde{V}$ is the subvariety of $\mathbf{P}^n$ defined by $\tilde{I}_V$.

We should point out that if $P \in V(\mathbf{Q})$, then $\tilde{P} \in \tilde{V}(\mathbf{F}_p)$, and this property is specific to closed (projective) varieties. For example, if V is a curve in $\mathbf{P}^2$ with equation $F(X, Y, Z) = 0$ and U is the affine curve $F(x, y, 1) = 0$ (seen as an open set in V), a point P which is in $U(\mathbf{Q})$ has reduction modulo p, denoted $\tilde{P}$, and there is no reason that this should be in $\tilde{U}$, the affine curve with equation $\tilde{F}(x, y, 1) = 0$. In fact, $\tilde{P} \in \tilde{U}$ if and only if x and y are p-integers, i.e., if P is a *p-integral* point of U.

1.22. Proposition. *Let V be a smooth subvariety of $\mathbf{P}^n(\mathbf{Q})$. For all p, except for a finite number, the subvariety $\tilde{V}$ is a smooth subvariety of $\mathbf{P}^n(\mathbf{F}_p)$.*

Proof. If V is a hypersurface given by $F(x_0, \ldots, x_n) = 0$, the hypothesis that the hypersurface $F(x) = 0$ is smooth means that the resultant, R, of the partial derivatives of F is non-zero; the latter R can be expressed as a polynomial in the coefficients of F. Therefore, the hypersurface remains smooth modulo p for all the primes numbers not dividing R. □

2. Intersection

The theorem below is fundamental to classical algebraic geometry. It shows, in particular, that there is a bijection between algebraic subsets and *reduced* ideals (i.e., such that if a power of an element is in the ideal, then the element itself is in the ideal).

2.1. Theorem. (Hilbert's Nullstellensatz) *Let $P_1, \dots, P_m$ be polynomials in $K[X_1, \dots, X_n]$ where K is algebraically closed. If Q is a polynomial which vanishes at the set of common zeros of the P_i, then a power of Q is in the ideal generated by the P_i. In other words, there exist $r \geqslant 1$ and $A_i \in K[X_1, \dots, X_n]$ such that*

$$Q^r = A_1 P_1 + \cdots + A_m P_m.$$

Proof. (Sketch) The key algebraic lemma is the fact that a finitely generated K-algebra which is a field must be algebraic over K, hence equal to K if K is algebraically closed (see for example [43]). We thus consider the polynomials $P_1, \dots, P_m, 1 - TQ \in K[X_1, \dots, X_m, T]$. According to the hypotheses these polynomials do not have any common zeros in K^{m+1}. We now prove that this implies that they generate (as an ideal) the ring $K[X_1, \dots, X_m, T]$. If this were not the case, they would be contained in a maximal ideal $\mathfrak{M}$, and the quotient $K[X_1, \dots, X_m, T]/\mathfrak{M}$ would be an algebraic extension of K and would thus, by the lemma recalled above, be isomorphic to K. If we let $x_i := X_i \bmod \mathfrak{M} \in K$, we have constructed a common zero $(x_1, \dots, x_m, t) \in K^{m+1}$ to all of the polynomials of $\mathfrak{M}$. We can deduce from this the existence of polynomials $U_i(X, T) \in K[X_1, \dots, X_m, T]$ such that

$$1 = U_1(X,T)P_1(X) + \cdots + U_m(X,T)P_m(X,T) + U_{m+1}(X,T)(1 - TQ(X)).$$

By interpreting this identity in $K(X)[T]$, substituting $T = 1/Q(X)$ and multiplying by $Q(X)^r$ where $r = \max \deg_T U_i(X,T)$, we obtain

$$Q^r(X) = A_1(X)P_1(X) + \cdots + A_m(X)P_m(X)$$

where $A_i(X) := Q(X)^r U_i(X, \dfrac{1}{Q(X)})$. □

2.2. Remarks. 1) In the course of the proof, we proved that a maximal ideal of $K[X_1, \dots, X_m]$ is of the form $(X_1 - a_1, \dots, X_m - a_m)$.

2) If $k \subset K$ and if $Q, P_1, \dots, P_m$ have coefficients in k, we can easily see that we can choose the A_i to have coefficients in k. Likewise, if $Q, P_1, \dots, P_m$ are homogeneous, we can choose the A_i to be homogeneous.

In the case where the polynomials define a finite set, we can estimate its cardinality by the following theorem (see [35] or [31]).

2.3. Theorem. *Let $Z_1, \ldots, Z_n$ be hypersurfaces with degrees $d_1, \ldots, d_n$ in $\mathbf{P}^n$. The intersection $Z_1 \cap \cdots \cap Z_n$ is non-empty, and if the intersection is finite, then the cardinality of this set satisfies*

$$\operatorname{card}(Z_1 \cap \cdots \cap Z_n) \leqslant d_1 \cdots d_n.$$

We can actually define multiplicities so as to obtain an equality in the previous assertion. We will do this for the case of two plane curve $C_1, C_2 \subset \mathbf{P}^2$ without any common components. To define the multiplicity at a point P, we can work in the affine plane and consider $f_i(x, y) = 0$ the affine equations of the C_i. We define the *local ring* $\mathscr{O}_P$ at $P = (a, b)$ as:

$$\mathscr{O}_P := \{F \in K(x, y) \mid \operatorname{ord}_P(F) \geqslant 0\} = S^{-1}K[x, y] \qquad \text{(B.1)}$$

where S is the multiplicative set of polynomials which do not vanish at P (or which do not appear in the ideal $(x - a, y - b)$). We then set

$$\operatorname{mult}(P; C_1, C_2) = \dim \mathscr{O}_P/(f_1, f_2)_P \qquad \text{(B.2)}$$

where $(f_1, f_2)_P$ is the ideal generated by f_1 and f_2 in $\mathscr{O}_P$. The dimension is well-defined whenever C_1 and C_2 do not have any common components containing P. The main properties of this notion of multiplicity are that it is positive, biadditive (meaning that if $C_1 = C + C'$, then $\operatorname{mult}(P; C_1, C_2) = \operatorname{mult}(P; C, C_2) + \operatorname{mult}(P; C', C_2)$) and equal to 1 whenever C_1 and C_2 intersect transversally at P (meaning that the tangents intersect transversally). In particular, we have

$$\operatorname{mult}(P; C_1, C_2) \geqslant 1 \Leftrightarrow P \in C_1 \cap C_2.$$

2.4. Theorem. (Bézout) *Let C_1 and C_2 be two plane curves of degree d_1 and d_2 in $\mathbf{P}^2$ without any common components. Then*

$$\sum_{P \in C_1 \cap C_2} \operatorname{mult}(P; C_1, C_2) = d_1 d_2.$$

Proof. We point out that the finiteness of $C_1 \cap C_2$ follows easily from the existence of non-zero polynomials such that $a(x)f_1(x, y) + b(x)f_2(x, y) = c(x)$ and $a'(y)f_1(x, y) + b'(y)f_2(x, y) = c'(y)$. Up to changing projective coordinates, we can thus assume that the line at infinity does not intersect $C_1 \cap C_2$. This condition translates to the fact that the homogeneous parts of largest degree, $f_1^{(d_1)}$ and $f_2^{(d_2)}$, are relatively prime. We will now prove, using this hypothesis, that

$$\dim k[x, y]/(f_1, f_2) = d_1 d_2. \qquad \text{(B.3)}$$

Let A_d be the set of polynomials of $k[x, y]$ of degree $\leqslant d$. It is a vector space of dimension $s(d) = \binom{d+2}{2} = (d + 1)(d + 2)/2$. The map $A_d \to$

$k[x,y]/(f_1, f_2)$ is surjective for large enough d; the kernel $B_d := A_d \cap (f_1, f_2)$ contains $I_d := A_{d-d_1} f_1 + A_{d-2} f_2$. We will prove that $I_d = B_d$ whenever $d \geqslant d_1 + d_2$. Let $f = g_1 f_1 + g_2 f_2 \in B_d$. We can assume that g_1 and g_2 have minimal degrees e_1 and e_2, and we want to show that $e_1 \leqslant d - d_1$ and $e_2 \leqslant d - d_2$. If we have, for example, $e_1 > d - d_1$, then we see, considering the homogeneous parts of largest degree, that $e_1 + d_1 = e_2 + d_2$ and that $g_1^{(e_1)} f_1^{(d_1)} + g_2^{(e_2)} f_2^{(d_2)} = 0$. Since $f_1^{(d_1)}$ and $f_2^{(d_2)}$ are relatively prime, we can deduce from this that $g_2^{(e_2)} = -f_1^{(d_1)} h$ and $g_1^{(e_1)} = f_2^{(d_2)} h$, but then $f = f_1(g_1 - f_2 h) + f_2(g_2 + f_1 h)$ allows us to write f with polynomials of degree $< e_1$ and e_2. Thus, for large enough d we have $A_d/I_d \cong k[x,y]/(f_1, f_2)$. We know that $A_{d-d_1} f_1 \cap A_{d-d_2} f_2 = A_{d-d_1-d_2} f_1 f_2$, hence $\dim(k[x,y]/(f_1, f_2))$ equals:

$$\begin{aligned}\dim A_d/I_d &= \dim A_d - \dim A_{d-d_1} f_1 - \dim A_{d-d_2} f_2 + \dim A_{d-d_1-d_2} f_1 f_2\\ &= s(d) - s(d-d_1) - s(d-d_2) + s(d-d_1-d_2)\\ &= d_1 d_2.\end{aligned}$$

The proof of the following equality finishes the proof of the theorem.

$$\dim k[x,y]/(f_1, f_2) = \sum_{P \in C_1 \cap C_2} \operatorname{mult}(P; C_1, C_2). \tag{B.4}$$

This equality is a special case of the decomposition of a module with finite length. Recall that an A-module has finite length if there exists a sequence of submodules $M = M_0 \supset M_1 \supset \cdots \supset M_\ell = 0$ such that each M_i/M_{i+1} is a simple A-module (i.e., a module of the type $A/\mathfrak{M}$, where $\mathfrak{M}$ is a maximal ideal). We therefore have the decomposition

$$M \cong \oplus_{\mathfrak{p}} M_{\mathfrak{p}}$$

where the (finite) sum is taken over the maximal ideals of A, and $M_{\mathfrak{p}}$ designates the localization of the module with respect to $\mathfrak{p}$. The proof of this last assertion can be done by induction on the length ℓ. If $\ell = 1$, then $M \cong A/\mathfrak{M}$ and $M_{\mathfrak{p}} = 0$ if $\mathfrak{p} \neq \mathfrak{M}$ whereas $(A/\mathfrak{M})_{\mathfrak{M}} = A/\mathfrak{M}$. Then, if we know the result for M_1 (which has length $\leqslant \ell - 1$) and M/M_1 (which is simple), we can deduce the result for M using this. □

2.5. Remark. We have seen that we can describe the points of $\mathbf{P}^n$ as lines through the origin of a vector space E of dimension $n+1$ or as the hyperplanes of E^*. We can generalize this by introducing the set $G(E,k)$ of vector subspaces of E with a given dimension k and by endowing this set with the structure of a projective variety which is called the *Grassmannian*. To do this, we can proceed as follows: for a subspace L of dimension k in a vector space E of dimension $n+1$, we choose $e_1, \ldots, e_k$ to be a basis for L, and we denote by $\phi(e_1, \ldots, e_k) = e_1 \wedge \cdots \wedge e_k \in \Lambda^k(E)$ (if we write the coordinates of the e_i in a fixed basis of E in a matrix, then the coordinates of $e_1 \wedge \cdots \wedge e_k$ are the $k \times k$ minors of this matrix). We can check that two bases of L generate the same line in $\Lambda^k(E)$. Hence, we can set $\Phi(L) := [\phi(e_1, \ldots, e_k)] \in \mathbf{P}\Lambda^k(E)$. We then show that the map $\Phi : G(E,k) \to \mathbf{P}\Lambda^k(E)$ is injective, and its image is a projective subvariety.

2.6. Remark. More generally, we can define the *intersection number* of two subvarieties W_1 and W_2 of V of complementary dimensions, i.e., such that $\dim W_1 + \dim W_2 = \dim V$. The construction of these intersection numbers is out of the scope of this text (see [35] and most of all [31]), but we can easily state some of its properties.

2.7. Definition. An *algebraic cycle* of codimension i on V is a linear combination with integer coefficients of subvarieties of codimension i. The group of cycles of codimension i is denoted

$$\mathscr{Z}^i(V) := \oplus_{\mathrm{codim}(W)=i} \mathbf{Z}[W],$$

where W ranges over the subvarieties of codimension i.

2.8. Proposition. *Let V be a smooth projective variety of dimension r. There exists a $\mathbf{Z}$-bilinear mapping, invariant under algebraic deformation,*

$$\begin{array}{ccc} \mathscr{Z}^i(V) \times \mathscr{Z}^{r-i}(V) & \to & \mathbf{Z} \\ (W, W') & \mapsto & W \cdot W' \end{array}$$

such that if W and W' intersect transversally at a finite number of points, then $W \cdot W' = \mathrm{card}(W \cap W')$. If we further impose functoriality: for every finite morphism $\phi : V' \to V$ we have $\phi^{-1}(W) \cdot \phi^{-1}(W') = \deg(\phi) W \cdot W'$, then the mapping is unique.

The notion of an algebraic deformation of cycles of V can be briefly described as follows. Let T be a variety and $Z \subset V \times T$ a subvariety such that for every point $t \in T$, we can define a cycle $Z_t := Z \cap V \times \{t\}$ on V. If t_1 and t_2 are in T, we say that Z_{t_1} can be deformed into Z_{t_2}. The invariance property of the proposition can be translated into the fact that for every cycle W of dimension complementary to Z_t, we have $Z_{t_1} \cdot W = Z_{t_2} \cdot W$.

2.9. Definition. Two cycles $W, W' \in \mathscr{Z}^i(V)$ are *numerically equivalent* if for every $Y \in \mathscr{Z}^{r-i}(V)$, we have $W \cdot Y = W' \cdot Y$. The quotient of $\mathscr{Z}^i(V)$ by this equivalence relation is denoted $\mathrm{Num}^i(V)$.

The numerical equivalence relation is of great importance in algebraic geometry: it is at the heart of the theory of Grothendieck motives. We know that $\operatorname{rank}_{\mathbf{Z}} \operatorname{Num}^i(V)$ is finite. The formal properties of the numerical equivalence relation resemble those of the homological equivalence relation and Grothendieck also conjectures that they coincide. We have, for example, $\operatorname{Num}^i(\mathbf{P}^n) = \mathbf{Z}$.

Appendix C

Galois Theory

"I'm Nobody! Who are you?
Are you nobody, too?
Then there's a pair of us! ? don't tell!
They'd banish us, you know!"
Emily Dickinson

By relying on some results from Galois theory (see, for example, [43]), we will more explicitly describe the decomposition law of prime ideals of rings in a number field in the case where the extension is Galois (see [7] for more details) before stating Chebotarev's theorem which connects this algebraic theory to analytic theory and provides an elegant generalization of the Dirichlet's theorem on arithmetic progressions. The last two sections present the beginnings of the theory of Galois representations, i.e., the study of the absolute Galois group $G_{\mathbf{Q}} := \mathrm{Gal}(\bar{\mathbf{Q}}/\mathbf{Q})$. *First, class field theory (see [44]) provides a description of abelian extensions and also representations of dimension 1 (which allows us to state a vast generalization of the quadratic reciprocity law). Then, we give some examples and basic properties of representations of dimension* > 1 *(see [27], [63] and [64]).*

1. Galois Theory and Number Fields

Let us briefly recall the fundamentals of Galois theory.

Notation. We denote by $\mathrm{Aut}(F)$ the group of automorphisms of a field. If K is a subfield of F, then we denote by $\mathrm{Aut}(F/K)$ the subgroup of $\mathrm{Aut}(F)$ which acts trivially on K. If G is a subgroup of $\mathrm{Aut}(F)$, we denote by $F^G := \{x \in F \;\; \forall g \in G,\; g(x) = x\}$ the subfield fixed by G.

An extension F/K is *Galois* if it is normal and separable. In this case, we call the group $\mathrm{Aut}(F/K)$ *the Galois group* of the extension, denoted by

M. Hindry, *Arithmetics*, Universitext,
DOI 10.1007/978-1-4471-2131-2,

Gal(F/K). If F/K is finite of degree $n=[F:K]$, this amounts to saying that card Aut$(F/K)=n$. To every subgroup H of Gal(F/K), we can associate the extension F^H, and to every extension L, we can associate the subgroup Aut$(F/L)=$ Gal(F/L). The fundamental theorem of Galois theory (for finite extensions) says that we have thus established a bijection between intermediate extensions $K\subset L\subset F$ and subgroups of Gal(F/K). The same theorem further states that the extension F^H is Galois over K if and only if H is normal in Gal(F/K), and if that is the case, then $\mathrm{Gal}(F^H/K)\cong \mathrm{Gal}(F/K)/H$. More generally, if $\sigma\in\mathrm{Gal}(F/K)$ and $K\subset L\subset F$, we have $\mathrm{Gal}(F/\sigma L)=\sigma\,\mathrm{Gal}(F/L)\sigma^{-1}$, and if $H\subset\mathrm{Gal}(F/K)$, we have $\sigma\left(F^H\right)=F^{\sigma H\sigma^{-1}}$.

To generalize this to possibly infinite extensions, we introduce the *Krull topology* on $G:=\mathrm{Gal}(F/K)$, where a basis of neighborhoods of the identity is given by finite index subgroups of G. The Galois correspondence is therefore a bijection between subextensions $K\subset L\subset F$ and *closed* subgroups of G, which are finite extensions of K corresponding to subgroups of G which are both closed and open (clopen).

1.1. Examples. 1) *(Finite fields)* The Galois group $\mathrm{Gal}(\mathbf{F}_{q^m}/\mathbf{F}_q)$ is canonically isomorphic to $\mathbf{Z}/m\mathbf{Z}$, the canonical generator being the "Frobenius" $\Phi(x)=x^q$. We can deduce from this a description of the absolute Galois group:

$$\mathrm{Gal}(\bar{\mathbf{F}}_q/\mathbf{F}_q)=\varprojlim_m \mathbf{Z}/m\mathbf{Z}=\prod_\ell \mathbf{Z}_\ell$$

where the product is over prime numbers ℓ and $\mathbf{Z}_\ell$ designates the ℓ-adic integers.

2) *(Cyclotomic extensions)* Let ζ be a primitive nth root of unity (for example $\zeta=\exp(2\pi i/n)$). The Galois group $\mathrm{Gal}(\mathbf{Q}(\zeta)/\mathbf{Q})$ is canonically isomorphic to $(\mathbf{Z}/n\mathbf{Z})^*$, and the isomorphism $\sigma\mapsto m(\sigma)$ is given by $\sigma(\zeta)=\zeta^{m(\sigma)}$ (we are using the irreducibility of cyclotomic polynomials here, Theorem 2-6.2.7). From this, we can deduce a description of the Galois group of the extension $\mathbf{Q}(\mu_{\ell^\infty})$ generated by all of the ℓ^nth roots of unity:

$$\mathrm{Gal}(\mathbf{Q}(\mu_{\ell^\infty})/\mathbf{Q})=\varprojlim_n(\mathbf{Z}/\ell^n\mathbf{Z})^*=\mathbf{Z}_\ell^*.$$

3) *(Kummer extensions)* Let K be a field containing the mth roots of unity, i.e., $\mu_m\subset K^*$, and let $\alpha\in K^*$ and $\beta\in\bar{K}$ such that $\beta^m=\alpha$, which will be (slightly abusively) denoted by $\beta=\sqrt[m]{\alpha}$. The extension $K(\beta)/K$ is therefore Galois, and its Galois group is isomorphic to a subgroup of μ_m, the injective homomorphism $\zeta:\mathrm{Gal}(K(\beta)/K)\to\mu_m$ being given by $\sigma(\beta)=\zeta(\sigma)\beta$.

4) *(Extensions generated by torsion points of an elliptic curve)* Let E be an elliptic curve defined over K. We denote by $K(E[N])$ the field generated

by the coordinates (in $\bar{K}$) of the torsion points killed by N. The extension $K(E[N])$ is Galois, and by remembering (cf. Theorem 5-5.6) that $E[N] := \mathrm{Ker}[N] \cong (\mathbf{Z}/N\mathbf{Z})^2$, we see that $\mathrm{Gal}(K(E[N])/K)$ can be identified with a subgroup of $\mathrm{Aut}(\mathbf{Z}/N\mathbf{Z} \times \mathbf{Z}/N\mathbf{Z}) = \mathrm{GL}_2(\mathbf{Z}/N\mathbf{Z})$.

We will now move on to describing the decomposition into prime ideals in Galois extensions (see, for example, [7] for the proofs).

Let $K/\mathbf{Q}$ be a Galois extension with Galois group $G := \mathrm{Gal}(K/\mathbf{Q})$. Let p be a prime number and $\mathfrak{p}$ a prime ideal of $\mathcal{O}_K$ over p. The *decomposition group* of $\mathfrak{p}$ is

$$D(\mathfrak{p}/p) := \{\sigma \in G \mid \sigma(\mathfrak{p}) = \mathfrak{p}\}.$$

If $\sigma \in D(\mathfrak{p}/p)$, we can define $\tilde{\sigma} : \mathcal{O}_K/\mathfrak{p} \to \mathcal{O}_K/\mathfrak{p}$ by the diagram

$$\begin{array}{ccc} \mathcal{O}_K & \xrightarrow{\sigma} & \mathcal{O}_K \\ \downarrow & & \downarrow \\ \mathcal{O}_K/\mathfrak{p} & \xrightarrow{\tilde{\sigma}} & \mathcal{O}_K/\mathfrak{p}. \end{array}$$

We set $\mathbf{F}_\mathfrak{p} = \mathcal{O}_K/\mathfrak{p}$. The map $\sigma \mapsto \tilde{\sigma}$ defines a homomorphism $r_\mathfrak{p} : D(\mathfrak{p}/p) \to \mathrm{Gal}(\mathbf{F}_\mathfrak{p}/\mathbf{F}_p)$. By definition, the kernel is called the *inertia group* of $\mathfrak{p}$, in other words

$$I(\mathfrak{p}/p) := \{\sigma \in D(\mathfrak{p}/p) \mid \forall x \in \mathcal{O}_K,\ \sigma(x) \equiv x \bmod \mathfrak{p}\}.$$

1.2. Lemma. *The Galois group $G = \mathrm{Gal}(K/\mathbf{Q})$ acts transitively on the set of prime ideals of $\mathcal{O}_K$ over p. The homomorphism $r_p : D(\mathfrak{p}/p) \to \mathrm{Gal}(\mathbf{F}_\mathfrak{p}/\mathbf{F}_p)$ is surjective.*

We will now do some calculations which allow us to see how the inertia groups and decomposition groups vary when we change the ideal $\mathfrak{p}$.

1.3. Lemma. *Let $\mathfrak{p}$ and $\mathfrak{p}'$ be prime ideals of $\mathcal{O}_K$ over p, and let $\sigma \in G$ such that $\sigma(\mathfrak{p}) = \mathfrak{p}'$. Then*

$$D(\mathfrak{p}'/p) = \sigma D(\mathfrak{p}/p)\sigma^{-1} \qquad \textit{and} \qquad I(\mathfrak{p}'/p) = \sigma I(\mathfrak{p}/p)\sigma^{-1}.$$

As we recalled, $\mathrm{Gal}(\mathbf{F}_\mathfrak{p}/\mathbf{F}_p)$ is a cyclic group whose canonical generator is given by the Frobenius homomorphism $\Phi(x) = x^p$.

1.4. Definition. A *Frobenius* of $\mathfrak{p}$ is an element $\sigma \in D(\mathfrak{p}/p)$ such that $r_\mathfrak{p}(\sigma) = \Phi$. That is to say, σ satisfies

$$\forall x \in \mathcal{O}_K,\ \sigma(x) \equiv x^p \bmod \mathfrak{p}.$$

We denote by $\mathrm{Frob}_\mathfrak{p}$ such an element (if we need to specify the field, we write $\mathrm{Frob}_{\mathfrak{p},K/\mathbf{Q}}$).

1.5. Remarks. 1) If we replace $\mathfrak{p}$ by $\mathfrak{p}' := \sigma(\mathfrak{p})$, we can easily see that

$$\mathrm{Frob}_{\mathfrak{p}'} = \sigma\,\mathrm{Frob}_{\mathfrak{p}}\,\sigma^{-1}.$$

In particular, if the extension is abelian, the element only depends on p, and we can denote it by Frob_p. Keep in mind, however, that the notation Frob_p in general only designates a "conjugacy class modulo the inertia group".

2) We will again look at the example $K = \mathbf{Q}(\mu_n)$. Then if p does not divide n, the element $\sigma_p = \mathrm{Frob}_p$ is well-defined and equals $\sigma_p(\zeta) = \zeta^p$. However, if, for example, $n = p^m$, then the extension $\mathbf{Q}(\mu_{p^m})/\mathbf{Q}$ is totally ramified at p, and every element of the Galois group is a Frobenius at p.

3) In the example $K = \mathbf{Q}(\sqrt{d})$, we can identify $\mathrm{Gal}(K/\mathbf{Q})$ with the group $\{+1, -1\}$, the nontrivial automorphism being given by $\sigma(a + b\sqrt{d}) = a - b\sqrt{d}$. Let p be an odd prime which does not divide d. Then we have

$$\left(\sqrt{d}\right)^p = d^{\frac{p-1}{2}}\sqrt{d} \equiv \left(\frac{d}{p}\right)\sqrt{d} \bmod p,$$

therefore Frob_p is nothing other than the Legendre symbol of d with respect to p (i.e., $+1$ if d is a square modulo p and -1 if not).

4) We can generalize these notions to Galois extensions of a number field L/K. For example, if $\mathfrak{q}$ is an ideal of L over $\mathfrak{p}$, an ideal of K, we denote by $\mathrm{Frob}_{\mathfrak{q}}$ an element of $\mathrm{Gal}(L/K)$ such that $\mathrm{Frob}_{\mathfrak{q}}(x) \equiv x^{\mathrm{N}\,\mathfrak{p}} \bmod \mathfrak{q}$.

5) We have defined the decomposition group, the inertia group and the Frobenius element for finite Galois extensions. No real difficulties arise when extending these definitions to infinite extensions. For example, if we consider $G_{\mathbf{Q}} := \mathrm{Gal}(\bar{\mathbf{Q}}/\mathbf{Q})$ and p prime, we denote by $\bar{\mathbf{Z}}$ the ring of algebraic integers and $\mathfrak{p}$ a maximal ideal of $\bar{\mathbf{Z}}$ over p. We set $D(\mathfrak{p}/p) = \{\sigma \in G_{\mathbf{Q}} \mid \sigma(\mathfrak{p}) = \mathfrak{p}\}$, then define the reduction homomorphism $r_{\mathfrak{p}} : D(\mathfrak{p}/p) \to \mathrm{Gal}(\bar{\mathbf{F}}_p/\mathbf{F}_p)$ and let $I(\mathfrak{p}/p) := \mathrm{Ker}(r_{\mathfrak{p}})$, and finally let $\mathrm{Frob}_{\mathfrak{p}}$ be an element of $D(\mathfrak{p}/p)$ whose image under $r_{\mathfrak{p}}$ is the Frobenius in characteristic p.

By the previous lemma, such a Frobenius element associated to a prime ideal of K always exists and is unique modulo the inertia subgroup. Let us see when the inertia subgroup is trivial.

1.6. Proposition. *Let $K/\mathbf{Q}$ be a finite Galois extension with Galois group $G := \mathrm{Gal}(K/\mathbf{Q})$, and let p be a prime number. The decomposition of p in $\mathcal{O}_K$ is written*

$$p\mathcal{O}_K = (\mathfrak{p}_1 \cdots \mathfrak{p}_g)^e$$

where $e = \mathrm{card}\, I(\mathfrak{p}_i/p)$, $\mathrm{N}\,\mathfrak{p}_i = p^f$, $ef = \mathrm{card}\, D(\mathfrak{p}_i/p)$ and $g = (\mathrm{Gal}(K/\mathbf{Q}) : D(\mathfrak{p}_i/p))$. Then we have $efg = [K : \mathbf{Q}]$.

1.7. Corollary. *With the same hypotheses and notations, we have:*

$$e = 1 \Longleftrightarrow p \text{ is unramified} \Longleftrightarrow p \text{ does not divide } \Delta_K.$$

We can introduce an often useful filtration of the inertia group by defining the *higher ramification groups*:

$$G_{i,\mathfrak{p}} := \left\{\sigma \in D(\mathfrak{p}/p) \mid \forall x \in \mathscr{O}_K,\ \sigma(x) \equiv x \bmod \mathfrak{p}^{i+1}\right\}.$$

Thus G_0 is the inertia group. We can see that the G_i are p-groups for $i \geqslant 1$. In fact, G_1 is the Sylow p-subgroup of G_0, and, in particular, the G_i are trivial whenever p does not divide card(G) and $i \geqslant 1$.

We could ask ourselves how $\mathrm{Frob}_\mathfrak{p}$ varies when we vary the prime ideal. The response is given by the following theorem.

1.8. Theorem. (Chebotarev) *If C is a conjugacy class of $G = \mathrm{Gal}(K/\mathbf{Q})$, then there exist infinitely many prime numbers p such that Frob_p is in C. To be more precise, the density of such p is exactly $|C|/|G|$.*

The previous theorem is a vast generalization of the theorem on arithmetic progressions (see [44] for the proof). To see why, if we choose $K = \mathbf{Q}(\zeta)$ where $\zeta = \exp(2\pi i/n)$ and an element $a \in (\mathbf{Z}/n\mathbf{Z})^* = \mathrm{Gal}(K/\mathbf{Q})$, then the equality $\mathrm{Frob}_p = a$ means $p \equiv a \bmod n$.

2. Abelian Extensions

If G is a group, we denote by G^{ab} the quotient of G by its commutator group (i.e., the largest abelian quotient of G). We will describe—briefly and without proof—the group $\mathrm{Gal}(\bar{K}/K)^{\mathrm{ab}}$ for a number field and indicate why this theory can be considered to have sprouted from the quadratic reciprocity law.

Let L/K be a number field extension with an abelian Galois group. If $\mathfrak{p}$ is a ideal in $\mathscr{O}_K$ which is unramified in L/K, we have seen that the element $\mathrm{Frob}_\mathfrak{p} \in \mathrm{Gal}(L/K)$ is well-defined. If we call S the set of prime ideals of K ramified in L/K and I_K^S the group of fractional ideals relatively prime to S, then we can define the homomorphism

$$\begin{array}{rccc} \psi_{L/K}: & I_K^S & \longrightarrow & \mathrm{Gal}(L/K) \\ & \prod_{\mathfrak{p}\notin S} \mathfrak{p}^{m_\mathfrak{p}} & \longmapsto & \prod_{\mathfrak{p}\notin S} \mathrm{Frob}_\mathfrak{p}^{m_\mathfrak{p}}, \end{array}$$

which we know to be surjective by Chebotarev's theorem (Theorem C-1.8).

The first step in analyzing the kernel is to see that norms of ideals in L are in $\mathrm{Ker}\,\psi_{L/K}$.

2.1. Lemma. *Let L/K be an abelian extension and F/K an extension. If $\mathfrak{q}$ is an ideal relatively prime to F over an ideal $\mathfrak{p}$ of K, we denote by $f := [\mathscr{O}_F/\mathfrak{q} : \mathscr{O}/\mathfrak{p}]$. Then*

$$\mathrm{Frob}_{\mathfrak{q},LF/F} = (\mathrm{Frob}_{\mathfrak{p},L/K})^f .$$

More generally, we have the following formula for an ideal $\mathfrak{A}$ of F,

$$\psi_{LF/F}(\mathfrak{A}) = \psi_{L/K}\left(\mathrm{N}_K^F \mathfrak{A}\right). \tag{C.1}$$

In particular, the norm of an ideal in L is in the kernel of $\psi_{L/K}$.

Proof. The statement of the lemma implies a natural identification of $\mathrm{Gal}(LF/F)$ with a subgroup of $\mathrm{Gal}(L/K)$. If $\tau := \mathrm{Frob}_{\mathfrak{q},LF/F}$, then we have $\tau(x) \equiv x^{\mathrm{N}\,\mathfrak{q}} \bmod \mathfrak{q}'$ (for $\mathfrak{q}'$ relatively prime to LF over $\mathfrak{q}$ and $x \in \mathscr{O}_{LF}$). If we restrict to $x \in \mathscr{O}_L$, we can thus write $\tau(x) \equiv x^{\mathrm{N}\,\mathfrak{p}^f} \bmod \mathfrak{q}' \cap \mathscr{O}_L$. We know that $\mathfrak{q}' \cap \mathscr{O}_L$ is a prime ideal of L over $\mathfrak{p}$, which indeed shows that the restriction of τ to L is equal to $\mathrm{Frob}_{\mathfrak{p}}^f$. The formula for the norms can be deduced by multiplicativity, and the last statement follows immediately from taking $F = L$. □

The second step in the analysis of the kernel of $\psi_{L/K}$ is much deeper and forms the core of class field theory. For the proof, you can refer, for example, to [44], but we should first introduce some vocabulary.

2.2. Definition. Let K be a number field. A *cycle* $\mathfrak{M}$ is given by an ideal $\mathscr{O}_K$ and a sign for every real place of K. Alternatively, we can write

$$\mathfrak{M} := \sum_{\mathfrak{p}} m_{\mathfrak{p}}\,[\mathfrak{p}] + \sum_{v\,|\,\infty} n_v[v],$$

where $m_{\mathfrak{p}} \in \mathbf{N}$ are almost all zero and $n_v = 0$ or 1 if v is Archimedean.

For $\alpha \in \mathscr{O}_K$, we write $\alpha \equiv 1\,[\mathfrak{M}]$ if $\alpha \equiv 1 \bmod \mathfrak{p}^{m_{\mathfrak{p}}}$ and $\sigma_v(\alpha) > 0$ if v is real and $n_v = 1$, and we set:

$$\mathscr{P}_{\mathfrak{M}} := \{\mathfrak{A} = \alpha\mathscr{O}_K \mid \alpha \equiv 1\,[\mathfrak{M}]\}.$$

2.3. Theorem. (Artin reciprocity law) *Let L/K be an abelian extension which is unramified outside of a set S of places of K. There exists a cycle $\mathfrak{M}$ with support in S such that $\mathscr{P}_{\mathfrak{M}} \subset \mathrm{Ker}\,\psi_{L/K}$ (such a cycle is called "admissible"). Moreover, we have the equality*

$$\mathrm{Ker}\,\psi_{L/K} = \mathscr{P}_{\mathfrak{M}}\,\mathrm{N}\left(I_L^{\mathfrak{M}}\right) \tag{C.2}$$

and consequently the following isomorphism

$$\psi_{L/K} : I_K^{\mathfrak{M}} / \mathscr{P}_{\mathfrak{M}}\,\mathrm{N}\left(I_L^{\mathfrak{M}}\right) \cong \mathrm{Gal}(L/K). \tag{C.3}$$

2.4. Remark. *(Hilbert class field)* We consider the case of an *unramified* (including at Archimedean places) abelian extension L/K. We then obtain a reciprocity homomorphism:

$$\psi_{L/K} : C\ell_K := I_K/\mathscr{P} \to \mathrm{Gal}(L/K).$$

Note that the condition of being unramified at Archimedean places signifies that the real places of K are uniquely extended to real places of L, and that if a homomorphism $\sigma : L \to \mathbf{C}$ satisfies $\sigma(K) \subset \mathbf{R}$, then $\sigma(L) \subset \mathbf{R}$. Since the composition of two unramified abelian extensions is unramified abelian, we see that there exists a *maximal* unramified abelian extension, called the *Hilbert class field* of K. In this case, we can specify the kernel of $\psi_{L/K}$.

2.5. Proposition. *Let H/K be the Hilbert class field of K (i.e., the maximal unramified abelian extension). Then*

$$\psi_{H/K} : C\ell_K \longrightarrow \mathrm{Gal}(H/K)$$

is an isomorphism. Furthermore, every ideal $\mathfrak{A}$ of K becomes principal in H (i.e., $\mathfrak{A}\mathscr{O}_H$ is a principal ideal).

Class field theory also allows us to "classify" abelian extensions, but for this, a more elegant method comes from Chevalley, namely to introduce J_K, the idele group of K (see the last part of the section on p-adic numbers in Chap. 6). To do this, we can reformulate the Artin reciprocity law using the following lemma.

2.6. Lemma. *Let L/K be an abelian extension of a number field and let $\mathfrak{M}$ be an admissible cycle (i.e., such that $\mathscr{P}_{\mathfrak{M}} \subset \operatorname{Ker}\psi_{L/K}$). We then have the following natural isomorphism:*

$$\mathrm{J}_K/K^*\,\mathrm{N}\,(\mathrm{J}_L) \longrightarrow I_K^{\mathfrak{M}}/\mathscr{P}_{\mathfrak{M}}\,\mathrm{N}\left(I_L^{\mathfrak{M}}\right). \tag{C.4}$$

The isomorphism is obtained by associating to an idele $a = (a_v)_{v\in M_K}$, whose coordinates are units at every finite v appearing in $\mathfrak{M}$ (resp. $a_v > 0$ if v is real Archimedean and $n_v = 1$), the ideal $\prod_{\mathfrak{p}} \mathfrak{p}^{\mathrm{ord}_{\mathfrak{p}}(a_{\mathfrak{p}})}$.

2.7. Corollary. *If L/K is a abelian extension, we have a surjective homomorphism:*

$$\mathrm{J}_K \longrightarrow \mathrm{Gal}(L/K),$$

whose kernel is $K^*\,\mathrm{N}\,(\mathrm{J}_L)$.

2.8. Theorem. (Class field theory) *The correspondence defined by*

$$L \longmapsto K^* \mathrm{N}(\mathrm{J}_L)$$

establishes a bijection between abelian extensions of K and closed subgroups of J_K containing K^* (i.e., closed subgroups of $\mathrm{C}_K := \mathrm{J}_K/K^*$). Finite extensions therefore correspond to subgroups which are both open and closed.

If H (resp H') is the subgroup associated to L/K (resp. to L'/K), then $L \subset L'$ if and only if $H' \subset H$. The subgroup associated to LL'/K is $H \cap H'$, and the subgroup associated to $L \cap L'/K$ is HH'. In other words, if $H = K^* \mathrm{N}(\mathrm{J}_L)$, then $(L : K) = (\mathrm{J}_K : H)$ and $\mathrm{Gal}(L/K) = \mathrm{J}_K/H$. Furthermore, if v is a place of K, the decomposition group (resp. the inertia group) is the image of $K_v^* \to \mathrm{J}_K \to \mathrm{J}_K/H$ (resp. the image of $\mathcal{O}_v^*$). In particular, L/K is unramified at v if and only if $\mathcal{O}_v^* \subset H$.

2.9. Examples. Let $\mathfrak{M} = \sum_{\mathfrak{p}} m_{\mathfrak{p}}\,[\mathfrak{p}] + \sum_{v\,|\,\infty} n_v[v]$ be a cycle of K. We denote by

$\mathrm{J}_K^{\mathfrak{M}} := \{(a_v)_{v \in M_K} \in \mathrm{J}_K \mid a_{\mathfrak{p}} \equiv 1 \bmod \mathfrak{p}^{m_{\mathfrak{p}}} \text{ and } \sigma_v(\alpha) > 0 \text{ if } v \text{ is real and } n_v = 1\}$.

The abelian extension $K^{\mathfrak{M}}$ of K associated to the group $K^*\mathrm{J}_K^{\mathfrak{M}}$ is called the *ray class field* modulo $\mathfrak{M}$. If we restrict ourselves to the case $K = \mathbf{Q}$ and $\mathfrak{M} = m(\infty) = \sum_p \mathrm{ord}_p(m)[p] + [\infty]$, we can prove that $\mathbf{Q}^{m(\infty)} = \mathbf{Q}(\exp(2\pi i/m))$. In particular, we thus obtain the following classical result.

2.10. Theorem. (Kronecker-Weber) *Every abelian extension of $\mathbf{Q}$ is contained in a cyclotomic extension (generated by the roots of unity). In particular,*

$$\mathrm{Gal}(\bar{\mathbf{Q}}/\mathbf{Q})^{\mathrm{ab}} = \mathrm{Gal}(\mathbf{Q}(\mu_\infty/\mathbf{Q}) \cong \prod_\ell \mathbf{Z}_\ell^* \,. \tag{C.5}$$

2.11. Remark. Let p and q be two distinct odd primes. Consider the field $K = \mathbf{Q}(\sqrt{q'})$ (where $q' = (-1)^{\frac{q-1}{2}} q$, so that $K/\mathbf{Q}$ is only ramified at q and, if $q \equiv 3 \bmod 4$, at ∞), and identify $\mathrm{Gal}(K/\mathbf{Q})$ with $\{+1, -1\}$. We have seen (Remark C-1.5, part 3) that $\mathrm{Frob}_p = \left(\frac{q'}{p}\right)$. The Artin reciprocity law (Theorem C-2.3) tells us that this element only depends on the congruence class of p modulo $\mathfrak{M} = q$, and it can be proven that it is equal to the identity, $+1$, if and only if p is a square modulo q. We thus obtain the quadratic reciprocity law in the form

$$\left(\frac{q'}{p}\right) = \left(\frac{p}{q}\right).$$

3. Galois Representations

The study of the group $G_{\mathbf{Q}} = \mathrm{Gal}(\bar{\mathbf{Q}}/\mathbf{Q})$, or more generally $G_L := \mathrm{Gal}(\bar{\mathbf{Q}}/L)$ for a number field L, is clearly a fundamental problem. One way to attack it is to study the *representations* of this group, i.e. the homomorphisms $\rho : G_{\mathbf{Q}} \to \mathrm{GL}(V) \cong \mathrm{GL}_n(K)$ where V is a vector space over K of dimension n. The three most interesting cases are $K = \mathbf{F}_q$ (a finite field), $K = \mathbf{Q}_\ell$ (a p-adic field) and $K = \mathbf{C}$.

The previous section essentially corresponds to one-dimensional representations, since a representation $G_L \to \mathrm{GL}_1(K) = K^*$ has an abelian image and can therefore be factored through G_L^{ab}.

3.1. Definition. An *Artin representation* of K is a continuous finite dimensional representation $\rho : \mathrm{Gal}(\bar{K}/K) \to \mathrm{GL}(V) \cong \mathrm{GL}_n(\mathbf{C})$.

Since G_K is compact and discontinuous, the image of ρ is necessarily finite and $L := \bar{K}^{\mathrm{Ker}(\rho)}$ is thus a finite extension of K. The representation is therefore factored though the representation of a finite group $\mathrm{Gal}(L/K) \to \mathrm{GL}_n(\mathbf{C})$. We say that ρ is *unramified* at p if $I(\mathfrak{p}/p) \subset \mathrm{Ker}(\rho)$. It is clear that ρ is unramified outside of a finite set of prime numbers.

3.2. Definition. The Artin *conductor* of $\rho : G \to \mathrm{GL}(V)$ is defined as:

$$N_\rho := \prod_p p^{n(p,\rho)}$$

where

$$n(p,\rho) := \sum_{i=0}^{\infty} \frac{\dim V/V^{G_{i,\mathfrak{p}}}}{(G_{0,\mathfrak{p}} : G_{i,\mathfrak{p}})}.$$

Here $\mathfrak{p}$ designates an ideal of K over p. It can be shown that this formula does not depend on the choice of $\mathfrak{p}$. It is clear that $n(p,\rho) = 0$ if ρ is not ramified over p. Next, if $G_{1,p} = \{1\}$ (no "wild" ramification), we have $n(p,\rho) = \dim V - \dim V^{I(\mathfrak{p}/p)}$. In the general case it is still true, but more tricky to prove, that $n(p,\rho)$ is an integer.

We can essentially identify one-dimensional Artin representations over $\mathbf{Q}$ with Dirichlet characters in the following sense. Given a Dirichlet character $\chi : (\mathbf{Z}/n\mathbf{Z})^* \to \mathbf{C}^*$, we associate an Artin representation to it by the following diagram:

$$\mathrm{Gal}(\bar{\mathbf{Q}}/\mathbf{Q}) \to \mathrm{Gal}(\mathbf{Q}(\mu_n)/\mathbf{Q}) \cong (\mathbf{Z}/n\mathbf{Z})^* \xrightarrow{\chi} \mathbf{C}^* = \mathrm{GL}_1(\mathbf{C}).$$

Furthermore, Weber's theorem tells us that we obtain all the one-dimensional Artin representations of $G_{\mathbf{Q}}$ this way.

Let $\rho : G_{\mathbf{Q}} \to \mathrm{GL}(V)$ be an Artin representation. We define the characteristic polynomial at p of ρ as follows: we choose $\mathfrak{p}$ to be a prime ideal over p and set

$$P_p(\rho; T) := \det\left(1 - \rho(\mathrm{Frob}_{\mathfrak{p}})T \mid V^{I(\mathfrak{p}/p)}\right). \tag{C.6}$$

$\mathrm{Frob}_{\mathfrak{p}}$ is only defined modulo $I(\mathfrak{p}/p)$, and the action on $V^{I(\mathfrak{p}/p)}$ only depends on the chosen representative. Finally, the determinant does not depend on the conjugacy class of $\mathrm{Frob}_{\mathfrak{p}}$, thus only depends on p. We know that (see, for example, [43] or [63]) a representation of a finite group is determined by its *character*, in other words, by the trace function $\chi_\rho = \mathrm{Tr} \circ \rho$. By using the elementary formula

$$\det{(I - TA)}^{-1} = \exp\left\{\sum_{m=1}^{\infty} \frac{\mathrm{Tr}(A^m)}{m} T^m\right\}$$

where A is a square matrix, we can write the previous definition as

$$P_p(\rho, T)^{-1} = \exp\left\{\sum_{m=1}^{\infty} \frac{\chi_\rho(\mathrm{Frob}_p^m)}{m} T^m\right\}, \tag{C.7}$$

where, in the ramified case, we restrict the representation to $V^{I(\mathfrak{p}/p)}$.

3.3. Definition. Let $\rho : G_{\mathbf{Q}} \to \mathrm{GL}(V)$ be an Artin representation. Its L-function is defined as

$$L(\rho, s) := \prod_p P_p(\rho; p^{-s})^{-1} = \prod_p \det\left(1 - \rho(\mathrm{Frob}_{\mathfrak{p}})p^{-s} \mid V^{I(\mathfrak{p}/p)}\right)^{-1}.$$

Since the eigenvalues have absolute value 1, we can easily see that the Euler product converges absolutely for $\mathrm{Re}(s) > 1$. This construction generalizes Dirichlet L-series, which we recover whenever the representation is one-dimensional. In fact, a famous theorem of Brauer on representations of finite groups (see, for example, [63]) shows that Artin $L(\rho, s)$ functions can be written in the form of a product $\prod L(\chi_i, s)^{m_i}$, where $m_i \in \mathbf{Z}$ (and where the χ_i are abelian characters which generalize Dirichlet characters). Since we know the analytic continuation of the series $L(\chi, s)$ and their functional equation, we can deduce from this a *meromorphic* continuation of $L(\rho, s)$ to the complex plane with a functional equation. Artin conjectured that in fact $L(\rho, s)$ is everywhere holomorphic, except for a possible pole at $s = 1$ with order equal to the multiplicity of the trivial representation in ρ.

To write the functional equation, we introduce the dimension $n = \dim V$ of the representation and the element $c \in \mathrm{Gal}(\bar{\mathbf{Q}}/\mathbf{Q})$ defined by *complex conjugation*, and we denote by $n^+ = \dim V^+$ and $n^- = \dim V^-$, where V^+

(resp. V^-) is the subspace with eigenvalue +1 (resp. −1) for $\rho(c)$. We then let $\Gamma_{\mathbf{R}}(s) := \pi^{-s/2}\Gamma(s/2)$ and

$$\Lambda(\rho, s) := N_\rho^{s/2}\Gamma_{\mathbf{R}}(s)^{n^+}\Gamma_{\mathbf{R}}(s+1)^{n^-}L(\rho, s).$$

The functional equation is therefore written as

$$\Lambda(\rho, s) = w_\rho \Lambda(\check{\rho}, 1-s), \tag{C.8}$$

where $|w_\rho| = 1$ and $\check{\rho}$ is the *dual* representation.[1]

3.4. Remark. We have introduced Artin representations over $\mathbf{Q}$, but no difficulties arise when generalizing to continuous representations $\rho : \mathrm{Gal}(\bar{\mathbf{Q}}/K) \to \mathrm{GL}(V)$.

3.5. Definition. An *ℓ-adic representation* is a continuous representation $\rho : \mathrm{Gal}(\bar{K}/K) \to \mathrm{GL}_n(\mathbf{Q}_\ell)$.

We often assume the following added condition: the representation is unramified outside of a finite set of primes $\mathfrak{p}$ of $\mathscr{O}_K$. This condition is automatically satisfied in the case of Artin representations, but not in the case of ℓ-adic representations.

3.6. Examples. 1) Let K_{ℓ^∞} be the field generated by the ℓ^nth roots of unity (for an arbitrary n). We can associate to it the following representation (christened "the cyclotomic character"):

$$\mathrm{Gal}(\bar{\mathbf{Q}}/\mathbf{Q}) \to \mathrm{Gal}(K_{\ell^\infty}/\mathbf{Q}) \cong (\mathbf{Z}_\ell)^* = \mathrm{GL}_1(\mathbf{Z}_\ell) \hookrightarrow \mathrm{GL}_1(\mathbf{Q}_\ell).$$

2) Let $E/\mathbf{Q}$ be an elliptic curve defined over $\mathbf{Q}$, and let

$$E[\ell^n] := \mathrm{Ker}\left\{[\ell^n] : E(\bar{\mathbf{Q}}) \to E(\bar{\mathbf{Q}})\right\}.$$

Recall the definition of a Tate module, $T_\ell(E) := \lim_n E[\ell^n]$. Since $T_\ell(E) \cong (\mathbf{Z}_\ell)^2$ (5-5.8), and since the Galois group acts $\mathbf{Z}$-linearly on $E[\ell^n]$, it acts $\mathbf{Z}_\ell$-linearly on $T_\ell(E)$, and we thus obtain a representation

$$\rho_{E,\ell} : \mathrm{Gal}(\bar{\mathbf{Q}}/\mathbf{Q}) \to \mathrm{GL}(T_\ell(E)) \cong \mathrm{GL}_2(\mathbf{Z}_\ell) \hookrightarrow \mathrm{GL}_2(\mathbf{Q}_\ell).$$

This representations happens to be unramified outside of ℓ and the places of bad reduction of the elliptic curve (see [70]). Moreover, by composing with the determinant, we obtain a representation $\det \circ \rho_{E,\ell} : \mathrm{Gal}(\bar{\mathbf{Q}}/\mathbf{Q}) \to \mathrm{GL}_1(\mathbf{Z}_\ell) = \mathbf{Z}_\ell^*$, which coincides with the cyclotomic character (cf. for example [70]). The conductor of the representation is defined as in C-3.2.

[1] If $\rho : G \to \mathrm{GL}(V)$ and if V^* is the vector space dual to V, the dual representation $\check{\rho} : G \to \mathrm{GL}(V^*)$ is given by $\langle \rho(g)(v), v^* \rangle = \langle v, \check{\rho}(g^{-1})(v^*) \rangle$.

In fact, it can be shown that the exponent $n(\rho, p)$ is, for $p \neq \ell$, independent of ℓ. This allows us to define, in an abstract manner, the *conductor* of E.

In general, by making use of the compactness of $G = \mathrm{Gal}(\bar{K}/K)$, we see that there exists in $V \cong \mathbf{Q}_\ell^n$ a lattice $\Lambda \cong \mathbf{Z}_\ell^n$ which is stable under $\rho(G)$. To do this, we choose a basis $v_1, \ldots, v_n$ of the $\mathbf{Q}_\ell$-vector space, we set $\Lambda_0 = \mathbf{Z}_\ell v_1 + \cdots + \mathbf{Z}_\ell v_n$ and $\Lambda := G \cdot \Lambda_0$, and we prove that $(\Lambda : \Lambda_0)$ is finite. Thus, up to a change of basis in V, we can always assume that ρ has values in $\mathrm{GL}_n(\mathbf{Z}_\ell)$. We can see however that the image $\rho(G)$ is not finite in general. The ℓ-adic representations are in this sense richer than complex representations.

3.7. Definition. A representation mod ℓ is a continuous representation $\rho : \mathrm{Gal}(\bar{K}/K) \to \mathrm{GL}_n(\mathbf{F}_\ell)$ (or more generally $\mathrm{GL}_n(\mathbf{F}_{\ell^m})$).

Such a representation obviously has a finite image. It therefore factors through the representation of a finite Galois group. One way to obtain such representations is to reduce an ℓ-adic representation modulo ℓ. In other words, starting with an ℓ-adic representation $\rho : \mathrm{Gal}(\bar{K}/K) \to \mathrm{GL}_n(\mathbf{Q}_\ell)$ which is normalized so that it has values in $\mathrm{GL}_n(\mathbf{Z}_\ell)$, we can compose with the reduction homomorphism $r_\ell : \mathrm{GL}_n(\mathbf{Z}_\ell) \to \mathrm{GL}_n(\mathbf{F}_\ell)$ and thus obtain a representation:

$$\bar{\rho} := r_\ell \circ \rho : \mathrm{Gal}(\bar{K}/K) \to \mathrm{GL}_n(\mathbf{Z}_\ell) \to \mathrm{GL}_n(\mathbf{F}_\ell).$$

More generally, if A is a ring and $r : A \to \mathbf{F}_q$ a homomorphism, we say that a representation $\bar{\rho} : \mathrm{Gal}(\bar{K}/K) \to \mathrm{GL}_n(\mathbf{F}_q)$ can be lifted to A if there exists $\rho : \mathrm{Gal}(\bar{K}/K) \to \mathrm{GL}_n(A)$ such that $\bar{\rho} = r \circ \rho$. This can be represented by the diagram

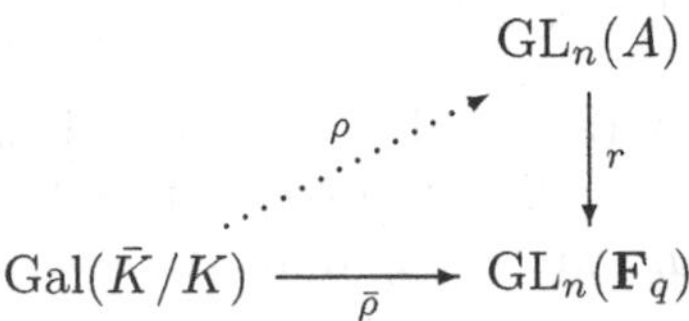

In the case $K = \mathbf{Q}$, $n = 2$ and $\bar{\rho}$ is irreducible and odd (i.e., the image of the complex conjugation is of determinant -1), a conjecture of Serre, for the statement of which we refer you to [66], describes these representations as coming from "modular" representations. This conjecture has just been proven by Khare and Wintenberger.

Bibliography

We will first give a short bibliography of works which are on a slightly higher level than this one. A series of more specialized references follows. The first works are cited with a brief commentary, in particular our three favorite books: those of Borevich & Shafarevich, Hardy & Wright and Serre.

[1] Baker, A.: Transcendental Number Theory. Cambridge University Press (1975) (*In the first few pages, they prove the transcendence of e and π; in the 2nd and 3rd chapters, they prove "Baker's theorem" on linear forms of logarithms of algebraic numbers; the rest of the book is more specialized.* An updated edition came out in 1990.)

[2] Borevich, Z., Shafarevich, I.: Number Theory (translated from the Russian Теория Чисел). Academic Press (1966) (*A very nice book that, even if it starts at an elementary level, is at a higher level than our text.*)

[3] Demazure, M.: Cours d'algèbre. Cassini, Paris (1997) (*Especially for the sections about "finite structures" and algorithms.*)

[4] Hardy, G.H., Wright, E.M.: An Introduction to the Theory of Numbers, 4th ed. Oxford University Press (1960) (*Presents most of the subjects in number theory at an elementary level, very inviting!*)

[5] Ireland, K., Rosen, M.: A Classical Introduction to Modern Number Theory, GTM **84**. Springer (1982) (*A classic, as the title indicates.*)

[6] Perrin, D.: Cours d'algèbre. Ellipses (1982) (*Excellent book, especially recommended for those studying for l'agrégation.*)

[7] Samuel, P.: Théorie algébrique des nombres. Hermann (1967) (*Covers Dedekind rings at a higher level than this book, very well written. A classic.*)

[8] Serre, J.-P.: Cours d'arithmétique. Presses Universitaires de France (1970) (*An unsurpassable classic, by one of the great names in mathematics of the 20th century.*)

[9] Silverman, J., Tate, J.: Rational Points on Elliptic Curves, UTM. Springer (1992) (*A nice, elementary introduction to elliptic curves, including many examples.*)

M. Hindry, *Arithmetics*, Universitext,
DOI 10.1007/978-1-4471-2131-2,

ARTICLES and ADVANCED BOOKS:

[10] Agrawal, M., Kayal, N., Saxena, N.: PRIMES is in P. Ann. of Math. **160**, 781–793 (2004)

[11] Alford, W., Granville, A., Pomerance, C.: There are infinitely many Carmichael numbers. Ann. of Math. **139**, 703–722 (1994)

[12] Bertrand, D. et al: Les nombres transcendants. Mém. Soc. Math. France no. 13 (1984)

[13] Birch, B., Swinnerton-Dyer, P.: The Hasse problem for rational surfaces. J. für die reine und angew. Math. **274/275**, 164–174 (1975)

[14] Birch, B., Swinnerton-Dyer, P.: Notes on elliptic curves I, II. J. für die reine und angew. Math. **212**, 7–25 (1963) and **218**, 79–108 (1965)

[15] Blake, I., Seroussi, G., Smart N.: Elliptic Curves in Cryptography, London Mathematical Society Lecture Note Series **265**. Cambridge University Press (1999)

[16] Bombieri, E., Gubler, W.: Heights in Diophantine Geometry, Cambridge University Press (2006)

[17] Boneh, D.: Twenty years of attacks on the RSA cryptosystem. Notices Am. Math. Soc. **46**, 203–213 (1999)

[18] Bordellès, O.: Thèmes d'arithmétique. Ellipses (2006)

[19] Buhler, J., Lenstra, H. Jr., Pomerance, C.: Factoring integers with the number field sieve. The development of the number field sieve, Lecture Notes in Math. **1554**, pp 50–94. Springer, Berlin (1993)

[20] Bump, D.: Automorphic Forms and Representations, CSAM **55**. Cambridge University Press (1997)

[21] Canfield, E., Erdös, P., Pomerance, C.: On a problem of Oppenheim concerning "factorisatio numerorum". J. Number Theory **17**, 1–28 (1983)

[22] Cohen, H.: A Course in Computational Algebraic Number Theory, GTM **138**. Springer-Verlag (1993)

[23] Daboussi, H.: Sur le théorème des nombres premiers. C. R. Acad. Sci., Paris, Sér. **I 298**, 161–164 (1984)

[24] Davis, M., Matijasevic, Y., Robinson J.: Hilbert's tenth problem: Diophantine equations: Positive aspects of a negative solution. Math. Dev. Hilbert Probl., Proc. Symp. Pure Math. **28**, 323–378 (1976)

[25] Delange, H.: Généralisation du théorème de Ikehara. Ann. Sci. ENS **71**, 213–242 (1954)

[26] Deligne, P., Serre, J.-P.: Formes modulaires de poids 1. Ann. Sci. ENS **7**, 507–530 (1974)

[27] Diamond, F., Shurman, J.: A First Course in Modular Forms, GTM **228**. Springer (2005)

[28] Dickson, L.: History of the Theory of Numbers. Carnegie Institution of Washington (1923)

[29] Elkies, N.: *ABC* implies Mordell. Int. Math. Res. Notices no. 7, 99–109 (1991)

[30] Faltings, G.: Endlichkeitssätze für abelsche Varietäten über Zahlkörpern. Invent. Math. **73**, 349–366 (1983)

[31] Fulton, W.: Intersection Theory, EMG 3. Springer-Verlag (1984)

[32] Gelbart, S., Miller, S.: Riemann's zeta function and beyond. Bull. Am. Math. Soc. **41**, 59–112 (2004)

[33] Granville, A.: It is easy to determine whether a given integer is prime. Bull. Am. Math. Soc. **42**, 3–38 (2005)

[34] Guy, R.: Unsolved Problems in Number Theory, 3rd ed. Springer-Verlag (2004)

[35] Hartshorne, R.: Algebraic Geometry, GTM **52**. Springer-Verlag (1977)

[36] Heath-Brown, D.: Cubic forms in ten variables. Proc. London Math. Soc. **47**, 225–257 (1983)

[37] Hellegouarch, Y.: Invitation aux mathématiques de Fermat-Wiles. Masson (1997)

[38] Hindry, M., Silverman, J.: Diophantine Geometry. An Introduction, GTM **201**. Springer-Verlag (2000)

[39] Hooley, C.: On nonary cubic forms. J. für die reine und angew. Math. **386**, 32–98 (1998)

[40] Ikehara, S.: An extension of Landau's theorem in the analytical theory of numbers. J. of Math. Massachusetts **10**, 1–12 (1931)

[41] Iwaniec, H., Kowalski, E.: Analytic Number Theory. American Mathematical Society, Providence, RI (2004)

[42] Knuth, D.: The Art of Computer Programing. Addison-Wesley (1969)

[43] Lang, S.: Algebra. Addison-Wesley (1965)

[44] Lang, S.: Algebraic Number Theory, GTM **110**. Springer-Verlag (1986)

[45] Lang, S.: Hyperbolic and diophantine analysis. Bull. Am. Math. Soc. **14**, 159–205 (1986)

[46] Lang, S.: Diophantine problems in complex hyperbolic analysis. Contemporary Math. **67**, 229–246 (1987)

[47] Lang, S., Weil, A.: Number of points of varieties in finite fields. Am. J. Math. **76**, 819–827 (1954)

[48] Lenstra, A.: Integer factoring. Towards a quarter-century of public key cryptography. Des. Codes Cryptogr. **19**, 101–128 (2000)

[49] Lenstra, H. Jr.: Factoring integers with elliptic curves. Ann. of Math. **126**, 649–673 (1987)

[50] Lenstra, H. Jr.: Solving the Pell equation. Notices Am. Math. Soc. **49**, 182–192 (2002)

[51] Levinson, N.: A motivated account of an elementary proof of the prime number theorem. Am. Math. Mon. **76**, 225–245 (1969)

[52] Lewis, D.J.: Cubic homogeneous polynomials over p-adic fields. Ann of Math. **56**, 473–478 (1952)

[53] Mendès-France, M., Tenenbaum, G.: Les nombres premiers, Que sais-je, no. 571. PUF (1997)

[54] Murty, M.: Bounds for congruence primes. In: Automorphic forms, automorphic representations, and arithmetic (ed. Doran et al.), pp. 177–192. Am. Math. Soc., Proc. Symp. Pure Math. **66** (1999)

[55] Newman, D.: Simple analytic proof of the prime number theorem. Am. Math. Monthly **87**, 693–696 (1980)

[56] Oesterlé, J.: Nouvelles approches du "théorème" de Fermat, Séminaire Bourbaki, Exposé no. 694. Astérisque **161/162**, 165–186 (1988)

[57] Perrin, D.: Géométrie algébrique. Une introduction, Savoirs Actuels, InterEditions. CNRS Éditions, Paris (1995)

[58] Ribenboim, P.: 13 Lectures on Fermat's Last Theorem. Springer (1979)

[59] Ribet, K.: On modular representations of $\mathrm{Gal}(\bar{\mathbf{Q}}/\mathbf{Q})$ arising from modular forms. Invent. Math. **100**, 431–476 (1990)

[60] Riemann, B.: Ueber die Anzahl der Primzahlen unter einer gegebenen Grösse. Monatsberichte der Berliner Akademie (1859) [*On the number of prime numbers smaller than a given size*]

[61] Rivest, R., Shamir, A., Adleman, L.: A method for obtaining digital signatures and public-key cryptosystems. Commun. ACM **21**, 120–126 (1978)

[62] Rosser, J., Schoenfeld, L.: Approximate formulas for some functions of prime numbers. Illinois J. Math. **6**, 64–94 (1962). See also: Sharper bounds for the Chebyshev functions $\theta(x)$ and $\psi(x)$. Math. Comp. **29**, 243–269 (1975)

[63] Serre, J.-P.: Représentations des groupes finis. Hermann, Paris (1967)

[64] Serre, J.-P.: Abelian l-Adic Representations and Elliptic Curves. Benjamin Inc., New York-Amsterdam (1968)

[65] Serre, J.-P.: Facteurs locaux des fonctions zêta des variétés algébriques (définitions et conjectures). Séminaire Delange-Pisot-Poitou, No 19 (1969/70)

[66] Serre, J.-P.: Sur les représentations modulaires de degré 2 de Gal($\bar{\mathbf{Q}}/\mathbf{Q}$). Duke Math. J. **54**, 179–230 (1987)

[67] Shannon, C.: A mathematical theory of communication. Bell System Tech. J. **27**, 379–423 and 623–656 (1948)

[68] Shimura, G.: Introduction to the Arithmetic Theory of Automorphic Functions, Publications of the Mathematical Society of Japan **11**. Iwanami Shoten, Publishers and Princeton University Press (1971)

[69] Shor, P.: Polynomial-time algorithms for prime factorization and discrete logarithms on a quantum computer. SIAM J. Comput. **26**, 1484–1509 (1997)

[70] Silverman, J.: The Arithmetic of Elliptic Curves, GTM **106**. Springer-Verlag (1986)

[71] Silverman, J.: Advanced Topics in the Arithmetic of Elliptic Curves, GTM **151**. Springer-Verlag (1994)

[72] Tenenbaum, G.: Introduction à la théorie analytique et probabiliste des nombres, Cours Spécialisés. Société Mathématique de France, Paris (1995)

[73] Tietäväinen, A.: On the nonexistence of perfect codes over finite fields. SIAM J. Appl. Math. **24**, 88–96 (1973)

[74] Vogel, P.: Fonctions analytiques: cours et exercices avec solutions. Dunod (1999)

[75] Vojta, P.: Diophantine Approximation and Value Distribution Theory, Lecture Notes in Mathematics **1239**. Springer (1987)

[76] Waldschmidt, M.: Nombres transcendants, Lecture Notes in Mathematics **402**. Springer-Verlag (1974)

[77] Washington, L.: Introduction to Cyclotomic Fields, GTM **83**. Springer-Verlag (1982)

[78] Weil, A.: Numbers of solutions of equations in finite fields. Bull. Amer. Math. Soc. **55**, 497–508 (1949)

[79] Weil, A.: Sur les sommes de trois et quatre carrés. Enseignement Math. **20**, 215–222 (1974)

[80] Wiles, A.: Modular elliptic curves and Fermat's last theorem. Ann. of Math. **141**, 443–551 (1995)

[81] Zagier, D.: Newman's short proof of the prime number theorem. Am. Math. Monthly **104**, 705–708 (1997)

[82] Zémor, G.: Cours de cryptographie. Cassini (2000)

List of Notations

M. Hindry, *Arithmetics*, Universitext,
DOI 10.1007/978-1-4471-2131-2,

Index

M. Hindry, *Arithmetics*, Universitext,
DOI 10.1007/978-1-4471-2131-2,

"Ai! que preguiça!..."
MÁRIO DE ANDRADE (MACUNAÍMA)